Louis Shapiro · Renzo Sprugnoli · Paul Barry · Gi-Sang Cheon · Tian-Xiao He · Donatella Merlini · Weiping Wang

AF609419

The Riordan Group and Applications

Springer

Louis Shapiro
Department of Mathematics
Howard University
Washington DC, USA

Paul Barry
Waterford Institute of Technology
Waterford, Ireland

Tian-Xiao He
Department of Mathematics
Illinois Wesleyan University
Bloomington, IL, USA

Weiping Wang
School of Science
Zhejiang Sci-Tech University
Hangzhou, China

Renzo Sprugnoli
Dipartimento di Statistica, Informatica,
Applicazioni "Giuseppe Parenti"
Università degli Studi di Firenze
Firenze, Italy

Gi-Sang Cheon
Department of Mathematics
Sungkyunkwan University
Suwon, Korea (Republic of)

Donatella Merlini
Dipartimento di Statistica, Informatica,
Applicazioni "Giuseppe Parenti"
Università degli Studi di Firenze
Firenze, Italy

ISSN 1439-7382 ISSN 2196-9922 (electronic)
Springer Monographs in Mathematics
ISBN 978-3-030-94153-6 ISBN 978-3-030-94151-2 (eBook)
https://doi.org/10.1007/978-3-030-94151-2

Mathematics Subject Classification: 05A15

© Springer Nature Switzerland AG 2022
This work is subject to copyright. All rights are solely and exclusively licensed by the Publisher, whether the whole or part of the material is concerned, specifically the rights of translation, reprinting, reuse of illustrations, recitation, broadcasting, reproduction on microfilms or in any other physical way, and transmission or information storage and retrieval, electronic adaptation, computer software, or by similar or dissimilar methodology now known or hereafter developed.
The use of general descriptive names, registered names, trademarks, service marks, etc. in this publication does not imply, even in the absence of a specific statement, that such names are exempt from the relevant protective laws and regulations and therefore free for general use.
The publisher, the authors and the editors are safe to assume that the advice and information in this book are believed to be true and accurate at the date of publication. Neither the publisher nor the authors or the editors give a warranty, expressed or implied, with respect to the material contained herein or for any errors or omissions that may have been made. The publisher remains neutral with regard to jurisdictional claims in published maps and institutional affiliations.

This Springer imprint is published by the registered company Springer Nature Switzerland AG
The registered company address is: Gewerbestrasse 11, 6330 Cham, Switzerland

Springer Monographs in Mathematics

Editors-in-Chief

Minhyong Kim, School of Mathematics, Korea Institute for Advanced Study, Seoul, South Korea, International Centre for Mathematical Sciences, Edinburgh, UK

Katrin Wendland, Research group for Mathematical Physics, Albert Ludwigs University of Freiburg, Freiburg, Germany

Series Editors

Sheldon Axler, Department of Mathematics, San Francisco State University, San Francisco, CA, USA

Mark Braverman, Department of Mathematics, Princeton University, Princeton, NY, USA

Maria Chudnovsky, Department of Mathematics, Princeton University, Princeton, NY, USA

Tadahisa Funaki, Department of Mathematics, University of Tokyo, Tokyo, Japan

Isabelle Gallagher, Département de Mathématiques et Applications, Ecole Normale Supérieure, Paris, France

Sinan Güntürk, Courant Institute of Mathematical Sciences, New York University, New York, NY, USA

Claude Le Bris, CERMICS, Ecole des Ponts ParisTech, Marne la Vallée, France

Pascal Massart, Département de Mathématiques, Université de Paris-Sud, Orsay, France

Alberto A. Pinto, Department of Mathematics, University of Porto, Porto, Portugal

Gabriella Pinzari, Department of Mathematics, University of Padova, Padova, Italy

Ken Ribet, Department of Mathematics, University of California, Berkeley, CA, USA

René Schilling, Institute for Mathematical Stochastics, Technical University Dresden, Dresden, Germany

Panagiotis Souganidis, Department of Mathematics, University of Chicago, Chicago, IL, USA

Endre Süli, Mathematical Institute, University of Oxford, Oxford, UK

Shmuel Weinberger, Department of Mathematics, University of Chicago, Chicago, IL, USA

Boris Zilber, Mathematical Institute, University of Oxford, Oxford, UK

This series publishes advanced monographs giving well-written presentations of the "state-of-the-art" in fields of mathematical research that have acquired the maturity needed for such a treatment. They are sufficiently self-contained to be accessible to more than just the intimate specialists of the subject, and sufficiently comprehensive to remain valuable references for many years. Besides the current state of knowledge in its field, an SMM volume should ideally describe its relevance to and interaction with neighbouring fields of mathematics, and give pointers to future directions of research.

More information about this series at https://link.springer.com/bookseries/3733

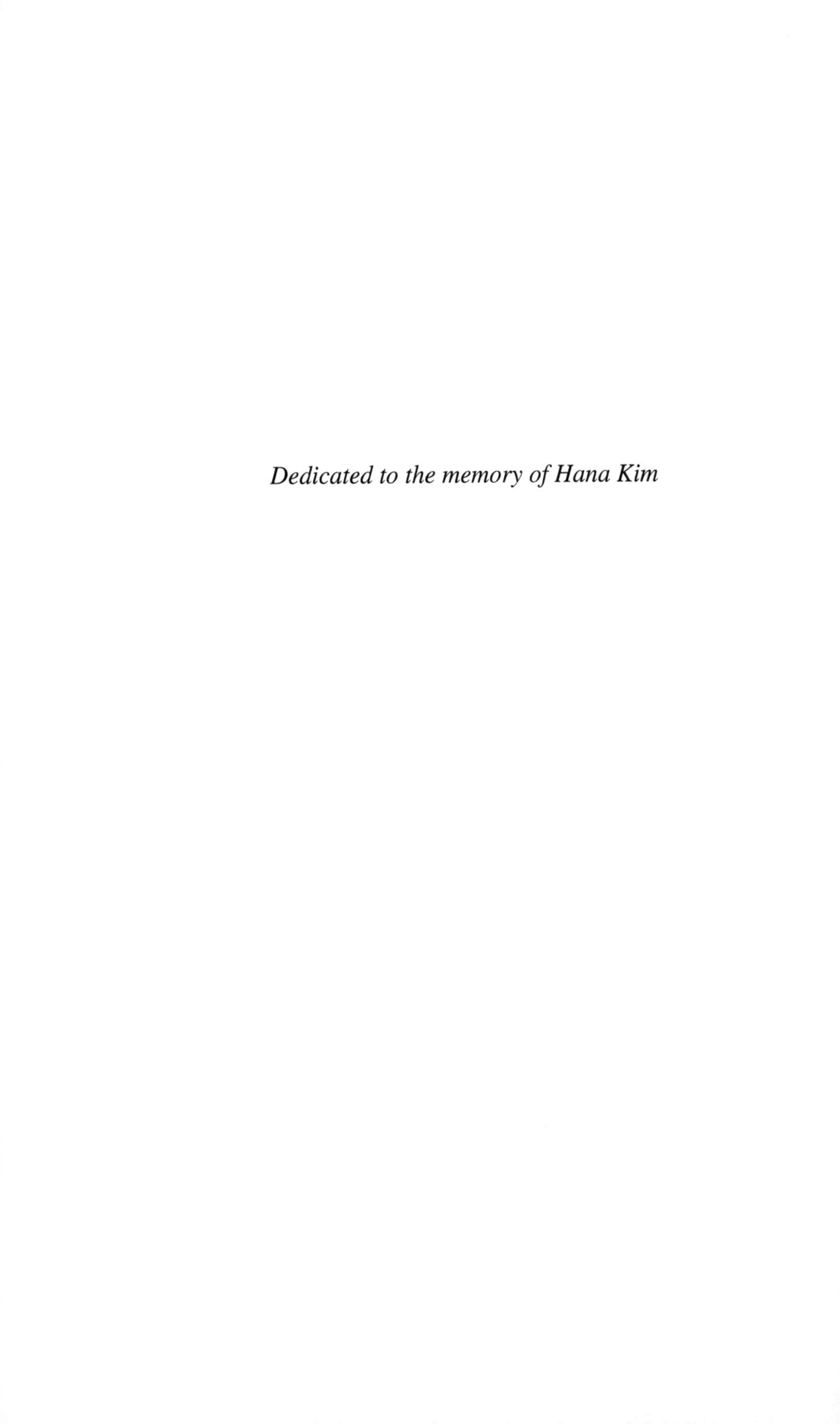

Dedicated to the memory of Hana Kim

Foreword

A Riordan array (or Riordan matrix) is an infinite lower triangular matrix encoding the coefficients of certain sequences of power series. Such an array can be regarded as a vast generalization of Pascal's triangle. The set of invertible Riordan arrays form the Riordan group. Riordan arrays and the Riordan group were introduced by Getu, Shapiro, Woan, and Woodson in 1991. These concepts lead to an elegant, unified, and fruitful approach for dealing with a host of enumerative problems.

I had some superficial acquaintance with Riordan arrays and the Riordan group prior to 2012. For 1 year beginning September 1, 2012, Hana Kim, to whom this book is dedicated, was a visiting Postdoctoral Associate at M.I.T. I learned from her many further aspects of Riordan arrays, leading both to a joint paper on hex trees and a heightened appreciation of the underlying theory.

It is a pleasure to see a comprehensive and clearly written treatise on Riordan arrays and the Riordan group arrive on the combinatorics scene. There are beautiful connections with such topics as Lagrange inversion; umbral calculus; and the finite operator calculus of Rota et al., continued fractions, and orthogonal polynomials. Numerous generalizations enhance the Riordan forest and plant seeds for further development. Overall this book admirably introduces and develops a fascinating new aspect of enumerative combinatorics.

October 2020

Richard Stanley
Professor Emeritus of Mathematics
Massachusetts Institute of Technology
Cambridge, USA

Preface

Combinatorial problems arise in many areas of pure mathematics, notably in algebra, probability theory, topology, and geometry, as well as in its many application areas ranging from logic to statistical physics, from evolutionary biology to computer science, etc. Although primarily concerned with finite systems, some combinatorial questions and techniques can be extended to an infinite (specifically, countable) but discrete setting.

This preface is structured into three parts: Parts I and II describe the points of view of Lou Shapiro and Renzo Sprugnoli; Part III is on behalf of all the authors.

Part I: Lou Shapiro's

The Riordan group is a group of infinite lower triangular matrices defined by a pair of generating functions. Let $g(z) = g_0 + g_1 z + g_2 z^2 + \cdots = \sum_{n\geq 0} g_n z^n$ and $f(z) = f_1 z + f_2 z^2 + \cdots = \sum_{n\geq 1} f_n z^n$ with $g_0 f_1 \neq 0$. The matrix M determined by g and f is given by $m_{n,k} = [z^n]g(z)(f(z))^k$ for $n, k \geq 0$. This can be thought of as a geometric series of generating functions with the coefficients of g in the leftmost column and then the multiplier function f. So the next column lists the coefficients of gf and then the next column uses gf^2 and so on.

The idea that matrices are related to composition of functions is far from new. Some early articles that discuss this are as follows:

- Albert A. Bennett, The iteration of functions of one variable, Ann. of Math., Ser. 2. 17 (1), 23–60 (1915).
- Eri Jabotinsky, Sur la representation de la composition de fonctions par un produit de matrices. Application a l'iteration de e^x et de $e^e - 1$, C. R. Acad. Sci. Paris. 224, 323–324 (1947).
- Eri Jabotinsky, Representation of functions by matrices. Application to Faber polynomials. Proc. Amer. Math. Soc. 4, 546–553 (1953).
- Peter Henrici, Applied and Computational Complex Analysis, vol. 1. Wiley, 13–65 (1988).

The article of Albert A. Bennett attributes this connection of power series to matrices to an earlier article of Helge von Koch appearing in 1900.

Here is the history of how the Riordan group got its name. Until Gian-Carlo Rota and computer science came along in the 1970s combinatorics was not highly regarded in the mathematical research community. The quick summary had combinatorics as graph theory (topology oversimplified) and combinatorial identities (a bag of tricks). There were, however, three masters of this bag of tricks: Leonard Carlitz, John Riordan, and Henry Gould. They published extensively and one could work diligently through their many results. One book which appeared that brought some order to the underlying concepts was "Combinatorial Identities" by John Riordan. One method that was often employed was the umbral calculus where many identities worked out nicely by changing subscripts to exponents, doing the calculation, and switching back. It was a powerful method which did however have the drawback of not having rules or proofs and could lead to false results.

Two factors made combinatorics much more highly regarded. Rota found deep connections with many parts of mathematics. He also developed the operator calculus to address these problems with the umbral calculus and a group structure emerged. Several examples of Riordan arrays were classical and well known. The most famous was Pascal's triangle and it's almost identical inverse. That the Stirling numbers of the first and second kind are essentially inverses was also known.

We were looking at various Catalan and Motzkin matrices and knew there was a group structure underlying our examples. When we found the fundamental theorem of Riordan arrays everything became transparent. One of the triumphs was that inverse relations, a major topic in Combinatorial Identities often were easy to show using the Riordan group inverse.

John Riordan, a kind and generous scholar, passed away at that time so we decided that we should name the group after him. Similarly the Bell subgroup was named to honor Eric Temple Bell. We do not know of anyone named checkerboard but that was the name we used.

We thought that we had written a nice little paper that used simple ingredients: matrix multiplication, elementary group theory, and generating functions, to prove and invert combinatorial identities. We were very pleased when we learned that Renzo Sprugnoli and his group from Italy had written significant papers using the Riordan group concept. We never imagined that it would evolve into hundreds of articles and even an annual conference on Riordan Arrays and Related Topics (RART). We are very grateful to the many talented researchers who have done so much starting with the simple combination of techniques in the first paper. We are also grateful for the development of a generous and supportive international community of Riordan array scholars.

Washington DC, USA
October 2020

Louis Shapiro

Part II: Renzo Sprugnoli's

During the summer of 1987 I decided to begin to study seriously Analysis of Algorithms (AofA). I thought that my degree in mathematics would be a sufficient introduction to this new course of studies. Actually, I had published (or was going to publish) some papers on data structures such as hash tables, k-ary trees, or heaps, without abandoning formal and programming languages, my traditional fields of interest.

I was lucky and AofA was just that branch of math known as combinatorics; Jean Dieudonné [Pour l'honneur de l'esprit humain, *Hachette Paris*, 1987] describes the situation with the following words:

> The methods connected to these problems have long been considered of little interest by many mathematicians, who willingly placed them in the sector that takes the name of *mathematical games*, to which many readers of scientific dissemination are passionate. But in a few years the situation has changed considerably: it has been realized that certain combinatorial methods could have important applications in completely respectable mathematics sectors, like group theory, the theory of algebras of Lie, algebraic geometry and algebraic topology; they have been used successfully even in various applications of Mathematics, including Computer Science.

In practice, combinatorics is the study of finite math or the mathematics of finite sets, such as counting problems and probability, but also group theory just to return to the list of Dieudonné. I was fascinated by the number and variety of arguments considered, but what struck my imagination most were the problems of combinatorial identity and inversion.

Suppose we have a formula solving a combinatorial problem:

$$s_n = \sum_{k=0}^{n} \binom{n}{k} \frac{(-1)^k}{2+k},$$

where n is a counter. When n is large, the computation of s_n may become difficult for a computer, as well; we need an equivalent formula in closed form, that is, an expression the computation of which requires a number of operations which is independent of n. In this example, we can prove that $s_n = \frac{1}{(n+1)(n+2)}$ and with this expression the combinatorial sum can be effortless computed for every n. The closed form in this case can be easily determined because the binomial coefficient $\binom{n}{k}$ inside the summation corresponds to a Riordan array. In general, if a combinatorial sum $\sum_{k=0}^{n} d_{n,k} f_k$ involves the elements $d_{n,k}$ of a Riordan array, a closed form s_n for it can be computed by operating a suitable transformation on a generating function and then by extracting the corresponding coefficient; this yields the combinatorial identity $s_n = \sum_{k=0}^{n} d_{n,k} f_k$.

Another interesting problem is combinatorial inversion: suppose we have a combinatorial identity as before, is it possible to derive f_n from s_n? In other words, is it possible to find the inverse combinatorial identity $f_n = \sum_{k=0}^{n} d^*_{n,k} s_k$? If $d_{n,k}$ is the generic element of a Riordan array, this is possible by computing $d^*_{n,k}$ as the generic element of the *inverse* Riordan array. Our original problem, for example, corresponds to following pair of inverse combinatorial identities, where we used the fact that the binomial coefficient $\binom{n}{k}(-1)^{n-k}$ is the generic element of the inverse of the Pascal triangle:

$$\sum_{k=0}^{n} \binom{n}{k} \frac{(-1)^k}{x+k} = \frac{1}{x\binom{x+n}{n}}, \quad \sum_{k=0}^{n} \binom{n}{k} \frac{(-1)^k}{x\binom{x+k}{n}} = \frac{1}{x+n}.$$

As we will see in Chap. 5, the concept of Riordan arrays can give a substantial help in solving these problems, together with a technique, known as the "method of coefficients", which is based on generating functions and the extraction of their coefficients; this method will be illustrated in Chap. 2.

The first occurrence of a Riordan array in my notes is dated in the summer of 1978. As soon as I realized the importance of the method, I performed a bibliographic research that I summarize below:

- About 1860–1870, the English mathematician John Blissard observed the analogy:

$$(x+y)^n = \sum_{k=0}^{n} \binom{n}{k} x^{n-k} y^k, \quad B_n(x+y) = \sum_{k=0}^{n} \binom{n}{k} B_{n-k} y^k;$$

  where $B_n(x)$ are Bernoulli polynomials. If you have a candle over the powers, indices become their shadows (umbrae, in Latin). Are shadows valid as the original formulas? In the nineteenth century, John Blissard, Édouard Lucas, and James Joseph Sylvester used this technique without any formal proof. This is classical umbral or symbolic calculus.

- About 1930–1940, Eric Temple Bell tried to justify umbral calculus without success.
- About 1960–1970, John Riordan used frequently the symbolic method.
- About 1970–1980, Gian-Carlo Rota and his co-workers (Roman, Taylor, Mullin, Kahaner et al.) finally found a justification of umbral calculus by means of linear functionals over the vector space of polynomials. This is the modern interpretation of the umbral calculus.
- I also like to mention a first study of the relationship between umbral calculus and Riordan arrays that appeared on 1999 in the Ph.D. Thesis of Yin Dongsheng, Dalian University of Technology, titled *Umbral calculus and Hsu-Riordan array.*

The connection between umbral calculus, the sets of polynomials that arise in this context, and Riordan arrays will be illustrated in Chap. 6. The question of which Riordan arrays correspond to sequences of orthogonal polynomials will be described in Chap. 9.

After the application of Riordan arrays to the computation of combinatorial sums and the bibliographic research, I also dealt with more algebraic questions, relating to the group structure of these arrays, and in particular to their characterization through special sequences, which is particularly useful in combinatorial problems. Riordan's group is illustrated in Chap. 3 while the characterization by special sequences is presented in Chap. 4.

Over the past 20 years the Riordan array concept has been extensively studied and today continues to attract great attention in the literature and, as a consequence, several generalizations and applications have been proposed. In addition to those already mentioned above, the group of three-dimensional Riordan arrays and the q-analogs of Riordan arrays are presented in Chaps. 7 and 8, respectively.

Since 2014 a series of symposiums on Riordan arrays have been organized which have made the topic even more popular. In particular, the first meeting on *Riordan Arrays and Related Topics* was held on August 6–9, 2014, at Sungkyunkwan University, in Seoul, South Korea, as one of the invited minisymposiums in the 19th International Linear Algebra Society Conference (ILAS2014); the second meeting was held on July 14–16, 2015, at Politecnico di Milano, Campus Lecco, (on lake Como), Italy; the third symposium was held on June 20–23, 2016, at the campus of Illinois Wesleyan University, Bloomington, Illinois, USA; the fourth symposium was held on July 17–20, 2017 at the Universidad Complutense de Madrid, Madrid, Spain; the fifth meeting was held on June 25–29, 2018 in Busan, South Korea as an international conference; and the sixth meeting was held on July 1–5, 2019, in Sanya, China, as an international workshop. The 2020 meeting was supposed to be held on June 22–26 in Galway, Ireland, as one of the invited minisymposiums in the 23rd International Linear Algebra Society Conference (ILAS2020), but unfortunately it was canceled under the circumstances of the global COVID-19 emergency.

Hoping that the Riordan family can be reunited soon, I wish long life to Riordan Arrays!

Firenze, Italy
October 2020

Renzo Sprugnoli

Part III: Overview of the book

The Riordan array concept, with its attendant group, crystallized into a coherent structure in the 1990s. Its applications were firstly in combinatorics, particularly in the field of combinatorial identities. As its applications grew, so did the realization that the Riordan group itself had a rich structure, and was worthy of study per se. A fruitful interaction between algebra and linear algebra ensued, with representations of power series as an intermediary. Thus, the Riordan group may be regarded as a matrix group and a Lie group. Inverses are constructed using multiplicative and compositional inverses of appropriate power series, or equivalently by matrix inversion. The researcher can choose the appropriate mechanism for their area of interest, or a combination of both. A richness of approaches follows. With an increase in research activity, an expanding literature, and a growing community of practitioners, it became evident that the area was ripe for the development of some books to introduce the concept to a wider audience, as well as to provide a baseline and springboard for further developments. The first book to appear was entitled Riordan arrays: A Primer. This book, introductory in nature, is aimed principally at an interested undergraduate audience, to be used for self-study, or as a support for senior project work. It can also serve as a quick introduction to Riordan arrays and the Riordan group for those starting in research, who may just want a rapid overview of the essentials. The current work, on the other hand, is more comprehensive in its treatment, both in terms of the structure of the Riordan group, and in terms of its generalizations and its applications. Its intended audience is practising mathematicians and postgraduate researchers who may wish to explore the use of the Riordan array concept in their work.

The Riordan group, or its close relatives, is now making inroads into areas far removed from the ideas that motivated its first proponents. An example, for instance, is an appearance in quantum field theory, where the Riordan pairing (g, f) can model massive and massless entities. Such unexpected links attest to the underlying significance and potential of the Riordan array concept. The authors envisage that this book will serve to introduce the Riordan group to a wider audience so that its applications can be carried into fields of research as yet to be discovered.

We now present a guided tour of the book. Chapter 1 provides a historical context for the development of Riordan arrays, using sequence-based examples, along with some tree-based applications. It is noted that many important sequences and Riordan arrays encountered in this book can be found in the On-Line Encyclopedia of Integer Sequences, `https://oeis.org`.

Chapter 2 gives an overview of some of the principal tools used in working with Riordan arrays. Power series provide a basis for the definition of a Riordan array, while coefficient extraction is the means whereby the elements of the matrix representation are found. Series inversion defines the inverse of a Riordan array. Thus this chapter explains the Lagrange inversion formula, which is routinely used to extract the elements of the inverse of Riordan arrays.

In Chap. 3, the formal definition of Riordan arrays and the Riordan group is addressed. Important subgroups of the Riordan group are discussed. Notably, the semidirect nature of the Riordan group is shown in this chapter. The Lie group nature of the Riordan group is also indicated.

Given a lower triangular matrix, how do we know if it is a Riordan array? In Chap. 4, useful characterizations of Riordan arrays are given. Sequence characterizations are introduced based on two sequences, the A-sequence and the Z-sequence. More generally, a matrix characterization can be given based on a matrix, the A-matrix. The Z-sequence and A-sequence make up a special matrix, the production matrix, which plays an important role in many considerations of Riordan arrays.

The application of Riordan array techniques to combinatorial sums and combinatorial inversions is explored in Chap. 5. This chapter gives us a solid insight into the power of Riordan array techniques in this context. At little expense, many important identities found in classical texts are reproduced. In fact, the unifying ability of Riordan array techniques is evident in this chapter.

As an active field of research, there have been many developments in the area of Riordan arrays since their original introduction. Chapter 6 begins the exposition of some of these, under the banner of generalized Riordan arrays. These are defined in terms of a sequence (c_n) of nonzero constants with $c_0 = 1$. The definition, structure, and sequence characterization of these generalized arrays is given, along with applications to combinatorial inversions. Importantly, the group of generalized Riordan arrays is shown to be isomorphic to the group of Sheffer sequences. Many interesting polynomial sequences are explored in this context, including polynomial sequences whose study is taken up again in Chap. 9. The chapter ends with a study of the double Riordan arrays, which are defined by a triple of power series.

The next chapter, Chap. 7, presents a further generalization of the Riordan group. The constituent arrays of this generalization are now three-dimensional arrays, and this new structure provides a group extension of the Riordan group. This notion naturally extends to the Riordan group in several variables. As with the case of one variable, an element in the group can be represented as an infinite, lower triangular block matrix, called a multivariate Riordan array.

Given the applications of Riordan arrays to combinatorial problems, it is natural to ask what a q-analog of a Riordan array would look like. This question, the q-analog of Riordan arrays, is studied in Chap. 8. Although no longer a group, the resulting structure still has significant applications, some of which are explored in this chapter.

The final chapter of the book, Chap. 9, is concerned with the question: what orthogonal polynomial sequences have Riordan arrays as their coefficient arrays? In the case of ordinary Riordan arrays, such polynomial sequences are essentially generalized Chebyshev polynomials of the second kind. In the case of exponential Riordan arrays, we encounter, for instance, Laguerre and Hermite polynomials. Many important combinatorial sequences are shown to be moment sequences for the families of orthogonal polynomials under discussion. Applications, such as the application of Riordan arrays to the Toda chain equation, are discussed.

Each chapter ends with a set of exercises, designed to help the reader develop a deeper understanding of the theory and methods of Riordan arrays, and a reference section. Solutions to selected exercises are provided at the end of the book.

Washington DC, USA — Louis Shapiro
Firenze, Italy — Renzo Sprugnoli
Waterford, Ireland — Paul Barry
Suwon, South Korea — Gi-Sang Cheon
Bloomington, Illinois, USA — Tian-Xiao He
Firenze, Italy — Donatella Merlini
Hangzhou, China — Weiping Wang

Acknowledgements

We wish to thank the many people who have contributed, beyond the authors of this work, to the enrichment of the theory and applications of Riordan arrays in the last two decades.

We wish also to thank Elizabeth Loew, the Executive Editor, Springer Mathematics, for her constant support since the spring of 2017, when we first submitted the book proposal.

Compiling a technical text across a range of international contributors is a demanding task that requires commitment, skill, and forbearance. The authors would like to thank one of their own, Donatella Merlini, who possesses all these qualities. The authors wish to record their debt of gratitude to Donatella for her steadfast work, from the initial discussions, to helping see this work through to its final stage.

Washington, DC, USA	Louis Shapiro
Firenze, Italy	Renzo Sprugnoli
Waterford, Ireland	Paul Barry
Suwon, South Korea	Gi-Sang Cheon
Bloomington, Illinois, USA	Tian-Xiao He
Firenze, Italy	Donatella Merlini
Hangzhou, China	Weiping Wang
October 2020	

Contents

Notation

$\mathbb{N}$	The set of natural numbers $0, 1, 2, \cdots$
$\mathbb{Z}$	The ring of integers
$\mathbb{Q}$	The field of rational numbers
$\mathbb{R}$	The field of real numbers
$\mathbb{C}$	The field of complex numbers
$\mathbb{F}[t]$	The ring of polynomials in variable t over a field $\mathbb{F}$
$\mathbb{F}[[t]]$	The ring of formal power series in variable t over a field $\mathbb{F}$
$[t^n]f(t)$	The coefficient extraction operator of $f(t)$
$\mathcal{G}(f_n)$	The generating function operator for f_n
$o(f(t))$	The order of a power series $f(t)$, the smallest integer k such that $[t^k]f(t) \neq 0$
$\mathcal{F}_k$	The set of formal power series $f(t)$ such that $o(f(t)) = k$
$g \circ f$	The composition of two functions $g, f \in \mathbb{F}[[t]]$, also $(g \circ f)(t)$ or $g(f(t))$
$\bar{f}(t)$	The compositional inverse of $f(t) \in \mathcal{F}_1$, i.e., $f \circ \bar{f} = \bar{f} \circ f = t$, also $\bar{f}$
(g, f)	A Riordan array or a Riordan matrix, also $(g(t), f(t))$ or $\mathcal{R}(g, f)$ or $T(f\|g)$; the pair (d, h) is also used
A	The A-sequence of a Riordan array
Z	The Z-sequence of a Riordan array
$\langle g, f\rangle$	An exponential Riordan matrix, also $\langle g(t), f(t)\rangle$ or $[g(t), f(t)]$ or $E(g, f)$ or $\mathcal{E}(g, f)$
$\mathcal{R}$	The Riordan group or $\mathcal{RG}$
$\mathcal{R}_e$	The exponential Riordan group
$\mathbb{F}^n$	The vector space over a field $\mathbb{F}$ of n-dimensional column vectors with entries from $\mathbb{F}$
$M_n(\mathbb{F})$	The set of all $n \times n$ matrices with entries from a field $\mathbb{F}$
$\|\alpha\|$	The cardinality of a set α
$X(\alpha\|\beta)$	The $(n - \|\alpha\|) \times (n - \|\beta\|)$ matrix obtained from $X \in M_n(\mathbb{F})$ by deleting the rows and the columns indexed by the sets α and β
$X[\alpha\|\beta]$	The complement of $X(\alpha\|\beta)$

$\bar{X}$	The matrix obtained from an infinite matrix X by deleting the top row
$(x)_n$	Falling factorials
$\langle x\rangle_n$	Rising factorials
$\left[{n \atop k}\right]$	Unsigned Stirling numbers of the first kind
$\left\{{n \atop k}\right\}$	Stirling numbers of the second kind
$(d;\, f_1, f_2)$	A double Riordan array
$(g, f)_q$	A q-Riordan array or q-Riordan matrix
(g, f, h)	Three-dimensional Riordan array or 3-D Riordan array
(g, f, zh)	A shifted three-dimensional Riordan array or shifted 3-D Riordan array
$\mathcal{R}^{\langle 3\rangle}$	The three-dimensional Riordan group
$\mathcal{R}^d$	The multivariate Riordan group in d variables
$\mathbb{F}[[z_1, \ldots, z_d]]$	The ring of formal power series in d variables $z_1, \ldots, z_d$ over a field $\mathbb{F}$, also $\mathbb{F}[[\mathbf{z}]]$ in $\mathbf{z} = (z_1, \ldots, z_d)$
$\mathbf{z}^{\mathbf{i}}$	The monomial $z_1^{i_1} \cdots z_d^{i_d}$ with $\mathbf{z} = (z_1, \ldots, z_d)$ and $i = (i_1, \ldots, i_d)$
$F(\mathbf{z})$	Formal power series in $\mathbb{F}[[\mathbf{z}]]$, also $F(z_1, \ldots, z_d)$ or F
$\mathbf{F}(\mathbf{z})$	$(F_1, \ldots, F_d)$ with $F_i \in \mathbb{F}[[\mathbf{z}]]$, also $\mathbf{F}$
$[\mathbf{z}^{\mathbf{i}}]F(\mathbf{z})$	The $\mathbf{z}^{\mathbf{i}}$-coefficient of $F(\mathbf{z})$
$\mathcal{M}(G, \mathbf{F})$	A multivariate Riordan array, also $(G, \mathbf{F})$
$J_{\mathbf{F}}(\mathbf{z})$	Jacobian matrix of $\mathbf{F}$ with respect to $z_1, \ldots, z_d$
$\mathbf{i} \prec \mathbf{j}$	The graded reverse lexicographic order

Chapter 1
Introduction

Abstract Here we insert some general considerations on the history, the properties and the applications of the Riordan group. The group of Riordan arrays was introduced in 1991 by Shapiro, Getu, Woan, and Woodson [6], with the aim of defining a class of infinite lower triangular arrays with properties analogous to those of the Pascal triangle. Soon, the concept became popular [5] and Sprugnoli in particular showed that these arrays constitute a practical device for solving combinatorial sums by means of the generating functions [7, 8].

1.1 What are Riordan Arrays?

Suppose a friend comes down the hall and says "we're supposed to evaluate the sum

$$\sum_{k=0}^{n} \binom{n}{k} (-1)^k (2k+1). \tag{1.1.1}$$

What do we do?" The first step is pretty intuitive, start with a few small but not tiny cases. For instance, $n = 3$ yields

$$\binom{3}{0} \cdot 1 - \binom{3}{1} \cdot 3 + \binom{3}{2} \cdot 5 - \binom{3}{3} \cdot 7 = 1 - 9 + 15 - 7 = 0.$$

The second step is less intuitive. Put things in the form of a matrix times a column vector equals a column vector after also computing the results for $n = 0, 1,$ and 2 :

$$\begin{bmatrix} \binom{0}{0} & 0 & 0 & 0 \\ \binom{1}{0} & \binom{1}{1} & 0 & 0 \\ \binom{2}{0} & \binom{2}{1} & \binom{2}{2} & 0 \\ \binom{3}{0} & \binom{3}{1} & \binom{3}{2} & \binom{3}{3} \end{bmatrix} \begin{bmatrix} 1 \\ -3 \\ 5 \\ -7 \end{bmatrix} = \begin{bmatrix} 1 \\ -2 \\ 0 \\ 0 \end{bmatrix} \quad \text{that can be written as } Ab = c.$$

© The Author(s), under exclusive license to Springer Nature Switzerland AG 2022

L. Shapiro et al., *The Riordan Group and Applications*, Springer Monographs in Mathematics, https://doi.org/10.1007/978-3-030-94151-2_1

The next step is to introduce generating functions for each of the columns. We will use a two-sided arrow to identify a sequence with its generating function. The leftmost column of the matrix (say A) gives us

$$1, 1, 1, 1, \cdots \Longleftrightarrow A^{(0)}(t) = 1 + t + t^2 + t^3 + \cdots = \sum_{n \geq 0} t^n = \frac{1}{1-t}.$$

Here are the next few columns

$$A^{(1)}(t) = 0 + t + 2t^2 + 3t^3 + \cdots = \frac{t}{(1-t)^2}$$

$$A^{(2)}(t) = 0 + 0 \cdot t + t^2 + 3t^3 + \cdots = \frac{t^2}{(1-t)^3}$$

and so on. Next we have

$$1, -3, 5, -7, \cdots \Longleftrightarrow b(t) = \frac{1-t}{(1+t)^2}.$$

Since

$$A^{(i)}(t) = \left(\frac{1}{1-t}\right)\left(\frac{t}{1-t}\right)^i$$

it immediately follows that

$$c(t) = \sum_{i=0}^{\infty} b_i A^{(i)}(t) = \left(\frac{1}{1-t}\right) b\left(\frac{t}{1-t}\right) = 1 - 2t.$$

Thus, we have that $\sum_{k=0}^{n} \binom{n}{k} (-1)^k (2k+1) = 0$ for $n \geq 2$.

The essential ideas appear in the evaluation of (1.1.1). One idea is to go to the matrix representation in terms of column generating functions. The second is that we move from one column to the next by multiplying by $\frac{t}{1-t}$. It is very convenient to state that the generating function of the leftmost column should have a nonzero constant term. We call this column generating function $g(t)$ and with very little loss of generality we can assume

$$g(t) = 1 + g_1 t + g_2 t^2 + g_3 t^3 + \cdots.$$

We then multiply by the generating function

$$f(t) = f_1 t + f_2 t^2 + f_3 t^3 + \cdots, \quad (f_1 \neq 0)$$

and the next column has the generating function $g(t) f(t)$. This continues, we multiply by $f(t)$ again and the generating function is $g(t) f(t)^2$. It is easier to usually abbreviate $g(t)$ to g, $f(t)$ to f, and so on. Thus, the generating function for the kth column is gf^k for $k = 0, 1, 2, \cdots$. In our proof, we had $g = \frac{1}{1-t}$ and $f = \frac{t}{1-t}$. The notation for such a matrix is the rather plebeian $(g(t), f(t))$ or (g, f). This leads us to the notion of *Riordan arrays*. With this notion, if $B = (b_0, b_1, b_2, \cdots)^T$ and $C = (c_0, c_1, c_2, \cdots)^T$ are column vectors and $b(t)$ and $c(t)$ are the corresponding generating functions, then the product $(g, f)B = C$ translates into the following expression in terms of generating functions

$$g(t)b(f(t)) = c(t) \quad \text{or} \quad (g(t), f(t)) * b(t) = c(t) \tag{1.1.2}$$

which is called the fundamental theorem of Riordan array (FTRA). This result will be formally proved in Theorem 3.1.

1.2 Origins and Motivation

There are many counting problems in combinatorics whose solutions are given by interesting numbers forming a sequence of natural numbers that occur in various combinatorial questions, often involving recursively defined objects.

For example, the *Catalan numbers*, $C_n = \frac{1}{n+1}\binom{2n}{n}$, can be recursively given by

$$C_{n+1} = \sum_{k=0}^{n} C_k C_{n-k} \text{ with } C_0 = 1.$$

In terms of generating functions this becomes

$$C = 1 + tC^2 = \frac{1}{1 - tC}$$

or more compactly

$$C = \sum_{n\geq 0} \frac{1}{n+1}\binom{2n}{n} t^n = 1 + t + 2t^2 + 5t^3 + 14t^4 + \cdots = \frac{1 - \sqrt{1-4t}}{2t}.$$

They count many things including

- triangulations of a regular $(n + 2)$-gon (Euler),
- parenthesizations of $n + 1$ letters (Catalan),
- the number of Dyck paths from $(0, 0)$ to $(2n, 0)$ using the steps $U = (1, 1)$ and $D = (1, -1)$ that never go below the x-axis.

For more information about these numbers, see the Catalan addendum of Richard Stanley [9], now expanded to a book [10]. The version we will use most is ordered trees (also known as plane trees, planar trees) with n edges.

A companion generating function is that for the *central binomial coefficients*, $B_n = \binom{2n}{n}$. This sequence has the generating function

$$B = \sum_{n\geq 0} \binom{2n}{n} t^n = 1 + 2t + 6t^2 + 20t^3 + \cdots = \frac{1}{\sqrt{1-4t}}.$$

Note that

$$B = 1 + 2tCB = \frac{C}{1 - tC^2}.$$

A third important sequence is that of the *Motzkin numbers*

$$1, 1, 2, 4, 9, 21, 51, 127, \cdots$$

which correspond to the number of Motzkin paths from $(0, 0)$ to $(n, 0)$ using steps $U = (1, 1)$, $D = (1, -1)$ and $L = (1, 0)$ with the extra provision that the paths never go below the x-axis. The sequence has the number A001006 in the OEIS, the On-Line Encyclopedia of Integer Sequences, https://oeis.org/. There is a tremendous amount of information at this reference. The generating function for this sequence is denoted $m(t)$ or more briefly m. We see that

$$m = 1 + tm + t^2m^2.$$

The three terms on the right side correspond to the empty path from $(0, 0)$ to $(0, 0)$ (the 1 term), paths starting with a level step (the tm term), and paths starting with an up step (the t^2m^2 term), where the first t is for that up step and the second t is for the first down step from height 1 back to height 0. If we remove the restriction about going below the x-axis then we have Euler's generating function $E = \frac{1}{\sqrt{1-2t-3t^2}}$ and then

$$E = 1 + tE + 2t^2mE = \frac{1}{1 - t - 2t^2m}.$$

We also have

$$m - 3tm = m\,(1 - 3t) = 1 - 2tm + t^2m^2 = (1 - tm)^2\,.$$

Counting numbers with a doubly indexed sequence can be expressed in triangular arrays. Very interestingly, in many cases, these arrays are represented in terms of two formal power series that generate the numbers in each column, that is, the Riordan arrays introduced in the previous section. As a result, many interesting results can be obtained by algebraic techniques.

For example, we look at every odd column of the Pascal triangle arranged as a lower triangular array:

$$\begin{bmatrix} 1 & 0 & 0 & 0 & 0 & 0 \\ 1 & 1 & 0 & 0 & 0 & 0 \\ 1 & 3 & 1 & 0 & 0 & 0 \\ 1 & 6 & 5 & 1 & 0 & 0 \\ 1 & 10 & 15 & 7 & 1 & 0 \\ 1 & 15 & 35 & 28 & 9 & 1 \end{bmatrix}$$

where the column generating functions from the leftmost column are

$$\frac{1}{1-t}, \frac{1}{1-t}\frac{t}{(1-t)^2}, \frac{1}{1-t}\frac{t^2}{(1-t)^4}, \ldots. \tag{1.2.1}$$

We could ask about the row sums or equivalently multiply by the vector $(1, 1, 1, \cdots)^T$. The product is the column vector $(1, 2, 5, 13, 34, 89, \cdots)^T$ and either a good memory or a visit to the OEIS, the On-Line Encyclopedia of Integer Sequences, tells us that we have the odd-indexed Fibonacci numbers. The generating function for the Fibonacci numbers is $F(t) = \frac{1}{1-t-t^2}$ and we get the alternating sum as

$$\frac{1}{2}(F(t) + F(-t)) = \frac{1-t^2}{1-3t^2+t^4} \Leftrightarrow 1, 0, 2, 0, 5, 0, 13, 0, \cdots.$$

The generating function we want is thus $\frac{1-t}{1-3t+t^2}$. For the proof, summing all in (1.2.1) yields

$$\frac{1}{1-t} \cdot \frac{1}{1-\frac{t}{(1-t)^2}} = \frac{1-t}{1-3t+t^2},$$

as we required.

We can use the same matrix for another surprise. It is in proving and inverting identities. For instance, we observe that

$$\begin{bmatrix} 1 & 0 & 0 & 0 & 0 & 0 \\ 1 & 1 & 0 & 0 & 0 & 0 \\ 1 & 3 & 1 & 0 & 0 & 0 \\ 1 & 6 & 5 & 1 & 0 & 0 \\ 1 & 10 & 15 & 7 & 1 & 0 \\ 1 & 15 & 35 & 28 & 9 & 1 \end{bmatrix} \begin{bmatrix} 1 \\ -4 \\ 4^2 \\ -4^3 \\ 4^4 \\ -4^5 \end{bmatrix} = \begin{bmatrix} 1 \\ -3 \\ 5 \\ -7 \\ 9 \\ -11 \end{bmatrix}$$

which corresponds to the identity:

$$\sum_{k=0}^{n} \binom{n+k}{n-k}(-4)^k = (2n+1)(-1)^n.$$

Here is an example linking the Fibonacci numbers with the powers of 5:

$$\begin{bmatrix} 1 & 0 & 0 & 0 & 0 & 0 \\ 2 & 0 & 0 & 0 & 0 & 0 \\ 3 & 1 & 0 & 0 & 0 & 0 \\ 4 & 4 & 0 & 0 & 0 & 0 \\ 5 & 10 & 1 & 0 & 0 & 0 \\ 6 & 20 & 6 & 0 & 0 & 0 \end{bmatrix} \begin{bmatrix} 1 \\ 5 \\ 5^2 \\ 5^3 \\ 5^4 \\ 5^5 \end{bmatrix} = \begin{bmatrix} 1 \\ 2 \\ 8 \\ 24 \\ 80 \\ 256 \end{bmatrix} = \begin{bmatrix} 1 \cdot 1 \\ 2 \cdot 1 \\ 4 \cdot 2 \\ 8 \cdot 3 \\ 16 \cdot 5 \\ 32 \cdot 8 \end{bmatrix} = \begin{bmatrix} 1 \cdot F_0 \\ 2 \cdot F_1 \\ 2^2 \cdot F_2 \\ 2^3 \cdot F_3 \\ 2^4 \cdot F_4 \\ 2^5 \cdot F_5 \end{bmatrix}$$

which corresponds to the identity

$$\sum_{k=0}^{n} \binom{n+1}{n-2k} 5^k = 2^n F_{n-1}.$$

A proof of this can be achieved by introducing generating functions and by employing the fundamental theorem of Riordan arrays.

Here is a variation

$$\begin{bmatrix} 1 & 0 & 0 & 0 & 0 & 0 \\ 1 & 1 & 0 & 0 & 0 & 0 \\ 1 & 2 & 1 & 0 & 0 & 0 \\ 1 & 3 & 3 & 1 & 0 & 0 \\ 1 & 4 & 6 & 4 & 1 & 0 \\ 1 & 5 & 10 & 10 & 5 & 1 \end{bmatrix} \begin{bmatrix} 0 \\ 1 \\ 0 \\ 5 \\ 0 \\ 25 \end{bmatrix} = \begin{bmatrix} 0 \\ 1 \\ 2 \\ 8 \\ 24 \\ 80 \end{bmatrix} = \begin{bmatrix} 0 \\ 1 \cdot F_0 \\ 2 \cdot F_1 \\ 4 \cdot F_2 \\ 8 \cdot F_3 \\ 16 \cdot F_4 \end{bmatrix}.$$

corresponding to

$$\sum_{k=0}^{\lfloor (n-1)/2 \rfloor} \binom{n}{2k+1} 5^k = 2^n F_{n-1}, \quad n > 0.$$

Going back to the example in Sect. 1.1 we had

$$\begin{bmatrix} \binom{0}{0} & 0 & 0 & 0 & 0 \\ \binom{1}{0} & \binom{1}{1} & 0 & 0 & 0 \\ \binom{2}{0} & \binom{2}{1} & \binom{2}{2} & 0 & 0 \\ \binom{3}{0} & \binom{3}{1} & \binom{3}{2} & \binom{3}{3} & 0 \\ \binom{4}{0} & \binom{4}{1} & \binom{4}{2} & \binom{4}{3} & \binom{4}{4} \end{bmatrix} \begin{bmatrix} 1 \\ -3 \\ 5 \\ -7 \\ 9 \end{bmatrix} = \begin{bmatrix} 1 \\ -2 \\ 0 \\ 0 \\ 0 \end{bmatrix}.$$

Inverting this gives us

$$\begin{bmatrix} \binom{0}{0} & 0 & 0 & 0 & 0 \\ -\binom{1}{0} & \binom{1}{1} & 0 & 0 & 0 \\ \binom{2}{0} & -\binom{2}{1} & \binom{2}{2} & 0 & 0 \\ -\binom{3}{0} & \binom{3}{1} & -\binom{3}{2} & \binom{3}{3} & 0 \\ \binom{4}{0} & -\binom{4}{1} & \binom{4}{2} & -\binom{4}{3} & \binom{4}{4} \end{bmatrix} \begin{bmatrix} 1 \\ -2 \\ 0 \\ 0 \\ 0 \end{bmatrix} = \begin{bmatrix} 1 \\ -3 \\ 5 \\ -7 \\ 9 \end{bmatrix}$$

and which corresponds to

$$(-1)^n\binom{n}{0} - 2(-1)^{n-1}\binom{n}{1} = (-1)^n(1+2n),\ n \geq 0.$$

Now there are two natural questions. The first is to use the other way to alternate zeros and see what happens:

$$\begin{bmatrix} 1 & 0 & 0 & 0 & 0 & 0 \\ 1 & 1 & 0 & 0 & 0 & 0 \\ 1 & 2 & 1 & 0 & 0 & 0 \\ 1 & 3 & 3 & 1 & 0 & 0 \\ 1 & 4 & 6 & 4 & 1 & 0 \\ 1 & 5 & 10 & 10 & 5 & 1 \end{bmatrix}\begin{bmatrix} 1 \\ 0 \\ 5 \\ 0 \\ 25 \\ 0 \end{bmatrix} = \begin{bmatrix} 1 \\ 1 \\ 6 \\ 16 \\ 56 \\ 176 \end{bmatrix} = \begin{bmatrix} 1 \\ 1 \\ 2\cdot 3 \\ 4\cdot 4 \\ 8\cdot 7 \\ 16\cdot 11 \end{bmatrix} = \begin{bmatrix} 1 \\ 1\cdot L_1 \\ 2\cdot L_2 \\ 4\cdot L_3 \\ 8\cdot L_4 \\ 16\cdot L_5 \end{bmatrix}$$

where we see the Lucas numbers given by

$$\sum_{n\geq 0} L_n t^n = \frac{1+t^2}{1-t-t^2} = 1 + 3t^2 + 4t^3 + 7t^4 + 11t^5 + \cdots.$$

The second thought is to invert the identity:

$$\begin{bmatrix} 1 & 0 & 0 & 0 & 0 & 0 \\ -1 & 1 & 0 & 0 & 0 & 0 \\ 1 & -2 & 1 & 0 & 0 & 0 \\ -1 & 3 & -3 & 1 & 0 & 0 \\ 1 & -4 & 6 & -4 & 1 & 0 \\ -1 & 5 & -10 & 10 & -5 & 1 \end{bmatrix}\begin{bmatrix} 1\cdot 1 \\ 2\cdot 1 \\ 4\cdot 2 \\ 8\cdot 3 \\ 16\cdot 5 \\ 32\cdot 8 \end{bmatrix} = \begin{bmatrix} 1 \\ 1 \\ 5 \\ 5 \\ 25 \\ 25 \end{bmatrix}.$$

This yields that

$$\sum_{k=0}^{n}\binom{n}{k}(-1)^k\cdot 2^k F_{k-1} = 5^{\lfloor n/2\rfloor}.$$

We conclude this introductory section with two items. The first is a solution to a recent problem from the American Mathematical Monthly.

[Problem 11897] Evaluate the following sum:

$$\sum_{k+l=n,\,k,\ell\geq 0}\frac{\binom{2k}{k}\binom{2\ell+2}{\ell+1}}{k+1} = \sum_{k=0}^{n}\frac{\binom{2k}{k}\binom{2n-2k+2}{n-k+1}}{k+1} = \sum_{k=0}^{n} C_k\binom{2n-2k+2}{n-k+1},$$

where $C_k = \frac{1}{k+1}\binom{2k}{k}$ is the kth Catalan number.

We go to generating functions with

$$C(t) = C = \sum_{k=0}^{\infty} C_k t^k = 1 + tC^2 \text{ and } B(t) = B = \sum_{k=0}^{\infty} \binom{2k}{k} t^k.$$

The sum is the convolution of C and $\sum_{k=0}^{\infty} \binom{2k+2}{k+1} t^k = \frac{B-1}{t}$. We recall two basic facts from enumerative combinatorics:

$$B = 1 + 2tCB, \quad [t^n]\, BC^s = \binom{2n+s}{n},$$

where $[t^n]$ is an operator which allows us to extract the coefficient of t^n from a generating function. Since

$$C \cdot \frac{B-1}{t} = C \cdot \frac{2tCB}{t} = 2BC^2$$

the sum is

$$[t^n]\, 2BC^2 = 2\binom{2n+2}{n},$$

solving the problem.

Remark To see this in matrix terms, we have $C = 1 + t + 2t^2 + 5t^3 + 14t^4 + \cdots$ while $\frac{B-1}{t} = 2 + 6t + 20t^2 + 70t^3 + 252t^4 + \cdots$ and

$$\begin{bmatrix} 2 & 0 & 0 & 0 & 0 \\ 6 & 2 & 0 & 0 & 0 \\ 20 & 6 & 2 & 0 & 0 \\ 70 & 20 & 6 & 2 & 0 \\ 252 & 70 & 20 & 6 & 2 \end{bmatrix} \begin{bmatrix} 1 \\ 1 \\ 2 \\ 5 \\ 14 \end{bmatrix} = 2 \cdot \begin{bmatrix} 1 \\ 4 \\ 15 \\ 56 \\ 210 \end{bmatrix}.$$

Using FTRA (1.1.2) with the Riordan array $\left(\frac{B-1}{t}, t\right)$ we have

$$\frac{B-1}{t} \cdot C(t) = 2BC^2$$

as above.

We finish this section with a second application. Two contestants A and B are in an election where the results are coming in one vote at a time. The contestants finish in a tie and during the counting A was never behind B. The possible vote sequences can be viewed as a Dyck path with a vote for A representing an up step $U = (1, 1)$ and a vote for B a down step $D = (1, -1)$. The question we want to consider is how many times, on average, are A and B tied during the contest if there are $2n$ voters?

Dyck paths are counted by the Catalan numbers and by shrinking the x-axis to a point we obtain ordered trees. Each tie corresponds to an edge at the root and if we classify ordered trees by root degree we have the Riordan array

$$(1, tC) = \begin{bmatrix} 1 & 0 & 0 & 0 & 0 & 0 & 0 & \dots \\ 0 & 1 & 0 & 0 & 0 & 0 & 0 & \dots \\ 0 & 1 & 1 & 0 & 0 & 0 & 0 & \dots \\ 0 & 2 & 2 & 1 & 0 & 0 & 0 & \dots \\ 0 & 5 & 5 & 3 & 1 & 0 & 0 & \dots \\ 0 & 14 & 14 & 9 & 4 & 1 & 0 & \dots \\ 0 & 42 & 42 & 28 & 14 & 5 & 1 & \dots \\ \vdots & \vdots & \vdots & \vdots & \vdots & \vdots & \vdots & \ddots \end{bmatrix}.$$

To get the expected number of returns, i.e., root degree, we multiply by the column vector $(0, 1, 2, 3, 4, \cdots)^T$ with generating function $\frac{t}{(1-t)^2}$. By the FTRA (1.1.2) with the Riordan array $(1, tC)$, we have

$$1 \cdot \frac{tC}{(1-tC)^2} = tC^3.$$

Since $[t^n]\, C^s = \frac{s}{2n+s}\binom{2n+s}{n}$ we have

$$\left[t^n\right] tC^3 = \left[t^{n-1}\right] C^3 = \frac{3}{2\,(n-1)+3}\binom{2\,(n-1)+3}{n-1} =$$

$$= \frac{3}{2n+1}\binom{2n+1}{n-1} = \frac{3n}{n+2} \cdot \frac{1}{n+1}\binom{2n}{n} = \frac{3n}{n+2} C_n.$$

For instance, if $n = 3$ there are $C_3 = 5$ paths and the $\frac{3 \cdot 3}{3+2} \cdot 5 = 9$ returns are marked by "." for the following paths

$$UD.UD.UD. \quad UUDD.UD. \quad UD.UUDD.$$

$$UUDUDD. \quad UUUDDD.$$

Thus, the average over all C_n paths is $\frac{3n}{n+2}$ approaching 3 in the limit as n gets large.

1.3 Elementary Applications

Generating functions are the central objects of study of the theory in this book. An infinite sequence $(a_0, a_1, a_2, \ldots)$ can conveniently be represented as a power series in a variable z,

$$A(z) = a_0 + a_1 z + a_2 z^2 + \cdots = \sum_{n\geq 0} a_n z^n.$$

For brevity, we shorten $A(z)$ to A, and the coefficient extraction operator $[z^n]$ is defined by $[z^n]A(z) = a_n$. The indeterminate z can be replaced by t or x, and so on. A generating function is useful because it is a single quantity that represents an entire infinite sequence. Generating functions and the extraction of their coefficient will be illustrated in depth in Chap. 2.

There are many counting problems in combinatorics whose solution is given by the Catalan numbers with generating function C that satisfies $C = 1 + zC^2$. More generally, there is a family of power series that satisfy the functional equation:

$$\mathfrak{B}_r = 1 + z\mathfrak{B}_r^r.$$

Solving the above equation, we obtain the *generalized binomial series* [3]:

$$\mathfrak{B}_r = \sum_{n\geq 0} \frac{1}{(r-1)n+1}\binom{rn}{n} z^n,$$

which is known as the generating function for r*-ary numbers*. In the special case, $\mathfrak{B}_2$ and $\mathfrak{B}_3$ are the generating functions of the Catalan numbers $C_n = \frac{1}{n+1}\binom{2n}{n}$ and the ternary numbers $T_n = \frac{1}{2n+1}\binom{3n}{n}$,, respectively. It is also known [3] that the following identity is valid for all real numbers k:

$$\mathfrak{B}_r^k = \sum_{n\geq 0} \frac{k}{rn+k}\binom{rn+k}{n} z^n. \tag{1.3.1}$$

This identity will be useful for finding a closed form for the entries of Riordan arrays in connection with r-ary numbers.

We now look at some elementary applications of the Riordan group (see [1, 2, 4]). We start with ordered trees and can ask various questions. As already observed, an ordered tree, or plane tree, is a rooted tree in which the children of each vertex are ordered. It is well known that the ordered trees with n edges are counted by the nth Catalan numbers C_n. We draw ordered trees going up and we sketch these trees for 3 edges, see Fig. 1.1.

Let us count vertices of ordered trees by height. Then we get the following matrix whose nth row gives the number of edges and kth column gives the height, where $n, k \geq 0$ and the root is a vertex of height 0. A few rows of the matrix can be displayed by

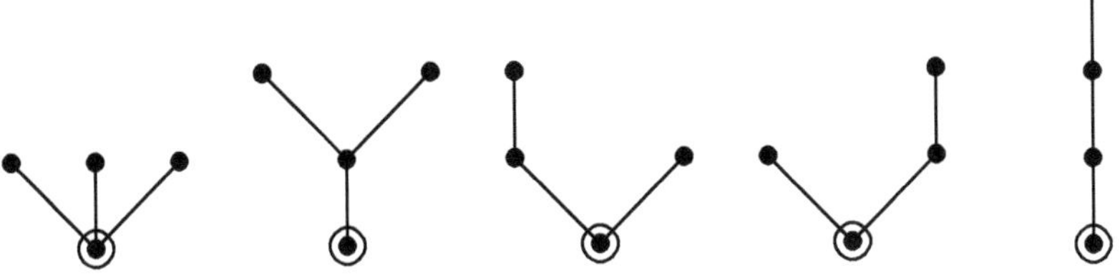

Fig. 1.1 Ordered trees with 3 edges

$$\mathbb{V} = \begin{bmatrix} 1 & 0 & 0 & 0 & 0 & 0 & 0 & 0 \\ 1 & 1 & 0 & 0 & 0 & 0 & 0 & 0 \\ 2 & 3 & 1 & 0 & 0 & 0 & 0 & 0 \\ 5 & 9 & 5 & 1 & 0 & 0 & 0 & 0 \\ 14 & 28 & 20 & 7 & 1 & 0 & 0 & 0 \\ 42 & 90 & 75 & 35 & 9 & 1 & 0 & 0 \\ 132 & 297 & 275 & 154 & 54 & 11 & 1 & 0 \\ 429 & 1001 & 1001 & 637 & 273 & 77 & 13 & 1 \end{bmatrix}. \quad (1.3.2)$$

Similarly, we can classify leaves of ordered trees by height where *leaf* is a vertex with no children, and we get a similar matrix:

$$\mathbb{L} = \begin{bmatrix} 1 & 0 & 0 & 0 & 0 & 0 & 0 & 0 \\ 0 & 1 & 0 & 0 & 0 & 0 & 0 & 0 \\ 0 & 2 & 1 & 0 & 0 & 0 & 0 & 0 \\ 0 & 5 & 4 & 1 & 0 & 0 & 0 & 0 \\ 0 & 14 & 14 & 6 & 1 & 0 & 0 & 0 \\ 0 & 42 & 48 & 27 & 8 & 1 & 0 & 0 \\ 0 & 132 & 165 & 110 & 44 & 10 & 1 & 0 \\ 0 & 429 & 572 & 429 & 208 & 65 & 12 & 1 \end{bmatrix}. \quad (1.3.3)$$

We can obtain similar results for many other types of *UUR trees* by generalizing an ordered tree to a tree with the same updegree possibilities at each vertex. This is called the *Uniform Updegree Requirement (UUR)*. The generating function for a class of the trees is

$$T = \sum_{n\geq 0} t_n z^n,$$

where t_n is the number of such trees with n edges. We essentially are assigning a z to each edge. The generating function V counts these same trees but with a marked vertex. Similarly L is the generating function for these trees but with a marked leaf. Then L_1 is the generating function for trees with a marked leaf at height 1. Also L_k counts trees with a marked leaf at height k.

The *A-sequence* of a family of trees is $(a_0, a_1, a_2, \ldots)$ where a_k is the number or weight for updegree k. Thus, the A-sequence gives the number of possibilities for the updegree of a vertex. Often it is just $a_k = 0$ meaning that no vertex has updegree k. Dually $a_k = 1$ means that a vertex of updegree k is permitted. The generating function for the A-sequence is called the *updegree function* and it is denoted by $A(z)$ or A. This is the same A-sequence of a proper Riordan array in Sect. 4.1, but in the context of ordered trees there is this combinatorial meaning. For instance, a *Motzkin tree* is one where each vertex can have updegree 0, 1, or 2. Thus, it has the A-sequence $(1, 1, 1, 0, 0, \ldots)$ and A-function $A(z) = A = 1 + z + z^2$. *Complete binary trees* have the A-sequence $(1, 0, 1, 0, 0, \ldots)$ and $A = 1 + z^2$. However, *incomplete binary trees* can have a right or a left single edge so the A-function is $1 + 2z + z^2$. Here is a short list for the A-sequences of trees with uniform conditions on updegrees.

Type of UUR tree	A-sequence
ordered tree	$(1, 1, 1, 1, 1, 1, \ldots)$
Motzkin tree	$(1, 1, 1, 0, 0, 0, \ldots)$
incomplete binary tree	$(1, 2, 1, 0, 0, 0, \ldots)$
complete binary tree	$(1, 0, 1, 0, 0, 0, \ldots)$
even tree	$(1, 0, 1, 0, 1, 0, \ldots)$
incomplete ternary tree	$(1, 3, 3, 1, 0, 0, \ldots)$
complete ternary tree	$(1, 0, 0, 1, 0, 0, \ldots)$
Schröder tree	$(1, 2, 2, 2, 2, 2, \ldots)$
spoiled child tree	$(1, 2, 1, 1, 1, 1, \ldots)$
Hex tree	$(1, 3, 1, 0, 0, 0, \ldots)$
Gamma tree	$(1, 0, 1, 1, 1, 1, \ldots)$
simple tree (path)	$(1, 1, 0, 0, 0, 0, \ldots)$

We have the following result:

Theorem 1.1 *There are five basic equations for counting UUR trees:*

$$\begin{aligned}
&\text{(a)}\quad V = TL;\\
&\text{(b)}\quad V = (zT)';\\
&\text{(c)}\quad L = \frac{1}{1 - L_1};\\
&\text{(d)}\quad L_1 = zT'/V;\\
&\text{(e)}\quad L_1 = zA'(zT).
\end{aligned}$$

Proof (a) Consider a tree with a marked vertex v. Then cut the tree at v to produce two smaller trees. The lower subtree is a tree with a marked leaf where v was. The upper tree was sitting atop v. This translates to $V = TL$.
(b) Since a tree with n edges has $n + 1$ vertices we have

$$V = \sum_{n\geq 0} (n + 1)\, t_n z^n = \left(\sum_{n\geq 0} t_n z^{n+1}\right)' = (zT)'.$$

(c) Since every leaf must be at some height, it is obvious that $L = 1 + L_1 + L_2 + \cdots$. But $L_2 = L_1^2$ and indeed $L_k = L_1^k$. Thus $L = \frac{1}{1-L_1}$.
(d) Inverting $L = 1/(1 - L_1)$ gives

$$L_1 = 1 - \frac{1}{L} = 1 - \frac{1}{V/T} = \frac{V - T}{V} = \frac{(zT)' - T}{V} = \frac{zT'}{V}.$$

(e) We leave the proof as an exercise. See Problem 1.1. □

If the number of vertices (leaves, resp.) at height k over UUR trees with n edges is denoted by $v_{n,k}$ ($\ell_{n,k}$, resp.), then the corresponding matrices can be expressed as Riordan matrices as follows:

$$\mathbb{V} = [v_{n,k}]_{n,k\geq 0} = (T, L_1), \text{ and } \mathbb{L} = [\ell_{n,k}]_{n,k\geq 0} = (1, L_1), \tag{1.3.4}$$

where $L_1 = zT'/V$. The matrix version of $V = TL$ is also given by the following relation:

$$\mathbb{V} = \mathbb{T}\mathbb{L}, \quad \mathbb{T} = (T, z).$$

It turns out that many statistics related to UUR trees can be expressed in terms of Riordan arrays.

Example 1.1 *(Ordered trees)*
For ordered trees we have $T = C$, the Catalan number generating function. Then

$$\begin{aligned} V &= \sum_{n\geq 0} (n+1)\, C_n z^n = \sum_{n\geq 0} (n+1) \frac{1}{n+1} \binom{2n}{n} z^n \\ &= \sum_{n\geq 0} \binom{2n}{n} z^n = B = \frac{1}{\sqrt{1-4z}}. \end{aligned}$$

By Theorem 1.1 (a), we have $L = \frac{B}{C} = \frac{B+1}{2}$. Moreover, counting vertices and leaves by height, respectively, we obtain the matrices $\mathbb{V}$ in (1.3.2) and $\mathbb{L}$ in (1.3.3). By (1.3.4) both matrices can be expressed by Riordan arrays as follows:

$$\mathbb{V} = \left(C, zC^2\right) \text{ and } \mathbb{L} = \left(1, zC^2\right).$$

These two are connected by the equation $\mathbb{T}\mathbb{L} = \mathbb{V}$ where

$$\mathbb{T}=\begin{bmatrix} 1 & 0 & 0 & 0 & 0 & 0 & 0 & 0 & \dots \\ 1 & 1 & 0 & 0 & 0 & 0 & 0 & 0 & \dots \\ 2 & 1 & 1 & 0 & 0 & 0 & 0 & 0 & \dots \\ 5 & 2 & 1 & 1 & 0 & 0 & 0 & 0 & \dots \\ 14 & 5 & 2 & 1 & 1 & 0 & 0 & 0 & \dots \\ 42 & 14 & 5 & 2 & 1 & 1 & 0 & 0 & \dots \\ 132 & 42 & 14 & 5 & 2 & 1 & 1 & 0 & \dots \\ 429 & 132 & 42 & 14 & 5 & 2 & 1 & 1 & \dots \\ \vdots & \vdots & \vdots & \vdots & \vdots & \vdots & \vdots & \vdots & \ddots \end{bmatrix} = (C, z)\,.$$

We can now prove several interesting results. The first is that for nontrivial ordered trees half of the vertices are leaves. The companion equation which follows from $B = 1 + 2zCB$ is $\frac{B-1}{2} = zBC = L - 1$.

Next we want to compute the total height of all the leaves of all the trees with n edges. A quick count of the trees with 2 or 3 edges yields a total height 4 for the two trees with 2 edges and 16 for the 5 trees with 3 edges. The key observation is that if we pick a leaf at height k then there are $k-1$ vertices between it and the root. Pick one of these $k-1$ vertices and cut the path at that point. The appropriate generating function is $L\,(L-1)$. But for ordered trees we then have

$$L\,(L-1) = \frac{B}{C}\cdot zBC = zB^2 = \frac{z}{1-4z} = \sum_{n\geq 1} 4^{n-1} z^n.$$

The total leaf height is 4^{n-1} for trees with n edges. This translates directly to all Dyck paths of $2n$ steps. The total height of all peaks is also 4^{n-1}.

We can obtain similar results for many other types of UUR trees. We will work through one example here and set a few of the others as exercises.

Example 1.2 *(Traffic light trees)* We consider ordered trees where the edges above a vertex can be either red or green. If both are present we require that all the green edges are to the left of all the red edges. Thus, a vertex of updegree k has $k+1$ possibilities for the edges above it ranging from all green to all red. Thus, the A-function is $A = \sum_{n\geq 0}(n+1)\,z^n = 1/\,(1-z)^2$. Differentiating gives

$$A'\,(z) = 2\,(1-z)^{-3}\,.$$

From $A(z) = 1/(1-z)^2$ it follows that $\overline{f} = z\,(1-z)^2$ (see Theorem 4.3) and hence $z = f\,(1-f)^2$. Looking at the first few values of L_1 which are $0, 2, 6, 24, 110, \cdots$, suggests that $L_1 = 2g - 2$ where

$$g = 1 + zg^3 = \frac{1}{1-zg^2} = \sum_{n\geq 0}\frac{1}{3n+1}\binom{3n+1}{n} z^n,$$

which is the generating function for the ternary numbers. It turns out that

$$T = g^2 = \sum_{n\geq 0} \frac{2}{3n+2}\binom{3n+2}{n} z^n.$$

Thus

$$V = (zT)' = \sum_{n\geq 0} \binom{3n+1}{n} z^n.$$

We can express L as either V/T or as $\frac{1}{1-L_1}$ where

$$\begin{aligned} L_1 &= zA'(zT) = z \cdot \frac{2}{(1-zT)^3} = \frac{2z}{\left(1-zg^2\right)^3} = 2zg^3 = 2(g-1) \\ &= 2z + 6z^2 + 24z^3 + 110z^4 + \cdots. \end{aligned}$$

For the Traffic Light trees, we thus have

$$\mathbb{V} = \begin{bmatrix} 1 & 0 & 0 & 0 & 0 & \dots \\ 2 & 2 & 0 & 0 & 0 & \dots \\ 7 & 10 & 4 & 0 & 0 & \dots \\ 30 & 50 & 32 & 8 & 0 & \dots \\ 143 & 260 & 208 & 88 & 16 & \dots \\ \vdots & \vdots & \vdots & \vdots & \vdots & \ddots \end{bmatrix} = \left(g^2, 2(g-1)\right),$$

$$\mathbb{L} = \begin{bmatrix} 1 & 0 & 0 & 0 & 0 & \dots \\ 0 & 2 & 0 & 0 & 0 & \dots \\ 0 & 6 & 4 & 0 & 0 & \dots \\ 0 & 24 & 24 & 8 & 0 & \dots \\ 0 & 110 & 132 & 72 & 16 & \dots \\ \vdots & \vdots & \vdots & \vdots & \vdots & \ddots \end{bmatrix} = (1, 2(g-1)).$$

We now go back to ordered trees and ask what is the average number of leaves at height 1. Since $L_1 = zC^2$ we have

$$\frac{[z^n]\,zC^2}{[z^n]\,C} = \frac{\frac{1}{n+1}\binom{2n}{n}}{\frac{1}{n+1}\binom{2n}{n}} = 1 \text{ for } n \geq 1 \text{ since } C = 1 + zC^2.$$

Thus, for all trees with n edges there is on average one leaf of height 1 per tree. In terms of Dyck paths the number of hills, subsequences UD starting and ending on the x-axis, is, on average, 1.

By way of contrast we have Motzkin trees where every vertex can have updegree 0, 1, or 2. The generating function counting Motzkin trees is denoted $m(z)$ or m. The defining equation for this generating function is

$$m = 1 + zm + z^2m^2.$$

The three terms correspond to the root having degree 0, 1, or 2. Thus

$$m = \frac{1 - z - \sqrt{1 - 2z - 3z^2}}{2z^2} = 1 + z + 2z^2 + 4z^3 + 9z^4 + \cdots = \sum_{n \geq 0} m_n z^n.$$

Note that $2z^2m = 1 - z - \sqrt{(1+z)(1-3z)}$, so the radius of convergence is $1/3$ where $1 - 3z = 0$ occurs. The ratio test then says that $\lim_{n\to\infty} \frac{m_{n+1}}{m_n} = 3$. Since the A-sequence is $(1, 1, 1, 0, 0, \ldots)$, we have $A(z) = 1 + z + z^2$ and $A'(z) = 1 + 2z$. With the L_1 equation, we get

$$L_1 = zA'(zm) = z(1 + 2zm).$$

Since $[z^n]z(1 + 2zm) = 2m_{n-2}$ for $n \geq 2$ it follows that

$$\frac{2m_{n-2}}{m_n} = 2 \cdot \frac{m_{n-2}}{m_{n-1}} \cdot \frac{m_{n-1}}{m_{n-2}} \to 2 \cdot \left(\frac{1}{3}\right)^2 = \frac{2}{9}.$$

To check this, we see that

$$\frac{2m_{10}}{m_{12}} = \frac{2 \cdot 2188}{15511} \simeq 0.282\,12,$$
$$\frac{2m_{20}}{m_{22}} = 2 \cdot \frac{50\,852\,019}{400\,763\,223} \simeq 0.253\,78$$

showing the convergence, albeit slowly, to $2/9 \simeq 0.222\,22$.

Exercises

1.1 Use identity (1.3.1) to find closed forms for (n, k)-entries of the following Riordan matrices:

(1) $\mathbb{V} = \left(C, zC^2\right)$ and $\mathbb{L} = \left(1, zC^2\right)$ in Example 1.1, where C is the generating function for Catalan numbers.

(2) $\mathbb{V} = \left(g^2, 2(g-1)\right)$ and $\mathbb{L} = (1, 2(g-1))$ in Example 1.2, where g is the generating function for ternary numbers.

1.2 Prove that $L_1 = zA'(zT)$. See (e) of Theorem 1.1.

1.3 Work out the equations for Traffic Light trees where there is at least one green edge for any vertex with positive updegree.

1.4 Show for the traffic light trees that

$$\frac{[z^n]\,L_1}{[z^n]\,T} = \frac{[z^n]\,2\,(g-1)}{[z^n]\,g^2} = \frac{2\cdot\frac{1}{3n+1}\binom{3n+1}{n}}{\frac{2}{3n+2}\binom{3n+2}{2}} = \frac{2n+2}{3n+1} \to \frac{2}{3}.$$

References

1. G.-S. Cheon, L.W. Shapiro, Protected points in ordered trees. Appl. Math. Lett. **21**(5), 516–520 (2008)
2. G.-S. Cheon, H. Kim, L.W. Shapiro, Profiles of ordered trees with mutation and associated Riordan matrices. Linear Algebr. Appl. **511**, 296–317 (2016)
3. R. Graham, D. Knuth, O. Patashnik, *Concrete Mathematics*, 2nd edn. (Addison-Wesley, 1994)
4. H. Kim, L.W. Shapiro, Pick two points in a tree. J. Korean Math. Soc. **56**, 1247–1263 (2019)
5. L.W. Shapiro, A survey of the Riordan Group. Talk at a meeting of the American Mathematical Society, Richmond, Virginia (1994)
6. L.W. Shapiro, S. Getu, W.-J. Woan, L. Woodson, The Riordan group. Discret. Appl. Math. **34**, 229–239 (1991)
7. R. Sprugnoli, Riordan arrays and combinatorial sums. Discret. Math. **132**(1–3), 267–290 (1994)
8. R. Sprugnoli, Riordan arrays and the Abel-Gould identity. Discret. Math. **142**(1–3), 213–233 (1995)
9. R. Stanley, Catalan addendum to Enumerative Combinatorics, vol. 2, http://www-math.mit.edu/~rstan/ec/catadd.pdf (1998)
10. R. Stanley, *Catalan Numbers* (Cambridge University Press, 2015)

Chapter 2
Extraction of Coefficients and Generating Functions

Abstract Generating functions have emerged as one of the most popular approaches to combinatorial problems, above all to problems arising in the analysis of algorithms (see, for example, D. E. Knuth [8] and R. Sedgewick and Ph. Flajolet [11] and Ph. Flajolet and R. Sedgewick [3]). A clear exposition of this concept is given in three books, namely, those of I. P. Goulden and D. M. Jackson [5], R. Stanley [13], and H. S. Wilf [14]; further discussion of this topic can be found in L. Comtet [1] and R.L. Graham, D.E. Knuth, and O. Patashnik [6]. Greene and Knuth [7, p. 7] show that combinatorial sums can be found in closed form by means of certain transformations on generating functions and the extraction of coefficients, attributing this elegant technique, the "method of coefficients", to G. P. Egorychev [2]. The method has been described in D. Merlini, R. Sprugnoli, and M. C. Verri [10]. The present chapter is devoted to formal power series and generating functions. For example, we will prove that the series $1 + t + t^2 + t^3 + \cdots$ can be conveniently abbreviated as $1/(1 - t)$, and from this fact we will be able to infer that the series has a formal power series inverse, which is $1 - t + 0t^2 + 0t^3 + \cdots$, or we will prove that the coefficient of t^n in the series expansion of $t/(1 - t - t^2)$ is the nth Fibonacci number F_n satisfying $F_n = F_{n-1} + F_n$, $F_0 = 0$, $F_1 = 1$ while the coefficient of t^n in $(1 - \sqrt{1 - 4t})/(2t)$ is the nth Catalan number $\binom{2n}{n}/(n + 1)$.

2.1 Formal Power Series

Let $\mathbb{R}$ be the field of real numbers and let t be any indeterminate over $\mathbb{R}$, i.e., a symbol different from any element in $\mathbb{R}$. A *formal power series* (f.p.s.) over $\mathbb{R}$ in the indeterminate t is an expression:

$$f(t) = f_0 + f_1 t + f_2 t^2 + f_3 t^3 + \cdots + f_n t^n + \cdots = \sum_{k=0}^{\infty} f_k t^k,$$

where $f_0, f_1, f_2, \ldots$ are all real numbers. The same definition applies to every set of numbers, in particular, to the field of rational numbers $\mathbb{Q}$ and to the field of complex numbers $\mathbb{C}$. The developments we are now going to see, and depending

© The Author(s), under exclusive license to Springer Nature Switzerland AG 2022

L. Shapiro et al., *The Riordan Group and Applications*, Springer Monographs in Mathematics, https://doi.org/10.1007/978-3-030-94151-2_2

on the field structure of the numeric set, can be easily extended to every field $\mathbb{F}$ of characteristic 0. The set of formal power series over $\mathbb{F}$ in the indeterminate t is denoted by $\mathbb{F}[[t]]$. The use of a particular indeterminate t is irrelevant, and there exists an obvious 1-1 correspondence between, say, $\mathbb{F}[[t]]$ and $\mathbb{F}[[y]]$; it is simple to prove that this correspondence is indeed an isomorphism. In order to stress that our results are substantially independent of the particular field $\mathbb{F}$ and of the particular indeterminate t, we denote $\mathbb{F}[[t]]$ by $\mathcal{F}$, but the reader can think of $\mathcal{F}$ as $\mathbb{R}[[t]]$. In fact, in combinatorial analysis and in the analysis of algorithms, the coefficients $f_0, f_1, f_2, \dots$ of a formal power series are mostly used to count objects, and therefore they are positive integer numbers, or, in some cases, positive rational numbers (e.g., when they are the coefficients of an exponential generating function).

If $f(t) \in \mathcal{F}$, the *order* of $f(t)$, denoted by $\operatorname{ord}(f(t))$, is the smallest index r for which $f_r \neq 0$. The set of all f.p.s. of order exactly r is denoted by $\mathcal{F}_r$ or by $\mathbb{F}_r[[t]]$. The formal power series $0 = 0 + 0t + 0t^2 + 0t^3 + \cdots$ has infinite order.

If $(f_0, f_1, f_2, \dots) = (f_k)_{k\in\mathbb{N}}$ is a sequence of (real) numbers, there is no substantial difference between the sequence and the f.p.s. $\sum_{k=0}^{\infty} f_k t^k$, which will be called the *(ordinary) generating function* of the sequence. The term *ordinary* is used to distinguish these functions from *exponential generating functions*, which will be introduced in a later chapter. The indeterminate t is used as a "place-marker", i.e., a symbol to denote the place of the element in the sequence. For example, in the f.p.s. $1 + t + t^2 + t^3 + \cdots$, corresponding to the sequence $(1, 1, 1, \dots)$, the term $t^5 = 1 \cdot t^5$ simply denotes that the element in position 5 (starting from 0) in the sequence is the number 1.

There are two main reasons why f.p.s. are more easily studied than sequences:

1. the algebraic structure of f.p.s. is very well understood and can be developed in a standard way;
2. many f.p.s. can be "abbreviated" by expressions easily manipulated by elementary algebra.

We conclude this section by defining the concept of a *formal Laurent (power) series* (f.L.s.), as an expression:

$$g(t) = g_{-m}t^{-m} + g_{-m+1}t^{-m+1} + \cdots + g_{-1}t^{-1} + g_0 + g_1 t + g_2 t^2 + \cdots = \sum_{k=-m}^{\infty} g_k t^k$$

with $m \in \mathbb{N}$. The set of f.L.s., which we denote by $\mathcal{L}$, strictly contains the set of f.p.s. For a f.L.s. $g(t)$ the order can be negative; when the order of $g(t)$ is non-negative, then $g(t)$ is actually a f.p.s. We observe explicitly that an expression such as $\sum_{k=-\infty}^{\infty} f_k t^k$ does *not* represent a f.L.s.

2.2 Coefficient Extraction

If $f(t) \in \mathcal{L}$, or in particular $f(t) \in \mathcal{F}$, the notation $[t^n]f(t)$ indicates the *extraction of the coefficient of* t^n from $f(t)$, and therefore we have $[t^n]f(t) = f_n$. In this sense, $[t^n]$ can be seen as a mapping: $[t^n] : \mathcal{L} \to \mathbb{R}$ or $[t^n] : \mathcal{L} \to \mathbb{C}$, depending on the underlying field of the set $\mathcal{L}$ or $\mathcal{F}$. Because of that, $[t^n]$ is called an *operator* and is in fact exactly the *"coefficient of" operator* or, more simply, the *coefficient operator*. We can state formally the main properties of this operator:

$$
\begin{array}{lr}
(K1)\ \text{(linearity)} & [t^n](\alpha f(t) + \beta g(t)) = \alpha[t^n]f(t) + \beta[t^n]g(t) \\
(K2)\ \text{(shifting)} & [t^n]tf(t) = [t^{n-1}]f(t) \\
(K3)\ \text{(differentiation)} & [t^n]f'(t) = (n+1)[t^{n+1}]f(t) \\
(K4)\ \text{(convolution)} & [t^n]f(t)g(t) = \sum_{k=0}^{n}[t^k]f(t)[t^{n-k}]g(t) \\
(K5)\ \text{(composition)} & [t^n]f(g(t)) = \sum_{k=0}^{\infty}([y^k]f(y))[t^n]g(t)^k \\
(K6)\ \text{(reversion)} & [t^n]\overline{f}(t) = \frac{1}{n}[t^{n-1}]\left(\frac{t}{f(t)}\right)^n.
\end{array}
$$

We observe that $\alpha, \beta \in \mathbb{R}$ or $\alpha, \beta \in \mathbb{C}$ are any constants; the use of the indeterminate y is only necessary to distinguish the action on different f.p.s. In rule $(K5)$, we require that $g(0) = 0$ in the composition, and so the sum in $(K5)$ is actually finite. In $(K6)$, the function $\overline{f}(t)$ is the *compositional inverse* of $f(t)$, such that $f(\overline{f}(t)) = \overline{f}(f(t)) = t$.

Some points require more lengthy comments. The property of shifting can be easily generalized to $[t^n]t^k f(t) = [t^{n-k}]f(t)$ and also to negative powers: $[t^n]f(t)/t^k = [t^{n+k}]f(t)$. These rules are very important and are often applied in the theory of f.p.s. and f.L.s. The property of differentiation for $n = -1$ gives $[t^{-1}]f'(t) = 0$; the operator $[t^{-1}]$ is also called the *residue* and is noted as "res"; so, for example, people write $\operatorname{res} f'(t) = 0$ and some authors use the notation $\operatorname{res} t^{-n-1}f(t)$ for $[t^n]f(t)$. The property of reversion corresponds to the Lagrange inversion formula and will be examined in detail in Sect. 2.3.

We shall have many occasions to apply the rules $(K1) - (K6)$ of coefficient extraction. However, just to give a meaningful example, let us find the coefficient of t^n in the series expansion of $(1 + \alpha t)^r$, when α and r are any two real numbers whatsoever. Rule $(K3)$ can be written in the form $[t^n]f(t) = \frac{1}{n}[t^{n-1}]f'(t)$ and we successively apply this to our case:

$$
[t^n](1+\alpha t)^r = \frac{r\alpha}{n}[t^{n-1}](1+\alpha t)^{r-1} = \frac{r\alpha}{n}\frac{(r-1)\alpha}{n-1}[t^{n-2}](1+\alpha t)^{r-2} = \cdots =
$$

$$
= \frac{r\alpha}{n}\frac{(r-1)\alpha}{n-1}\cdots\frac{(r-n+1)\alpha}{1}[t^0](1+\alpha t)^{r-n} = \alpha^n\binom{r}{n}[t^0](1+\alpha t)^{r-n}.
$$

We now observe that $[t^0](1+\alpha t)^{r-n} = 1$ because of our observations on f.p.s. operations. Therefore, we conclude with the so-called *Newton's rule*:

$$[t^n](1+\alpha t)^r = \binom{r}{n}\alpha^n$$

which is one of the most frequently used results in coefficient extraction. Let us remark explicitly that when $r = -1$, we have

$$[t^n]\frac{1}{1+\alpha t} = \binom{-1}{n}\alpha^n = \binom{1+n-1}{n}(-1)^n\alpha^n = (-\alpha)^n.$$

A simple, but important use of Newton's rule concerns the extraction of the coefficient of t^n from the inverse of a trinomial at^2+bt+c, in the case it is reducible, i.e., it can be written $(1+\alpha t)(1+\beta t)$; obviously, we can always reduce the constant c to 1; by the linearity rule, it can be taken outside the "coefficient of" operator. Therefore, our aim is to compute

$$[t^n]\frac{1}{(1+\alpha t)(1+\beta t)}$$

with $\alpha \neq \beta$, otherwise Newton's rule would be immediately applicable. The problem can be solved by using the technique of *partial fraction expansion*. We look for two constants A and B such that

$$\frac{1}{(1+\alpha t)(1+\beta t)} = \frac{A}{1+\alpha t} + \frac{B}{1+\beta t} = \frac{A + A\beta t + B + B\alpha t}{(1+\alpha t)(1+\beta t)};$$

if two such constants exist, the numerator in the first expression should equal the numerator in the last one, independently of t, or, if one so prefers, for every value of t. Therefore, the term $A+B$ should be equal to 1, while the term $(A\beta + B\alpha)t$ should always be 0. The values for A and B are therefore the solution of the linear system:

$$\begin{cases} A+B=1 \\ A\beta + B\alpha = 0 \end{cases}$$

The discriminant of this system is $\alpha - \beta$, which is always different from 0, because of our hypothesis $\alpha \neq \beta$. The system has therefore only one solution, which is $A = \alpha/(\alpha-\beta)$ and $B = -\beta/(\alpha-\beta)$. We can now substitute these values in the expression above:

$$[t^n]\frac{1}{(1+\alpha t)(1+\beta t)} = [t^n]\frac{1}{\alpha-\beta}\left(\frac{\alpha}{1+\alpha t} - \frac{\beta}{1+\beta t}\right) =$$

$$= \frac{1}{\alpha-\beta}\left([t^n]\frac{\alpha}{1+\alpha t} - [t^n]\frac{\beta}{1+\beta t}\right) = \frac{\alpha^{n+1}-\beta^{n+1}}{\alpha-\beta}(-1)^n.$$

Let us now consider a trinomial $1 + bt + ct^2$ for which $\Delta = b^2 - 4c < 0$ and $b \neq 0$. The trinomial is irreducible, but we can write

$$[t^n]\frac{1}{1+bt+ct^2} = [t^n]\frac{1}{\left(1 - \frac{-b+i\sqrt{|\Delta|}}{2}t\right)\left(1 - \frac{-b-i\sqrt{|\Delta|}}{2}t\right)}.$$

This time, a partial fraction expansion does not give a simple closed form for the coefficients; however, we can apply the formula above in the form:

$$[t^n]\frac{1}{(1-\alpha t)(1-\beta t)} = \frac{\alpha^{n+1}-\beta^{n+1}}{\alpha-\beta}.$$

Since α and β are complex numbers, the resulting expression is not very appealing. We can try to give it a better form. Let us set $\alpha = \left(-b + i\sqrt{|\Delta|}\right)/2$, so α is always contained in the positive imaginary half plane. This implies $0 < \arg(\alpha) < \pi$ and we have

$$\alpha - \beta = -\frac{b}{2} + i\frac{\sqrt{|\Delta|}}{2} + \frac{b}{2} + i\frac{\sqrt{|\Delta|}}{2} = i\sqrt{|\Delta|} = i\sqrt{4c-b^2}.$$

If $\theta = \arg(\alpha)$ and:

$$\rho = |\alpha| = \sqrt{\frac{b^2}{4} - \frac{4c-b^2}{4}} = \sqrt{c}$$

we can set $\alpha = \rho e^{i\theta}$ and $\beta = \rho e^{-i\theta}$. Consequently:

$$\alpha^{n+1} - \beta^{n+1} = \rho^{n+1}\left(e^{i(n+1)\theta} - e^{-i(n+1)\theta}\right) = 2i\rho^{n+1}\sin(n+1)\theta$$

and therefore:

$$[t^n]\frac{1}{1+bt+ct^2} = \frac{2(\sqrt{c})^{n+1}\sin(n+1)\theta}{\sqrt{4c-b^2}}.$$

At this point, we only have to find the value of θ. Obviously:

$$\theta = \arctan\left(\frac{\sqrt{|\Delta|}}{2}\Big/\frac{-b}{2}\right) + k\pi = \arctan\frac{\sqrt{4c-b^2}}{-b} + k\pi.$$

When $b < 0$, we have $0 < \arctan\left(-\sqrt{4c-b^2}\right)/2 < \pi/2$, and this is the correct value for θ. However, when $b > 0$, the principal branch of arctan is negative, and we should set $\theta = \pi + \arctan\left(-\sqrt{4c-b^2}\right)/2$. As a consequence, we have

$$\theta = \arctan\frac{\sqrt{4c-b^2}}{-b} + C,$$

where $C = \pi$ if $b > 0$ and $C = 0$ if $b < 0$.

An interesting and nontrivial example is given by

$$\sigma_n = [t^n]\frac{1}{1-3t+3t^2} = \frac{2(\sqrt{3})^{n+1}\sin((n+1)\arctan(\sqrt{3}/3))}{\sqrt{3}}$$
$$= 2(\sqrt{3})^n \sin\frac{(n+1)\pi}{6}.$$

2.3 Lagrange Inversion Theorem

The celebrated *Lagrange Inversion Formula* (LIF) states that if $\overline{f}(t)$ is the compositional inverse of $f(t)$, then we have

$$\overline{f}_n = [t^n]\overline{f}(t) = \frac{1}{n}[t^{n-1}]\left(\frac{t}{f(t)}\right)^n,$$

and, more generally:

$$[t^n]\overline{f}(t)^k = \frac{k}{n}[t^{n-k}]\left(\frac{t}{f(t)}\right)^n.$$

There are many proofs of Lagrange's inversion formula, both of algebraic and combinatorial types, in one or more variables (see, e.g., [2, 4, 9, 12, 13]). In Chap. 3, Theorem 3.3 and Corollary 3.1, we will see one based on the concept of Riordan arrays.

Many times, there is another way for applying the LIF. Suppose we have a functional equation $w = t\phi(w)$, where $\phi(t) \in \mathcal{F}_0$, and we wish to find the f.p.s. $w = w(t)$ satisfying this functional equation. Clearly $w(t) \in \mathcal{F}_1$ and if we set $f(y) = y/\phi(y)$, we also have $f(t) \in \mathcal{F}_1$. However, the functional equation can be written $f(w(t)) = t$, and this shows that $w(t)$ is the compositional inverse of $f(t)$. We therefore know that $w(t)$ is uniquely determined and the LIF gives us

$$[t^n]w(t) = \frac{1}{n}[t^{n-1}]\left(\frac{t}{f(t)}\right)^n = \frac{1}{n}[t^{n-1}]\phi(t)^n.$$

The LIF can also give us the coefficients of the powers $w(t)^k$, but we can obtain a still more general result. Let $F(t) \in \mathcal{F}$ and let us consider the composition $F(w(t))$ where $w = w(t)$ is, as before, the solution to the functional equation $w = t\phi(w)$, with $\phi(w) \in \mathcal{F}_0$. For the coefficient of t^n in $F(w(t))$, we have

$$[t^n]F(w(t)) = [t^n]\sum_{k=0}^{\infty} F_k w(t)^k = \sum_{k=0}^{\infty} F_k [t^n]w(t)^k = \sum_{k=0}^{\infty} F_k \frac{k}{n}[t^{n-k}]\phi(t)^n =$$

$$= \frac{1}{n}[t^{n-1}]\left(\sum_{k=0}^{\infty} kF_k t^{k-1}\right)\phi(t)^n = \frac{1}{n}[t^{n-1}]F'(t)\phi(t)^n.$$

We can obtain a different formulation of the LIF by applying to the previous result the differentiation rule $\frac{1}{n}[t^{n-1}]f'(t) = [t^n]f(t)$. We have

$$\frac{1}{n}[t^{n-1}]F'(t)\phi(t)^n = [t^n]\int F'(t)\phi(t)^n \delta t$$

and by using integration by parts we find

$$[t^n]\left(F(t)\phi(t)^n - n\int F(t)\phi(t)^{n-1}\phi'(t)\delta t\right) =$$

$$= [t^n]F(t)\phi(t)^n - \frac{n}{n}[t^{n-1}]F(t)\phi(t)^{n-1}\phi'(t),$$

hence

$$[t^n]F(w(t)) = [t^n]F(t)\phi(t)^{n-1}(\phi(t) - t\phi'(t)).$$

The formulas are equivalent but sometimes one is more convenient than the other. Note that $[t^0]F(w(t)) = F_0$, and this formula can be generalized to every $F(t) \in \mathcal{L}$.

As a meaningful example of an application of the LIF, let us consider the number b_n of binary trees with n nodes which, due to their recursive structure, satisfy the recurrence relation: $b_{n+1} = \sum_{k=0}^{n} b_k b_{n-k}$. Let us consider the f.p.s. $b(t) = \sum_{k=0}^{\infty} b_k t^k$; if we multiply the recurrence relation by t^{n+1} and sum for n from 0 to infinity, we find

$$\sum_{n=0}^{\infty} b_{n+1}t^{n+1} = \sum_{n=0}^{\infty} t^{n+1}\left(\sum_{k=0}^{n} b_k b_{n-k}\right).$$

Since $b_0 = 1$, we can add and subtract $1 = b_0 t^0$ from the left-hand member and can take t outside the summation sign in the right-hand member:

$$\sum_{n=0}^{\infty} b_n t^n - 1 = t\sum_{n=0}^{\infty}\left(\sum_{k=0}^{n} b_k b_{n-k}\right)t^n.$$

In the r.h.s. we recognize a convolution and substituting $b(t)$ for the corresponding f.p.s., we obtain

$$b(t) - 1 = tb(t)^2.$$

We are interested in evaluating $b_n = [t^n]b(t)$; let us therefore set $w = w(t) = b(t) - 1$, so that $w(t) \in \mathcal{F}_1$ and $w_n = b_n, \forall n > 0$. The previous relation becomes $w = t(1+w)^2$ and we see that the LIF can be applied (in the form relative to the functional equation) with $\phi(t) = (1+t)^2$. Therefore, we have

$$b_n = [t^n]w(t) = \frac{1}{n}[t^{n-1}](1+t)^{2n} = \frac{1}{n}\binom{2n}{n-1} =$$

$$= \frac{1}{n}\frac{(2n)!}{(n-1)!(n+1)!} = \frac{1}{n+1}\frac{(2n)!}{n!n!} = \frac{1}{n+1}\binom{2n}{n}.$$

As we said in the previous section, b_n is called the nth Catalan number and, under this name, it is often denoted by C_n. Now we have its form:

$$C_n = \frac{1}{n+1}\binom{2n}{n}$$

also valid in the case $n = 0$ when $C_0 = 1$.

In the same way, we can compute the number of *p-ary trees* with n nodes. A p-ary tree is a tree in which all the nodes have arity p, except for leaves, which have arity 0. A non-empty p-ary tree can be decomposed in the following way:

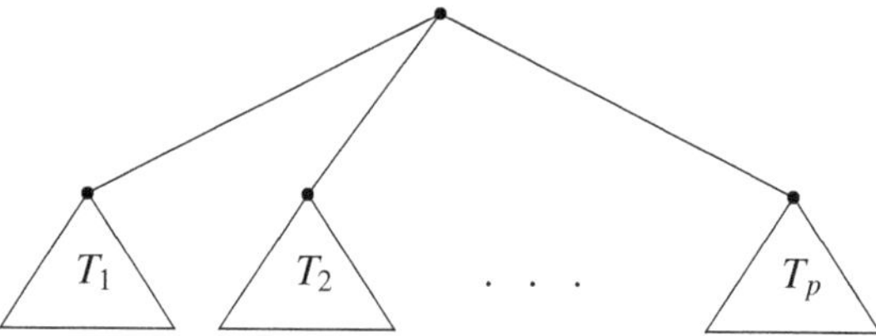

which proves that $T_{n+1} = \sum T_{i_1}T_{i_2}\cdots T_{i_p}$, where the sum is extended to all the p-uples $(i_1, i_2, \ldots, i_p)$ such that $i_1 + i_2 + \cdots + i_p = n$. As before, we can multiply by t^{n+1} the two members of the recurrence relation and sum for n from 0 to infinity. We find:

$$T(t) - 1 = tT(t)^p.$$

This time we have a p degree equation, which cannot be directly solved. However, if we set $w(t) = T(t) - 1$, so that $w(t) \in \mathcal{F}_1$, we have

$$w = t(1+w)^p$$

and the LIF gives

$$T_n = [t^n]w(t) = \frac{1}{n}[t^{n-1}](1+t)^{pn} = \frac{1}{n}\binom{pn}{n-1} =$$

$$= \frac{1}{n}\frac{(pn)!}{(n-1)!((p-1)n+1)!} = \frac{1}{(p-1)n+1}\binom{pn}{n}$$

which generalizes the formula for the Catalan numbers.

As another example, let us find the solution of the functional equation $w = te^w$. The LIF gives

$$w_n = \frac{1}{n}[t^{n-1}]e^{nt} = \frac{1}{n}\frac{n^{n-1}}{(n-1)!} = \frac{n^{n-1}}{n!}.$$

Therefore, the solution we are looking for is the f.p.s.:

$$w(t) = \sum_{n=1}^{\infty} \frac{n^{n-1}}{n!}t^n = t + t^2 + \frac{3}{2}t^3 + \frac{8}{3}t^4 + \frac{125}{24}t^5 + \cdots.$$

As noticed, $w(t)$ is the compositional inverse of

$$f(t) = t/\phi(t) = te^{-t} = t - t^2 + \frac{t^3}{2!} - \frac{t^4}{3!} + \frac{t^5}{4!} - \cdots.$$

It is a useful exercise to perform the necessary computations to show that $f(w(t)) = t$, for example, up to the term of degree 5 or 6, and verify that $w(f(t)) = t$ as well.

2.4 Generating Functions

Let us consider a sequence of numbers $F = (f_0, f_1, f_2, \ldots) = (f_k)_{k\in\mathbb{N}}$; the *generating function* for the sequence F is defined as the f.p.s. $f(t) = f_0 + f_1t + f_2t^2 + \cdots$, where the indeterminate t is arbitrary. Given the sequence $(f_k)_{k\in\mathbb{N}}$, we introduce the *generating function operator* $\mathcal{G}$, which applied to $(f_k)_{k\in\mathbb{N}}$ produces the generating function for the sequence, i.e., $\mathcal{G}(f_k)_{k\in\mathbb{N}} = f(t)$. A more accurate notation would be $\mathcal{G}_t(f_k)_{k\in\mathbb{N}} = f(t)$, this notation is essential when $(f_k)_{k\in\mathbb{N}}$ depends on some parameter or when we consider multivariate generating functions. In this latter case, for example, we should write $\mathcal{G}_{t,w}\left(f_{n,k}\right)_{n,k\in\mathbb{N}} = f(t,w)$ to indicate the fact that $f_{n,k}$ in the double sequence becomes the coefficient of t^nw^k in the function $f(t,w)$. However, whenever no ambiguity can arise, we will use the notation $\mathcal{G}(f_k) = f(t)$, leaving the binding for the variable t understood.

The operator $\mathcal{G}$ is clearly linear. The function $f(t)$ can be shifted or differentiated. Two functions $f(t)$ and $g(t)$ can be multiplied and composed. This leads to the following properties for the operator $\mathcal{G}$:

$$
\begin{array}{ll}
(G1)\ \text{(linearity)} & \mathcal{G}\left(\alpha f_k + \beta g_k\right) = \alpha \mathcal{G}(f_k) + \beta \mathcal{G}(g_k) \\
(G2)\ \text{(shifting)} & \mathcal{G}\left(f_{k+1}\right) = \dfrac{\mathcal{G}(f_k) - f_0}{t} \\
(G3)\ \text{(differentiation)} & \mathcal{G}\left(k f_k\right) = t D \mathcal{G}(f_k) \\
(G4)\ \text{(convolution)} & \mathcal{G}\left(\sum_{k=0}^{n} f_k g_{n-k}\right) = \mathcal{G}(f_k) \cdot \mathcal{G}(g_k) \\
(G5)\ \text{(composition)} & \sum_{n=0}^{\infty} f_n \left(\mathcal{G}(g_k)\right)^n = \mathcal{G}(f_k) \circ \mathcal{G}(g_k) \\
(G6)\ \text{(diagonalization)} & \mathcal{G}\left([t^n]F(t)\phi(t)^n\right) = \left[\dfrac{F(w)}{1 - t\phi'(w)} \;\middle|\; w = t\phi(w)\right].
\end{array}
$$

Note that formula $(G5)$ requires $g_0 = 0$. The first five formulas are easily verified by using the intended interpretation of the operator $\mathcal{G}$; the last formula can be proved by means of the LIF, in the form relative to the composition $F(w(t))$. In fact, we have

$$
[t^n]F(t)\phi(t)^n = [t^{n-1}]\frac{F(t)}{t}\phi(t)^n = n[t^n]\left[\int \frac{F(y)}{y} dy \;\middle|\; y = w(t)\right];
$$

in the last passage, we have applied backward the formula:

$$
[t^n]F(w(t)) = \frac{1}{n}[t^{n-1}]F'(t)\phi(t)^n \quad (w = t\phi(w))
$$

and therefore $w = w(t) \in \mathcal{F}_1$ is the unique solution of the functional equation $w = t\phi(w)$. Now applying the rule of differentiation for the “coefficient of” operator, we can go on:

$$
[t^n]F(t)\phi(t)^n = [t^{n-1}]\frac{d}{dt}\left[\int \frac{F(y)}{y} dy \;\middle|\; y = w(t)\right] =
$$

$$
= [t^{n-1}]\left[\frac{F(w)}{w} \;\middle|\; w = t\phi(w)\right]\frac{dw}{dt}.
$$

We have applied the chain rule for differentiation, and from $w = t\phi(w)$ we have

$$
\frac{dw}{dt} = \phi(w) + t\left[\frac{d\phi}{dw} \;\middle|\; w = t\phi(w)\right]\frac{dw}{dt}.
$$

We can therefore compute the derivative of $w(t)$:

$$
\frac{dw}{dt} = \left[\frac{\phi(w)}{1 - t\phi'(w)} \;\middle|\; w = t\phi(w)\right],
$$

where $\phi'(w)$ denotes the derivative of $\phi(w)$ with respect to w. We can substitute this expression in the formula above and observe that $w/\phi(w) = t$ can be taken outside of the substitution symbol:

$$[t^n]F(t)\phi(t)^n = [t^{n-1}]\left[\frac{F(w)}{w}\frac{\phi(w)}{1-t\phi'(w)} \;\middle|\; w = t\phi(w)\right] =$$

$$= [t^{n-1}]\frac{1}{t}\left[\frac{F(w)}{1-t\phi'(w)} \;\middle|\; w = t\phi(w)\right] = [t^n]\left[\frac{F(w)}{1-t\phi'(w)} \;\middle|\; w = t\phi(w)\right]$$

which is our diagonalization rule $(G6)$.

The name “diagonalization” is due to the fact that if we imagine the coefficients of $F(t)\phi(t)^n$ as constituting the row n in an infinite matrix, then $[t^n]F(t)\phi(t)^n$ are just the elements in the main diagonal of this array.

The rules $(G1) - (G6)$ can also be used to derive general theorems as well as specific functions for particular sequences. In the next subsections, we will prove a number of properties of the generating function operator. The proofs rely on the following fundamental *principle of identity*:

Given two sequences $(f_k)_{k\in\mathbb{N}}$ *and* $(g_k)_{k\in\mathbb{N}}$, *then* $\mathcal{G}(f_k) = \mathcal{G}(g_k)$ *if and only if for every* $k \in \mathbb{N}$ $f_k = g_k$.

The principle is rather obvious from the very definition of the concept of generating functions; however, it is important, because it states the condition under which we can pass from an identity about elements to the corresponding identity about generating functions. In fact, for two generating functions to be unequal it is sufficient that their coefficient sequences differ by a single element.

Some Theorems on Generating Functions

We are now going to prove a series of properties of generating functions; many others are given as exercises at the end of this section.

If we consider right instead of left shifting we have to be careful:

Theorem 2.1 *Let* $f(t) = \mathcal{G}(f_k)$ *be as above and let* $g_k = \begin{cases} 0, & k \le j-1 \\ f_{k-j}, & k \ge j \end{cases}$, *then*

$$\mathcal{G}(g_k) = t^j \mathcal{G}(f_k), \tag{2.4.1}$$

and we will write for simplicity $\mathcal{G}(g_k) = \mathcal{G}\left(f_{k-j}\right)$.

Proof We have $\mathcal{G}(f_k) = \mathcal{G}\left(f_{(k-1)+1}\right) = t^{-1}\left(\mathcal{G}(f_{k-1}) - f_{-1}\right)$ where f_{-1} denotes the coefficient of t^{-1} in $f(t)$. Since $f(t) \in \mathcal{F}$, $f_{-1} = 0$ and $\mathcal{G}(f_{k-1}) = t\mathcal{G}(f_k)$. The theorem then follows by mathematical induction. □

Theorem 2.2 *Let* $f(t) = \mathcal{G}(f_k)$ *be as above, then*

$$\mathcal{G}\left(k^j f_k\right) = S_j(tD)\mathcal{G}(f_k), \tag{2.4.2}$$

where $\mathcal{S}_j(w) = \sum_{r=1}^{j} \left\{ {j \atop r} \right\} w^r$ *is the* j*th Stirling polynomial of the second kind and* $\mathcal{S}_j(tD) = \sum_{r=1}^{j} \left\{ {j \atop r} \right\} t^r D^r$.

Proof Formula (2.4.2) is to be understood in the operator sense; so, for example, as $\mathcal{S}_3(w) = w + 3w^2 + w^3$, we have

$$\mathcal{G}\left(k^3 f_k\right) = tD\mathcal{G}\left(f_k\right) + 3t^2 D^2 \mathcal{G}\left(f_k\right) + t^3 D^3 \mathcal{G}\left(f_k\right).$$

The proof proceeds by induction, as $(G3)$ and

$$\mathcal{G}\left(k^2 f_k\right) = tD\mathcal{G}\left(f_k\right) + t^2 D^2 \mathcal{G}\left(f_k\right)$$

are the first two instances. Now:

$$\mathcal{G}\left(k^{j+1} f_k\right) = \mathcal{G}\left(k(k^j) f_k\right) = tD\mathcal{G}\left(k^j f_k\right)$$

that is:

$$\mathcal{S}_{j+1}(tD) = tD\mathcal{S}_j(tD) = tD\sum_{r=1}^{j} S(j,r) t^r D^r =$$

$$= \sum_{r=1}^{j} S(j,r) r t^r D^r + \sum_{r=1}^{j} S(j,r) t^{r+1} D^{r+1}.$$

By equating like coefficients we find $S(j+1,r) = rS(j,r) + S(j,r-1)$, which is the classical recurrence for the Stirling numbers of the second kind. Since initial conditions also coincide, we can conclude $S(j,r) = \left\{ {j \atop r} \right\}$. □

Let us now come to integration:

Theorem 2.3 *Let* $f(t) = \mathcal{G}\left(f_k\right)$ *be as above and let us define* $g_k = f_k/k, \forall k \neq 0$ *and* $g_0 = 0$*, then*

$$\mathcal{G}\left(\frac{1}{k} f_k\right) = \mathcal{G}\left(g_k\right) = \int_0^t \left(\mathcal{G}\left(f_k\right) - f_0\right) \frac{dz}{z}. \tag{2.4.3}$$

Proof Clearly, $kg_k = f_k$, except for $k = 0$. Hence we have $\mathcal{G}\left(kg_k\right) = \mathcal{G}\left(f_k\right) - f_0$. By using $(G3)$, we find $tD\mathcal{G}\left(g_k\right) = \mathcal{G}\left(f_k\right) - f_0$, from which (2.4.3) follows by integration and the condition $g_0 = 0$. □

Another classical formula is

Theorem 2.4 *Let* $f(t) = \mathcal{G}\left(f_k\right)$ *be as above or, equivalently, let* $f(t)$ *be a f.p.s. but not a f.L.s., then*

$$\mathcal{G}\left(\frac{1}{k+1}f_k\right) = \frac{1}{t}\int_0^t \mathcal{G}(f_k)\,dz = \frac{1}{t}\int_0^t f(z)\,dz. \tag{2.4.4}$$

Proof Let us consider the sequence $(g_k)_{k\in\mathbb{N}}$, where $g_{k+1} = f_k$ and $g_0 = 0$. So we have: $\mathcal{G}(g_{k+1}) = \mathcal{G}(f_k) = t^{-1}(\mathcal{G}(g_k) - g_0)$. Finally:

$$\mathcal{G}\left(\frac{1}{k+1}f_k\right) = \mathcal{G}\left(\frac{1}{k+1}g_{k+1}\right) = \frac{1}{t}\mathcal{G}\left(\frac{1}{k}g_k\right) =$$
$$= \frac{1}{t}\int_0^t (\mathcal{G}(g_k) - g_0)\frac{dz}{z} = \frac{1}{t}\int_0^t \mathcal{G}(f_k)\,dz.$$

□

More Advanced Results

The results obtained until now can be considered as simple generalizations of the axioms. They are very useful and will be used in many circumstances. However, we can also obtain more advanced results, concerning sequences derived from a given sequence by manipulating its elements in various ways. For example, let us begin by proving the well-known bisection formulas:

Theorem 2.5 *Let $f(t) = \mathcal{G}(f_k)$ denote the generating function of the sequence $(f_k)_{k\in\mathbb{N}}$; then:*

$$\mathcal{G}(f_{2k}) = \frac{f(\sqrt{t}) + f(-\sqrt{t})}{2}, \tag{2.4.5}$$

$$\mathcal{G}(f_{2k+1}) = \frac{f(\sqrt{t}) - f(-\sqrt{t})}{2\sqrt{t}}, \tag{2.4.6}$$

where $\sqrt{t}$ is a symbol with the property that $(\sqrt{t})^2 = t$.

Proof By $(G5)$ we have

$$\frac{f(\sqrt{t}) + f(-\sqrt{t})}{2} = \frac{\sum_{n=0}^{\infty} f_n\sqrt{t}^n + \sum_{n=0}^{\infty} f_n(-\sqrt{t})^n}{2} = \sum_{n=0}^{\infty} f_n\frac{(\sqrt{t})^n + (-\sqrt{t})^n}{2}.$$

For n odd $(\sqrt{t})^n + (-\sqrt{t})^n = 0$; hence, by setting $n = 2k$, we have

$$\sum_{k=0}^{\infty} f_{2k}(t^k + t^k)/2 = \sum_{k=0}^{\infty} f_{2k}t^k = \mathcal{G}(f_{2k}).$$

The proof of the second formula is analogous. □

The following proof is typical and introduces the use of ordinary differential equations in the calculus of generating functions:

Theorem 2.6 *Let $f(t) = \mathcal{G}(f_k)$ denote the generating function of the sequence $(f_k)_{k\in\mathbb{N}}$; then:*

$$\mathcal{G}\left(\frac{f_k}{2k+1}\right) = \frac{1}{2\sqrt{t}}\int_0^t \frac{\mathcal{G}(f_k)}{\sqrt{z}}\,dz. \tag{2.4.7}$$

Proof Let us set $g_k = f_k/(2k+1)$ or $2kg_k + g_k = f_k$. If $g(t) = \mathcal{G}(g_k)$, by applying $(G3)$ we have the differential equation $2tg'(t) + g(t) = f(t)$, whose solution satisfying $g(0) = f_0$ is just formula (2.4.7). □

We conclude with two general theorems on sums:

Theorem 2.7 (Partial Sum Theorem) *Let $f(t) = \mathcal{G}(f_k)$ denote the generating function of the sequence $(f_k)_{k\in\mathbb{N}}$; then:*

$$\mathcal{G}\left(\sum_{k=0}^{n} f_k\right) = \frac{1}{1-t}\mathcal{G}(f_k). \tag{2.4.8}$$

Proof If we set $s_n = \sum_{k=0}^{n} f_k$, then we have $s_{n+1} = s_n + f_{n+1}$ for every $n \in \mathbb{N}$ and we can apply the operator $\mathcal{G}$ to both members: $\mathcal{G}(s_{n+1}) = \mathcal{G}(s_n) + \mathcal{G}(f_{n+1})$, i.e.:

$$\frac{\mathcal{G}(s_n) - s_0}{t} = \mathcal{G}(s_n) + \frac{\mathcal{G}(f_n) - f_0}{t}.$$

Since $s_0 = f_0$, we find $\mathcal{G}(s_n) = t\mathcal{G}(s_n) + \mathcal{G}(f_n)$ and from this (2.4.8) follows directly. □

The following result is known as *Euler transformation* or *Binomial transformation*:

Theorem 2.8 *Let $f(t) = \mathcal{G}(f_k)$ denote the generating function of the sequence $(f_k)_{k\in\mathbb{N}}$; then:*

$$\mathcal{G}\left(\sum_{k=0}^{n}\binom{n}{k} f_k\right) = \frac{1}{1-t} f\left(\frac{t}{1-t}\right). \tag{2.4.9}$$

Proof By well-known properties of binomial coefficients, we have

$$\binom{n}{k} = \binom{n}{n-k} = \binom{-n+n-k-1}{n-k}(-1)^{n-k} = \binom{-k-1}{n-k}(-1)^{n-k}$$

and this is the coefficient of t^{n-k} in $(1-t)^{-k-1}$. We now observe that the sum in (2.4.9) can be extended to infinity, and by $(G5)$ we have

$$\sum_{k=0}^{n}\binom{n}{k}f_k=\sum_{k=0}^{n}\binom{-k-1}{n-k}(-1)^{n-k}f_k=\sum_{k=0}^{\infty}[t^{n-k}](1-t)^{-k-1}[y^k]f(y)=$$

$$=[t^n]\frac{1}{1-t}\sum_{k=0}^{\infty}[y^k]f(y)\left(\frac{t}{1-t}\right)^k=[t^n]\frac{1}{1-t}f\left(\frac{t}{1-t}\right).$$

Since the last expression does not depend on n, it represents the generating function of the sum. □

We observe explicitly that by $(K4)$ we have: $\sum_{k=0}^{n}\binom{n}{k}f_k=[t^n](1+t)^n f(t)$, but this expression does not represent a generating function because it depends on n. The Euler transformation can be generalized in several ways, as we shall see when dealing with Riordan arrays.

Common Generating Functions

The aim of the present subsection is to derive the most common generating functions by using the apparatus of the previous subsections. As a first example, let us consider the constant sequence $F=(1,1,1,\ldots)$, for which we have $f_{k+1}=f_k$ for every $k\in\mathbb{N}$. By applying the principle of identity, we find: $\mathcal{G}(f_{k+1})=\mathcal{G}(f_k)$, that is by $(G2)$: $\mathcal{G}(f_k)-f_0=t\mathcal{G}(f_k)$. Since $f_0=1$, we have immediately:

$$\mathcal{G}(1)=\frac{1}{1-t}.$$

For any constant sequence $F=(c,c,c,\ldots)$, by $(G1)$ we find that $\mathcal{G}(c)=c(1-t)^{-1}$. Similarly, by using the basic rules and the theorems of the previous subsections, we have

$$\mathcal{G}(n)=\mathcal{G}(n\cdot 1)=tD\frac{1}{1-t}=\frac{t}{(1-t)^2}$$

$$\mathcal{G}\left(n^2\right)=tD\mathcal{G}(n)=tD\frac{1}{(1-t)^2}=\frac{t+t^2}{(1-t)^3}$$

$$\mathcal{G}\left((-1)^n\right)=\mathcal{G}(1)\circ(-t)=\frac{1}{1+t}$$

$$\mathcal{G}\left(\frac{1}{n}\right)=\mathcal{G}\left(\frac{1}{n}\cdot 1\right)=\int_0^t\left(\frac{1}{1-z}-1\right)\frac{dz}{z}=\int_0^t\frac{dz}{1-z}=\ln\frac{1}{1-t}$$

$$\mathcal{G}(H_n)=\mathcal{G}\left(\sum_{k=0}^{n}\frac{1}{k}\right)=\frac{1}{1-t}\mathcal{G}\left(\frac{1}{n}\right)=\frac{1}{1-t}\ln\frac{1}{1-t},$$

where we consider $f_0=0$ in the sequence $f_n=1/n$ and H_n is the nth harmonic number.

We have

$$\mathcal{G}\left(\delta_{0,n}\right) = \mathcal{G}\left(1\right) - t\mathcal{G}\left(1\right) = \frac{1-t}{1-t} = 1,$$

where $\delta_{n,m}$ is Kronecker's delta. This last relation can be readily generalized to $\mathcal{G}\left(\delta_{n,m}\right) = t^m$.

An interesting example is given by $f_n = \frac{1}{n(n+1)}$ with $f_0 = 0$. Since $\frac{1}{n(n+1)} = \frac{1}{n} - \frac{1}{n+1}$, it is tempting to apply the operator $\mathcal{G}$ to both members. However, this relation is not valid for $n = 0$. In order to apply the principle of identity, we must define

$$\frac{1}{n(n+1)} = \frac{1}{n} - \frac{1}{n+1} + \delta_{n,0}$$

in accordance with the fact that the first element of the sequence is zero. We thus arrive to the correct generating function:

$$\mathcal{G}\left(\frac{1}{n(n+1)}\right) = 1 - \frac{1-t}{t}\ln\frac{1}{1-t}.$$

Let us now come to binomial coefficients. In order to find $\mathcal{G}\left(\binom{p}{k}\right)$ we observe that, from the definition:

$$\binom{p}{k+1} = \frac{p(p-1)\cdots(p-k+1)(p-k)}{(k+1)!} = \frac{p-k}{k+1}\binom{p}{k}.$$

Hence, by denoting $\binom{p}{k}$ as f_k, we have $\mathcal{G}\left((k+1)f_{k+1}\right) = \mathcal{G}\left((p-k)f_k\right) = p\mathcal{G}\left(f_k\right) - \mathcal{G}\left(kf_k\right)$. By $(G3)$ and using the fact that $\mathcal{G}\left((k+1)f_{k+1}\right) = D\mathcal{G}\left(f_k\right)$ we have $D\mathcal{G}\left(f_k\right) = p\mathcal{G}\left(f_k\right) - tD\mathcal{G}\left(f_k\right)$, i.e., the differential equation $f'(t) = pf(t) - tf'(t)$. By separating the variables and integrating, we find $\ln f(t) = p\ln(1+t) + c$, or $f(t) = c(1+t)^p$. For $t = 0$ we should have $f(0) = \binom{p}{0} = 1$, and this implies $c = 1$. Consequently:

$$\mathcal{G}\left(\binom{p}{k}\right) = (1+t)^p \qquad p \in \mathbb{R}.$$

We are now in a position to derive the recurrence relation for binomial coefficients. By using $(K1) \div (K5)$ we find easily:

$$\binom{p}{k} = [t^k](1+t)^p = [t^k](1+t)(1+t)^{p-1} =$$

$$= [t^k](1+t)^{p-1} + [t^{k-1}](1+t)^{p-1} = \binom{p-1}{k} + \binom{p-1}{k-1}.$$

By $(K4)$, we have the well-known *Vandermonde convolution*:

$$\binom{m+p}{n} = [t^n](1+t)^{m+p} = [t^n](1+t)^m(1+t)^p = \sum_{k=0}^{n}\binom{m}{k}\binom{p}{n-k}$$

which for $m = p = n$ becomes $\sum_{k=0}^{n}\binom{n}{k}^2 = \binom{2n}{n}$.

We can also find $\mathcal{G}\left(\binom{k}{p}\right)$, where p is a parameter. The derivation is purely algebraic and makes use of the generating functions already found and of various properties considered in the previous subsection:

$$\begin{aligned}\mathcal{G}\left(\binom{k}{p}\right) &= \mathcal{G}\left(\binom{k}{k-p}\right) = \mathcal{G}\left(\binom{-k+k-p+1}{k-p}(-1)^{k-p}\right) \\ &= \mathcal{G}\left(\binom{-p-1}{k-p}(-1)^{k-p}\right) = \mathcal{G}\left([t^{k-p}]\frac{1}{(1-t)^{p+1}}\right) = \mathcal{G}\left([t^k]\frac{t^p}{(1-t)^{p+1}}\right) \\ &= \frac{t^p}{(1-t)^{p+1}}.\end{aligned}$$

Several generating functions for different forms of binomial coefficients can be found by means of this method. They are summarized as follows, where p and m are two parameters and can also be zero:

$$\begin{aligned}\mathcal{G}\left(\binom{p}{m+k}\right) &= \frac{(1+t)^p}{t^m} \\ \mathcal{G}\left(\binom{p+k}{m}\right) &= \frac{t^{m-p}}{(1-t)^{m+1}} \\ \mathcal{G}\left(\binom{p+k}{m+k}\right) &= \frac{1}{t^m(1-t)^{p+1-m}}.\end{aligned}$$

These functions can make sense even when they are f.L.s. and not simply f.p.s.

The Method of Shifting

When the elements of a sequence F are given by an explicit formula, we can try to find the generating function for F by using the technique of *shifting*: we consider the element f_{n+1} and try to express it in terms of f_n. This can produce a relation to which we apply the principle of identity deriving an equation in $\mathcal{G}(f_n)$, the solution of which is the generating function. In practice, we find a recurrence for the elements $f_n \in F$ and then try to solve it by using the rules $(G1) \div (G5)$ and their consequences. It can happen that the recurrence involves several elements in F and/or that the resulting equation is indeed a differential equation. Whatever the case, the method of shifting allows us to find the generating function of many sequences.

Let us consider the geometric sequence $\left(1, p, p^2, p^3, \ldots\right)$; we have $p^{k+1} = pp^k, \forall k \in \mathbb{N}$ or, by applying the operator $\mathcal{G}$, $\mathcal{G}\left(p^{k+1}\right) = p\mathcal{G}\left(p^k\right)$. By $(G2)$ we have $t^{-1}\left(\mathcal{G}\left(p^k\right) - 1\right) = p\mathcal{G}\left(p^k\right)$, that is:

$$\mathcal{G}\left(p^k\right) = \frac{1}{1-pt}.$$

From this we obtain other generating functions:

$$\mathcal{G}\left(kp^k\right) = tD\frac{1}{1-pt} = \frac{pt}{(1-pt)^2}$$

$$\mathcal{G}\left(k^2p^k\right) = tD\frac{pt}{(1-pt)^2} = \frac{pt+p^2t^2}{(1-pt)^3}$$

$$\mathcal{G}\left(\frac{1}{k}p^k\right) = \int_0^t\left(\frac{1}{1-pz}-1\right)\frac{dz}{z} =$$

$$= \int\frac{p\,dz}{(1-pz)} = \ln\frac{1}{1-pt}$$

$$\mathcal{G}\left(\sum_{k=0}^{n}p^k\right) = \frac{1}{1-t}\frac{1}{1-pt} = \frac{1}{(p-1)t}\left(\frac{1}{1-pt}-\frac{1}{1-t}\right).$$

The last relation has been obtained by partial fraction expansion. By using the operator $[t^k]$ we easily find:

$$\sum_{k=0}^{n}p^k = [t^n]\frac{1}{1-t}\frac{1}{1-pt} = \frac{1}{(p-1)}[t^{n+1}]\left(\frac{1}{1-pt}-\frac{1}{1-t}\right) = \frac{p^{n+1}-1}{p-1}$$

the well-known formula for the sum of a geometric progression. We observe explicitly that the formulas above could have been obtained from formulas of the previous subsection and the general formula $\mathcal{G}\left(p^k f_k\right) = f(pt)$. In a similar way, we also have

$$\mathcal{G}\left(\binom{m}{k}p^k\right) = (1+pt)^m$$

$$\mathcal{G}\left(\binom{k}{m}p^k\right) = \mathcal{G}\left(\binom{k}{k-m}p^{k-m}p^m\right) = p^m\mathcal{G}\left(\binom{-m-1}{k-m}(-p)^{k-m}\right) = \frac{p^mt^m}{(1-pt)^{m+1}}.$$

As a very simple application of the shifting method, let us observe that

$$\frac{1}{(n+1)!} = \frac{1}{n+1}\frac{1}{n!},$$

that is, $(n+1)f_{n+1} = f_n$, where $f_n = 1/n!$. Therefore we have $f'(t) = f(t)$ or:

$$\mathcal{G}\left(\frac{1}{n!}\right) = e^t$$

$$\mathcal{G}\left(\frac{1}{n \cdot n!}\right) = \int_0^t \frac{e^z - 1}{z} dz$$

$$\mathcal{G}\left(\frac{n}{(n+1)!}\right) = tD\mathcal{G}\left(\frac{1}{(n+1)!}\right) = tD\frac{1}{t}(e^t - 1) = \frac{te^t - e^t + 1}{t}.$$

Let us now observe that $\binom{2n+2}{n+1} = \frac{2(2n+1)}{n+1}\binom{2n}{n}$; by setting $f_n = \binom{2n}{n}$, we have the recurrence $(n+1)f_{n+1} = 2(2n+1)f_n$. By using $(G1)$ and $(G3)$ we obtain the differential equation: $f'(t) = 4tf'(t) + 2f(t)$, the simple solution of which is

$$\mathcal{G}\left(\binom{2n}{n}\right) = \frac{1}{\sqrt{1-4t}}. \tag{2.4.10}$$

By using (2.4.4) we also have

$$\mathcal{G}\left(\frac{1}{n+1}\binom{2n}{n}\right) = \frac{1}{t}\int_0^t \frac{dz}{\sqrt{1-4z}} = \frac{1-\sqrt{1-4t}}{2t}$$

which is the well-known generating function for Catalan numbers $\binom{2n}{n}/(n+1)$.

Diagonalization

The technique of shifting is a rather general method for obtaining generating functions which produces first-order recurrence relations. Not every sequence can be defined by a first-order recurrence relation, and other methods are often necessary to find out generating functions. Sometimes, the rule of diagonalization can be used very conveniently. One of the most simple examples is how to determine the generating function of the central binomial coefficients, without having to pass through the solution of a differential equation. In fact, we have

$$\binom{2n}{n} = [t^n](1+t)^{2n}$$

and $(G6)$ can be applied with $F(t) = 1$ and $\phi(t) = (1+t)^2$. In this case, the function $w = w(t)$ is easily determined by solving the functional equation $w = t(1+w)^2$. By expanding, we find $tw^2 - (1-2t)w + t = 0$ or:

$$w = w(t) = \frac{1 - t \pm \sqrt{1-4t}}{2t}.$$

Since $w = w(t)$ should belong to $\mathcal{F}_1$, we must eliminate the solution with the $+$ sign; consequently, we have

$$\mathcal{G}\left(\binom{2n}{n}\right) = \left[\frac{1}{1-2t(1+w)} \;\middle|\; w = \frac{1-t-\sqrt{1-4t}}{2t}\right] = \frac{1}{\sqrt{1-4t}}$$

as we already know.

The function $\phi(t) = (1+t)^2$ gives rise to a second degree equation. More in general, let us study the sequence:

$$c_n = [t^n](1+\alpha t+\beta t^2)^n$$

and look for its generating function $C(t) = \mathcal{G}(c_n)$. In this case again we have $F(t) = 1$ and $\phi(t) = 1+\alpha t+\beta t^2$, and therefore we should solve the functional equation $w = t(1+\alpha w+\beta w^2)$ or $\beta t w^2 - (1-\alpha t)w + t = 0$. This gives

$$w = w(t) = \frac{1-\alpha t \pm \sqrt{(1-\alpha t)^2 - 4\beta t^2}}{2\beta t}$$

and again we have to eliminate the solution with the $+$ sign. By performing the necessary computations, we find

$$C(t) = \left[\frac{1}{1-t(\alpha+2\beta w)} \;\middle|\; w = \frac{1-\alpha t-\sqrt{1-2\alpha t+(\alpha^2-4\beta)t^2}}{2\beta t}\right] =$$

$$= \frac{1}{\sqrt{1-2\alpha t+(\alpha^2-4\beta)t^2}}$$

and for $\alpha = 2$, $\beta = 1$ we obtain again the generating function for the central binomial coefficients.

The coefficients of $(1+t+t^2)^n$ are called *trinomial coefficients*, in analogy with the binomial coefficients. They constitute an infinite array in which every row has two more elements, different from 0, with respect to the previous row (see Table 2.1).

Table 2.1 Trinomial coefficients

$n\backslash k$	0	1	2	3	4	5	6	7	8
0	**1**								
1	1	**1**	1						
2	1	2	**3**	2	1				
3	1	3	6	**7**	6	3	1		
4	1	4	10	16	**19**	16	10	4	1

If $T_{n,k}$ is $[t^k](1+t+t^2)^n$, a trinomial coefficient, by the obvious property $(1+t+t^2)^{n+1} = (1+t+t^2)(1+t+t^2)^n$, we immediately deduce the recurrence relation:

$$T_{n+1,k+1} = T_{n,k-1} + T_{n,k} + T_{n,k+1}$$

from which the array can be built, once we start from the initial conditions $T_{n,0} = 1$ and $T_{n,2n} = 1$, for every $n \in \mathbb{N}$. The elements $T_{n,n}$, marked in the table, are called the *central trinomial coefficients*; their sequence begins

n	0	1	2	3	4	5	6	7	8
T_n	1	1	3	7	19	51	141	393	1107

and by the formula above their generating function is

$$\mathcal{G}\left(T_{n,n}\right) = \frac{1}{\sqrt{1-2t-3t^2}} = \frac{1}{\sqrt{(1+t)(1-3t)}}.$$

Linear Recurrences with Constant Coefficients

If $(f_k)_{k\in\mathbb{N}}$ is a sequence, it may be defined by means of a *recurrence relation*, i.e., a relation relating the generic element f_n to other elements f_k having $k < n$. Usually, the first elements of the sequence must be given explicitly, in order to allow the computation of the successive values; they constitute the *initial conditions* and the sequence is well defined if and only if every element can be computed by starting with the initial conditions and going on with the other elements by means of the recurrence relation. For example, the constant sequence $(1, 1, 1, \ldots)$ can be defined by the recurrence relation $x_n = x_{n-1}$ and the initial condition $x_0 = 1$. By changing the initial conditions, the sequence can radically change; if we consider the same relation $x_n = x_{n-1}$ together with the initial condition $x_0 = 2$, we obtain the constant sequence $\{2, 2, 2, \ldots\}$.

In general, a recurrence relation can be written $f_n = F(f_{n-1}, f_{n-2}, \ldots)$; when F depends on all the values $f_{n-1}, f_{n-2}, \ldots, f_1, f_0$, then the relation is called a *full history recurrence*. If F only depends on a fixed number of elements $f_{n-1}, f_{n-2}, \ldots, f_{n-p}$, then the relation is called a *partial history recurrence* and p is called the *order* of the relation. Besides, if F is linear, we have a *linear recurrence*. Linear recurrences are surely the most common and important type of recurrence relations; if all the coefficients appearing in F are constant, we have a linear recurrence *with constant coefficients*, and if the coefficients are polynomials in n, we have a linear recurrence *with polynomial coefficients*. As we are now going to see, the method of generating functions allows us to find the solution of any linear recurrence with constant coefficients, in the sense that we find a function $f(t)$ such that $[t^n]f(t) = f_n$, $\forall n \in \mathbb{N}$. For linear recurrences with polynomial coefficients, the same method allows us to find a solution on many occasions, but the success is not guaranteed. On the other hand, no method is known that solves all the recurrences of this kind, and surely generating functions are the method giving the highest number of positive results. We will discuss this case in the next subsection.

The Fibonacci recurrence $F_n = F_{n-1} + F_{n-2}$ is an example of a recurrence relation with constant coefficients. When we have a recurrence of this kind, we begin by expressing it in such a way that the relation is valid for every $n \in \mathbb{N}$. In the example of Fibonacci numbers, this is not the case, because for $n = 0$ we have $F_0 = F_{-1} + F_{-2}$, and we do not know the values for the two elements in the r.h.s., which have no combinatorial meaning. However, if we write the recurrence as $F_{n+2} = F_{n+1} + F_n$ we have fulfilled the requirement. This first step has a great importance, because it allows us to apply the operator $\mathcal{G}$ to both members of the relation; this was not possible beforehand because of the principle of identity for generating functions.

The recurrence being linear with constant coefficients, we can apply the axiom of linearity to the recurrence:

$$f_{n+p} = \alpha_1 f_{n+p-1} + \alpha_2 f_{n+p-2} + \cdots + \alpha_p f_n$$

and obtain the relation:

$$\mathcal{G}(f_{n+p}) = \alpha_1 \mathcal{G}(f_{n+p-1}) + \alpha_2 \mathcal{G}(f_{n+p-2}) + \cdots + \alpha_p \mathcal{G}(f_n).$$

By using the rule $\mathcal{G}\left(f_{k+j}\right) = \frac{\mathcal{G}(f_k) - f_0 - f_1 t - \cdots - f_{j-1} t^{j-1}}{t^j}$ we can now express every $\mathcal{G}(f_{n+p-j})$ in terms of $f(t) = \mathcal{G}(f_n)$ and obtain a linear relation in $f(t)$, from which an explicit expression for $f(t)$ is immediately obtained. This is the solution of the recurrence relation. We observe explicitly that in writing the expressions for $\mathcal{G}(f_{n+p-j})$ we make use of the initial conditions for the sequence.

Let us go on with the example of the Fibonacci sequence $(F_k)_{k \in \mathbb{N}}$. We have

$$\mathcal{G}(F_{n+2}) = \mathcal{G}(F_{n+1}) + \mathcal{G}(F_n)$$

and by setting $F(t) = \mathcal{G}(F_n)$ we find

$$\frac{F(t) - F_0 - F_1 t}{t^2} = \frac{F(t) - F_0}{t} + F(t).$$

Because we know that $F_0 = 0$, $F_1 = 1$, we have

$$F(t) - t = tF(t) + t^2 F(t)$$

and by solving in $F(t)$ we have the explicit expression:

$$F(t) = \frac{t}{1 - t - t^2}.$$

This is the generating function for the Fibonacci numbers. We can now find an explicit expression for F_n in the following way. The denominator of $F(t)$ can be written $1 - t - t^2 = (1 - \phi t)(1 - \widehat{\phi} t)$ where

$$\phi = \frac{1+\sqrt{5}}{2} \approx 1.618033989$$

$$\widehat{\phi} = \frac{1-\sqrt{5}}{2} \approx -0.618033989.$$

The constant $\phi \approx 1.618033989$ is known as the *golden ratio*. By applying the method of *partial fraction expansion* we find

$$F(t) = \frac{t}{(1-\phi t)(1-\widehat{\phi} t)} = \frac{A}{1-\phi t} + \frac{B}{1-\widehat{\phi} t} = \frac{A - A\widehat{\phi} t + B - B\phi t}{1-t-t^2}.$$

We determine the two constants A and B by equating the coefficients in the first and last expression for $F(t)$:

$$\begin{cases} A + B = 0 \\ -A\widehat{\phi} - B\phi = 1 \end{cases} \qquad \begin{cases} A = 1/(\phi - \widehat{\phi}) = 1/\sqrt{5} \\ B = -A = -1/\sqrt{5} \end{cases}.$$

The value of F_n is now obtained by extracting the coefficient of t^n:

$$F_n = [t^n]F(t) = [t^n]\frac{1}{\sqrt{5}}\left(\frac{1}{1-\phi t} - \frac{1}{1-\widehat{\phi} t}\right) =$$

$$= \frac{1}{\sqrt{5}}\left([t^n]\frac{1}{1-\phi t} - [t^n]\frac{1}{1-\widehat{\phi} t}\right) = \frac{\phi^n - \widehat{\phi}^n}{\sqrt{5}}.$$

Linear Recurrences with Polynomial Coefficients

When a recurrence relation has polynomial coefficients, the method of generating functions does not guarantee a solution, but no other method is available to solve those recurrences which cannot be solved by a generating function approach. Usually, the rule of differentiation introduces a derivative in the relation for the generating function and a differential equation has to be solved. This is the actual problem of this approach, because the main difficulty just consists in dealing with the differential equation. We wish to present a case arising from an actual combinatorial problem.

A permutation is an arrangement of a set of objects in any order. For example, three objects, denoted by a, b, and c, can be arranged into six different ways:

$$(a, b, c),\ (a, c, b),\ (b, a, c),\ (b, c, a),\ (c, a, b),\ (c, b, a).$$

If we abstract from the particular nature of the objects and denote $\{1, 2, \ldots, n\} = \mathbb{N}_n$, we can define a *permutation* of n objects as a 1-1 mapping $\pi : \mathbb{N}_n \to \mathbb{N}_n$. By identifying a and 1, b and 2, c and 3, the six permutations of 3 objects are written $(1, 2, 3)$, $(1, 3, 2)$, $(2, 1, 3)$, $(2, 3, 1)$, $(3, 1, 2)$, $(3, 2, 1)$, by using the common *vector representation* for permutations. This consists in writing the images of the map in the form of a vector.

Let us examine the permutation $\pi = (3, 2, 1)$ for which we have $\pi(1) = 3$, $\pi(2) = 2$, and $\pi(3) = 1$. If we start with the element 1 and successively apply the mapping π, we have $\pi(1) = 3$, $\pi(\pi(1)) = \pi(3) = 1$, $\pi(\pi(\pi(1))) = \pi(1) = 3$ and so on. Since the elements in $\mathbb{N}_n$ are finite, by starting with any $k \in \mathbb{N}_n$ we must obtain a finite chain of numbers, which will repeat always in the same order. These numbers are said to form a *cycle* and the permutation $(3, 2, 1)$ is formed by two cycles, the first one composed of 1 and 3, the second one only composed of 2. We write $(3, 2, 1) = (1\ 3)(2)$, where every cycle is written between parentheses and numbers are separated by blanks, to distinguish a cycle from a vector. Conventionally, a cycle is written with the smallest element first and the various cycles are arranged according to their first element. Therefore, in this *cycle representation*, the six permutations are

$$(1)(2)(3),\ (1)(2\ 3),\ (1\ 2)(3),\ (1\ 2\ 3),\ (1\ 3\ 2),\ (1\ 3)(2).$$

Let P_n denote the set of all the permutations of n elements. If $\pi, \rho \in P_n$, we can perform their composition, i.e., a new permutation σ defined as $\sigma(k) = \pi(\rho(k)) = (\phi \circ \rho)(k)$. For example, if $\pi = (1, 5, 6, 7, 4, 2, 3)$ and $\rho = (4, 5, 2, 1, 7, 6, 3)$ then $\pi \circ \rho = (4, 7, 6, 3, 1, 5, 2)$.

An *involution* is a permutation $\pi \in P_n$ such that $\pi^2 = \pi \circ \pi = (1, 2, \cdots, n)$, and for the number I_n of involutions in P_n we have the recurrence relation:

$$I_n = I_{n-1} + (n-1)I_{n-2}$$

which has polynomial coefficients. This result depends on the property that an involution can only be composed of fixed points, that is, numbers k such that $\pi(k) = k$, and transpositions, that is, cycles with only two elements.

The number of involutions grows very fast and it can be a good idea to consider the quantity $i_n = I_n/n!$. Therefore, let us begin by changing the recurrence in such a way that the principle of identity can be applied, and then divide everything by $(n+2)!$:

$$\begin{aligned} I_{n+2} &= I_{n+1} + (n+1)I_n \\ \frac{I_{n+2}}{(n+2)!} &= \frac{1}{n+2}\frac{I_{n+1}}{(n+1)!} + \frac{1}{n+2}\frac{I_n}{n!}. \end{aligned}$$

The recurrence relation for i_n is

$$(n+2)i_{n+2} = i_{n+1} + i_n$$

and we can pass to generating functions. The term $\mathcal{G}((n+2)i_{n+2})$ can be seen as the shifting of $\mathcal{G}((n+1)i_{n+1}) = i'(t)$, if with $i(t)$ we denote the generating function of

$(i_k)_{k\in\mathbb{N}} = (I_k/k!)_{k\in N}$, which corresponds to the exponential generating function for $(I_k)_{k\in\mathbb{N}}$, as defined in Chap. 6. We have

$$\frac{i'(t)-1}{t} = \frac{i(t)-1}{t} + i(t)$$

because of the initial conditions $i_0 = i_1 = 1$, and so

$$i'(t) = (1+t)i(t).$$

This is a simple differential equation with separable variables and by solving it we find

$$\ln i(t) = t + \frac{t^2}{2} + C \quad \text{or} \quad i(t) = \exp\left(t + \frac{t^2}{2} + C\right),$$

where C is an integration constant. Because $i(0) = e^C = 1$, we have $C = 0$ and we conclude with the formula:

$$I_n = n![t^n]\exp\left(t + \frac{t^2}{2}\right).$$

Exercises

2.1 Prove that if $f(t) = \mathcal{G}(f_k)$ is the generating function of the sequence $(f_k)_{k\in\mathbb{N}}$, then

$$\mathcal{G}(f_{k+2}) = \frac{\mathcal{G}(f_k) - f_0 - f_1 t}{t^2}$$

and, more generally,

$$\mathcal{G}\left(f_{k+j}\right) = \frac{\mathcal{G}(f_k) - f_0 - f_1 t - \cdots - f_{j-1}t^{j-1}}{t^j}.$$

2.2 Prove that if $f(t) = \mathcal{G}(f_k)$ is the generating function of the sequence $(f_k)_{k\in\mathbb{N}}$, then

$$\mathcal{G}((k+1)f_{k+1}) = D\mathcal{G}(f_k).$$

2.3 Prove that if $f(t) = \mathcal{G}(f_k)$ is the generating function of the sequence $(f_k)_{k\in\mathbb{N}}$, then

$$\mathcal{G}\left(k^2 f_k\right) = tD\mathcal{G}(f_k) + t^2D^2\mathcal{G}(f_k).$$

2.4 Prove that if $f(t) = \mathcal{G}(f_k)$ is the generating function of the sequence $(f_k)_{k\in\mathbb{N}}$, then

$$\mathcal{G}\left(k^{\underline{r}} f_k\right) = t^r D^r \mathcal{G}(f_k),$$

where $k^{\underline{r}} = k(k-1)\cdots(k-r+1)$ is the falling factorial.

2.5 Prove that if $f(t) = \mathcal{G}(f_k)$ is the generating function of the sequence $(f_k)_{k\in\mathbb{N}}$, then

$$\mathcal{G}\left(p^k f_k\right) = f(pt).$$

2.6 Prove that if H_n is the nth harmonic number

$$\mathcal{G}(nH_n) = \frac{t}{(1-t)^2}\left(\ln\frac{1}{1-t} + 1\right)$$

$$\mathcal{G}\left(\frac{1}{n+1}H_n\right) = \frac{1}{2t}\left(\ln\frac{1}{1-t}\right)^2.$$

2.7 Prove the following identities:

$$\begin{aligned}
\mathcal{G}\left(k\binom{p}{k}\right) &= pt(1+t)^{p-1}\\
\mathcal{G}\left(k^2\binom{p}{k}\right) &= pt(1+pt)(1+t)^{p-2}\\
\mathcal{G}\left(\frac{1}{k+1}\binom{p}{k}\right) &= \frac{(1+t)^{p+1}-1}{(p+1)t}\\
\mathcal{G}\left(k\binom{k}{m}\right) &= \frac{mt^m + t^{m+1}}{(1-t)^{m+2}}\\
\mathcal{G}\left(k^2\binom{k}{m}\right) &= \frac{m^2t^m + (3m+1)t^{m+1} + t^{m+2}}{(1-t)^{m+3}}\\
\mathcal{G}\left(\frac{1}{k}\binom{k}{m}\right) &= \frac{t^m}{m(1-t)^m} \quad \text{for} m > 0.
\end{aligned}$$

2.8 Prove that

$$\sum_{k=0}^{n}\frac{k}{(k+1)!} = 1 - \frac{1}{(n+1)!}.$$

2.9 Prove the following identities:

$$\mathcal{G}\left(\frac{1}{n}\binom{2n}{n}\right) = 2\ln\frac{1-\sqrt{1-4t}}{2t}$$
$$\mathcal{G}\left(n\binom{2n}{n}\right) = \frac{2t}{(1-4t)\sqrt{1-4t}}$$
$$\mathcal{G}\left(\frac{1}{2n+1}\binom{2n}{n}\right) = \frac{1}{\sqrt{4t}}\arctan\sqrt{\frac{4t}{1-4t}}.$$

2.10 Prove the following identities:

$$\mathcal{G}\left(\frac{1}{2n}\frac{4^n}{2n+1}\binom{2n}{n}^{-1}\right) = 1-\sqrt{\frac{1-t}{t}}\arctan\sqrt{\frac{t}{1-t}}$$
$$\mathcal{G}\left(\frac{4^n}{2n+1}\binom{2n}{n}^{-1}\right) = \frac{1}{\sqrt{t(1-t)}}\arctan\sqrt{\frac{t}{1-t}}$$
$$\mathcal{G}\left(\frac{4^n}{2n^2}\binom{2n}{n}^{-1}\right) = \left(\arctan\sqrt{\frac{t}{1-t}}\right)^2.$$

2.11 Given the generating function of the nth Fibonacci number F_n,

$$\mathcal{G}(F_n) = \frac{t}{1-t-t^2},$$

find $\mathcal{G}(F_{2n})$ and $\mathcal{G}(F_{2n+1})$.

2.12 The Catalan numbers C_n satisfy the recurrence relation

$$C_{n+1} = \sum_{k=0}^{n} C_k C_{n-k}.$$

Use the properties of $\mathcal{G}$ operator to find $\mathcal{G}(C_n)$.

2.13 Find $\mathcal{G}(f_n)$ where

$$f_n = \delta_{n,1} + \frac{2}{n}\sum_{k=0}^{n-1} f_k, \quad f_0 = 0$$

where $\delta_{n,k}$ is a Kronecker's delta which is 1 if the variables are equal, and 0 otherwise.

References

1. L. Comtet, *Advanced Combinatorics* (Reidel, Dordrecht, 1974)
2. G.P. Egorychev, *Integral Representation and the Computation of Combinatorial Sums* (trans. H. H. McFadden), vol. 59 (American Mathematical Society, Providence, 1984)
3. P. Flajolet, R. Sedgewick, *Analytic Combinatorics* (Cambridge University Press, 2009)
4. I.M. Gessel, A combinatorial proof of the multivariable Lagrange inversion formula. J. Comb. Theory A **45**, 178–195 (1987)
5. I.P. Goulden, D.M. Jackson, *Combinatorial Enumeration* (Wiley, New York, 1983)
6. R.L. Graham, D.E. Knuth, O. Patashnik, *Concrete Mathematics* (Addison-Wesley, New York, 1989)
7. D.H. Greene, D.E. Knuth, *Mathematics for the Analysis of Algorithms* (Birkäuser, Boston, 1982)
8. D.E. Knuth, *The Art of Computer Programming*, vol. 1–3 (Addison-Wesley, Reading, MA, 1973)
9. D. Merlini, R. Sprugnoli, M.C. Verri, Lagrange inversion: when and how. Acta Appl. Math. **94**(3), 233–249 (2006)
10. D. Merlini, R. Sprugnoli, M.C. Verri, The method of coefficients. Am. Math. Mon. **114**(1), 40–57 (2007)
11. R. Sedgewick, P. Flajolet, *An Introduction to the Analysis of Algorithms* (Addison-Wesley, 1996)
12. A.D. Sokal, A ridiculously simple and explicit implicit function theorem. Preprint. https://arxiv.org/abs/0902.0069 (2009)
13. R.P. Stanley, *Enumerative Combinatorics*, vol. 1 (Wadsworth, Cambridge, 1986)
14. H. Wilf, *Generatingfunctionology* (Academic Press, San Diego, 1990)

Chapter 3
The Riordan Group

Abstract The group of Riordan arrays was introduced in 1991 by Shapiro, Getu, Woan, and Woodson [15], with the aim of defining a class of infinite lower triangular arrays with properties analogous to those of the Pascal triangle. A previous generalization of the Pascal, Catalan, and Motzkin triangles can be found in Rogers [13] who introduces the concept of renewal array. Sprugnoli [16, 17] also investigated these Riordan arrays and showed that they constitute a practical device for solving combinatorial sums by means of generating functions and the Lagrange inversion formula. Some of the structural properties of Riordan arrays were studied by Merlini, Rogers, Sprugnoli, and Verri [10]. Since then, these arrays attracted, and continue to attract, a lot of attention in the literature and many works on the Riordan arrays have been done, for example [3–6, 8, 12, 14, 18, 19].

3.1 Riordan Arrays and the Riordan Group

We begin with our first definition of a Riordan array. Recall that for the real field $\mathbb{R}$

$$\begin{aligned}\mathcal{F} &= \{c_0 + c_1 t + c_2 t^2 + \cdots \mid c_i \in \mathbb{R}\}\\ \mathcal{F}_0 &= \{g \in \mathcal{F} \mid g(0) \neq 0\}\\ \mathcal{F}_1 &= \{g \in \mathcal{F} \mid g(0) = 0, g'(0) \neq 0\} = t\mathcal{F}_0.\end{aligned}$$

Definition 3.1 *(Riordan array)* Let $L = [d_{n,k}]$ be an infinite matrix over $\mathbb{R}$. If there exists a pair of formal power series $(g(t), f(t)) \in \mathcal{F} \times \mathcal{F}_1$ such that

$$d_{n,k} = [t^n]g(t)f(t)^k, \tag{3.1.1}$$

the matrix L is called a *Riordan array* (R.a.) or *Riordan matrix* generated by $g(t)$ and $f(t)$. We write $L = (g(t), f(t))$, $L = \mathcal{R}(g(t), f(t))$, or $L = (g, f)$. In particular, if $g(t) \in \mathcal{F}_0$ and $f(t) \in \mathcal{F}_1$ then L is called a *proper* Riordan array (p.R.a.).

© The Author(s), under exclusive license to Springer Nature Switzerland AG 2022

L. Shapiro et al., *The Riordan Group and Applications*, Springer Monographs in Mathematics, https://doi.org/10.1007/978-3-030-94151-2_3

By definition, a Riordan array is lower triangular and a proper Riordan array is invertible.

Remark 3.1 A proper Riordan array is characterized by a pair of formal power series $g(t) \in \mathcal{F}_0$ and $f(t) \in \mathcal{F}_1$; $g(t)$ is the generating function of column 0 and $f(t)$ is the ratio between any two consecutive columns.

Given an infinite lower triangular array, how do we know whether it is a Riordan array or not?

Example 3.1 We start with the coefficients of $1 + t + t^2$ arranged in triangular form and compute the successive powers

$$\begin{array}{ccccccccccc} & & & & & \mathbf{1} & & & & & \\ & & & & 1 & \mathbf{1} & 1 & & & & \\ & & & 1 & 2 & \mathbf{3} & 2 & 1 & & & \\ & & 1 & 3 & 6 & \mathbf{7} & 6 & 3 & 1 & & \\ & 1 & 4 & 10 & 16 & \mathbf{19} & 16 & 10 & 4 & 1 & \\ 1 & 5 & 14 & 30 & 45 & \mathbf{51} & 45 & 30 & 14 & 5 & 1 \\ & & & & & \cdots & & & & & \end{array}$$

The center numbers in boldface are called the *central trinomial numbers* and Euler found their generating function

$$1 + t + 3t^2 + 7t^3 + 19t^4 + \cdots = \frac{1}{\sqrt{1 - 2t - 3t^2}}.$$

Cutting this triangle in half gives us the matrix that starts

$$T = \begin{bmatrix} 1 & 0 & 0 & 0 & 0 & 0 \\ 1 & 1 & 0 & 0 & 0 & 0 \\ 3 & 2 & 1 & 0 & 0 & 0 \\ 7 & 6 & 3 & 1 & 0 & 0 \\ 19 & 16 & 10 & 4 & 1 & 0 \\ 51 & 45 & 30 & 15 & 5 & 1 \end{bmatrix}.$$

If T is a Riordan array, we have to find f such that $T = \left(\frac{1}{\sqrt{1-2t-3t^2}}, f\right)$ assuming that it exists. Here is a naive approach. We know that

$$gf \Leftrightarrow 0, 1, 2, 6, 16, 45, \cdots .$$

Thus, we need

$$\left(1 + t + 3t^2 + 7t^3 + 19t^4 + \cdots\right)\left(f_1 t + f_2 t^2 + f_3 t^3 + f_4 t^4 + \cdots\right) =$$

$$= t + 2t^2 + 6t^3 + 16t^4 + 45t^5 + \cdots.$$

This forces first that $f_1 = 1$, then $f_2 = 1$, $f_3 = 2$, $f_4 = 4$, and $f_5 = 9$. Then a trip to the website OEIS, the On-Line Encyclopedia of Integer Sequences, gives the strong possibility that $f = tm(t)$ where

$$m = m(t) = \frac{1 - t - \sqrt{1 - 2t - 3t^2}}{2t^2}$$

is the generating function for the Motzkin numbers.

We have not proved anything but we do know what we would like to prove. By way of more evidence that we have a Riordan array we see that

$$f^2 = t^2 + 2t^3 + 5t^4 + 12t^5 + \cdots$$

so that

$$gf^2 = t^2 + 3t^3 + 10t^4 + 30t^5 + \cdots$$

agreeing with the second column of our matrix.

Example 3.2 The most important example of a Riordan array is $P = (1/(1 - t), t/(1 - t)) = [d_{n,k}]$, the elements of which are

$$d_{n,k} = [t^n]\frac{1}{1-t}\left(\frac{t}{1-t}\right)^k = [t^{n-k}]\frac{1}{(1-t)^{k+1}} = \binom{-k-1}{n-k}(-1)^{n-k} =$$
$$= \binom{k+1+n-k-1}{n-k} = \binom{n}{k}.$$

Therefore, P is the Pascal triangle (see Table 3.1), and we call P the *Pascal matrix*.

Example 3.3 A second example is

$$C = \left(\frac{1}{\sqrt{1-4t}}, \frac{1 - \sqrt{1-4t}}{2}\right),$$

the upper left part of which is given in Table 3.1. As we will see, $d_{n,k} = \binom{2n-k}{n-k}$ and it is a type of Catalan triangle.

Example 3.4 The third example is $S = (1, e^t - 1)$, so that

$$d_{n,k} = [t^n](e^t - 1)^k = \frac{k!}{n!}\left\{ n \atop k \right\};$$

Table 3.1 The Pascal (left) and a Catalan type triangle (right)

	0	1	2	3	4	5
0	1					
1	1	1				
2	1	2	1			
3	1	3	3	1		
4	1	4	6	4	1	
5	1	5	10	10	5	1

	0	1	2	3	4	5
0	1					
1	2	1				
2	6	3	1			
3	20	10	4	1		
4	70	35	15	5	1	
5	252	126	56	21	6	1

Table 3.2 Stirling numbers of the second kind and the related triangles

	0	1	2	3	4	5
0	1					
1	0	1				
2	0	1	1			
3	0	1	3	1		
4	0	1	7	6	1	
5	0	1	15	25	10	1

	0	1	2	3	4	5
0	1					
1	0	1				
2	0	1/2	1			
3	0	1/6	1	1		
4	0	1/24	7/12	3/2	1	
5	0	1/120	1/4	5/4	2	1

this is related to the triangle of Stirling numbers of the second kind (see Table 3.2), and can be used to prove properties of these numbers.

The important point about Riordan arrays is that their product corresponding to the usual row-by-column product between infinite matrices is actually a Riordan array.

Theorem 3.1 (Fundamental Theorem of Riordan Arrays)
Let $A = (a_0, a_1, a_2, \cdots)^T$ and $B = (b_0, b_1, b_2, \cdots)^T$ be column vectors and let $A(t)$ and $B(t)$ be the corresponding generating functions. If $(g, f)A = B$ then $B(t) = g(t)A(f(t))$.

Proof Multiplying out

$$(g, f)\,(a_0, a_1, a_2, \cdots)^T$$

we get, in terms of generating functions,

$$\begin{aligned} a_0 g + a_1 g f + a_2 g f^2 + \cdots &= g\left(a_0 + a_1 f + a_2 f^2 + \cdots\right) \\ &= g(t)A(f(t)) = B(t), \end{aligned}$$

which proves the result. □

This simple result is the backbone of everything that follows and is called the *Fundamental Theorem of Riordan Arrays (FTRA)*, and we write more briefly the FTRA by

$$(g, f)A = gA(f).$$

The first application is that we can define the multiplication of two Riordan arrays, (g, f) and (d, h). We apply the FTRA and find

$$(g, f)\, dh^k = g(t)d(f(t))(h(f(t)))^k$$

and this is the kth column of the matrix $(g(t)d(f(t)), h(f(t)))$. Thus, we have

$$(g, f)\,(d, h) = (gd(f), h(f))\,. \tag{3.1.2}$$

We will sometimes use $*$ to highlight the multiplication between matrices.

Example 3.5 Let us first consider the product $P * P = P^2$, we have

$$\left(\frac{1}{1-t}, \frac{t}{1-t}\right)\left(\frac{1}{1-t}, \frac{t}{1-t}\right) = \left(\frac{1}{1-t}\frac{1-t}{1-2t}, \frac{t}{1-t}\frac{1-t}{1-2t}\right) = \left(\frac{1}{1-2t}, \frac{t}{1-2t}\right).$$

The generic element is

$$[t^n]\frac{1}{1-2t}\left(\frac{t}{1-2t}\right)^k = [t^{n-k}]\frac{1}{(1-2t)^{k+1}} = \binom{-k-1}{n-k}(-2)^{n-k} = \binom{n}{k}2^{n-k}.$$

This result can be iterated, as many times as we wish.

Example 3.6 Let us show what $P * C$ is

$$\begin{aligned} P * C &= \left(\frac{1}{1-t}, \frac{t}{1-t}\right)\left(\frac{1}{\sqrt{1-4t}}, \frac{1-\sqrt{1-4t}}{2}\right) \\ &= \left(\frac{1}{1-t}\sqrt{\frac{1-t}{1-5t}}, \frac{1}{1-t}\frac{1-t}{2t}\left(1-\sqrt{1-\frac{4t}{1-t}}\right)\right) \\ &= \left(\frac{1}{\sqrt{1-6t+5t^2}}, \frac{1}{2t}\left(1-\sqrt{\frac{1-5t}{1-t}}\right)\right). \end{aligned}$$

For the first rows of corresponding infinite matrices, we have

$$\begin{bmatrix} 1 & & & & \\ 1 & 1 & & & \\ 1 & 2 & 1 & & \\ 1 & 3 & 3 & 1 & \\ 1 & 4 & 6 & 4 & 1 \end{bmatrix} \begin{bmatrix} 1 & & & & \\ 2 & 1 & & & \\ 6 & 3 & 1 & & \\ 20 & 10 & 4 & 1 & \\ 70 & 35 & 15 & 5 & 1 \end{bmatrix} = \begin{bmatrix} 1 & & & & \\ 3 & 1 & & & \\ 10 & 5 & 1 & & \\ 45 & 22 & 7 & 1 & \\ 195 & 97 & 37 & 9 & 1 \end{bmatrix}$$

and the generic element is

$$d_{n,k} = \sum_{j=0}^{n} \binom{n}{j} \binom{2j-k}{j-k}.$$

The first set of applications for the Riordan array concept is in proving and inverting identities. For instance, we observe that

$$\begin{bmatrix} 1 & 0 & 0 & 0 & 0 & 0 \\ 1 & 1 & 0 & 0 & 0 & 0 \\ 3 & 2 & 1 & 0 & 0 & 0 \\ 7 & 6 & 3 & 1 & 0 & 0 \\ 19 & 16 & 10 & 4 & 1 & 0 \\ 51 & 45 & 30 & 15 & 5 & 1 \end{bmatrix} \begin{bmatrix} 1 \\ 2 \\ 2 \\ 2 \\ 2 \\ 2 \end{bmatrix} = \begin{bmatrix} 1 \\ 3 \\ 9 \\ 27 \\ 81 \\ 243 \end{bmatrix}.$$

To actually prove that the right-hand column gives the powers of three we apply the FTRA. Since $1, 2, 2, 2, 2, \cdots \Leftrightarrow \frac{1+t}{1-t}$ it can be shown that

$$\left(\frac{1}{\sqrt{1-2t-3t^2}}, tm \right) \frac{1+t}{1-t} = \frac{1}{\sqrt{1-2t-3t^2}} \cdot \frac{1+tm}{1-tm} = \frac{1}{1-3t}. \quad (3.1.3)$$

Not every Riordan matrix is invertible and recall that an invertible Riordan matrix is called a proper Riordan array.

Theorem 3.2 (Riordan group) *Let $\mathcal{R}$ be the set of all proper Riordan arrays. Then $\mathcal{R}$ forms a group under the binary operation (3.1.2).*

Proof Let (g, f) and (d, h) be in the set $\mathcal{R}$. Since $g, d \in \mathcal{F}_0$ and $f, h \in \mathcal{F}_1$ it follows that $gd(f) \in \mathcal{F}_0$ and $h(f) \in \mathcal{F}_1$. Thus, the binary operation (3.1.2) is closed. Since matrix multiplication is associative so is the operation. Further, it can be easily shown that the identity is $I = (1, t)$, the usual identity matrix, and $(g, f)^{-1} = \left(\frac{1}{g(\overline{f})}, \overline{f} \right)$ where $\overline{f}$ is the compositional inverse of f, i.e., $f\left(\overline{f}\right) = \overline{f}(f) = t$. □

Finally, let us introduce the following concept due to Rogers: a *renewal array* is a p.R.a. $D = (d(t), h(t))$ such that $d(t) = h(t)/t$. The Pascal triangle is a renewal array and the paper of Rogers [13] is the first appearance in the literature of a concept similar to that of a p.R.a.

Given a f.p.s. $f(t) \in \mathcal{F}_1$, determining its compositional inverse poses some problems. Obviously, if in $f(\overline{f}(t)) = t$ we set $y = \overline{f}(t)$, we have $\overline{f}(t) = [y \mid f(y) = t]$,

Table 3.3 Stirling numbers of the first kind and the related Riordan array

	0	1	2	3	4	5
0	1					
1	0	1				
2	0	1	1			
3	0	2	3	1		
4	0	6	11	6	1	
5	0	24	50	35	10	1

	0	1	2	3	4	5
0	1					
1	0	1				
2	0	−1/2	1			
3	0	1/3	−1	1		
4	0	−1/4	11/12	−3/2	1	
5	0	1/5	−5/6	7/4	−2	1

so $\overline{f}(t)$ can be found by solving in y the equation $f(y) = t$. For example, let us consider our p.R.a. S in Example 3.4, having $f(t) = e^t - 1$. By solving in y the equation $e^y - 1 = t$ we find $y = \ln(1+t)$, and so $S^{-1} = (1, \ln(1+t))$ and its generic element is

$$d_{n,k} = [t^n](\ln(1+t))^k = \frac{k!}{n!}\begin{bmatrix} n \\ k \end{bmatrix}(-1)^{n-k},$$

related to the Stirling numbers of the first kind (see Table 3.3).

However, it is not always possible to solve the functional equation $f(y) = t$, and obtain the f.p.s. $\overline{f}(t)$; in these cases, we can look for a formula giving the single coefficient $\overline{f}_n = [t^n]\overline{f}(t)$. To this purpose, we begin by finding the p.R.a. inverse of the Lagrange matrix $(1, f(t))$, that is, $(1, \overline{f}(t))$.

Theorem 3.3 *Let* $D = (d_{n,k})_{n,k\in N} = (1, f(t))^{-1}$, *then we have*

$$d_{n,k} = \frac{k}{n}[t^{n-k}]\left(\frac{t}{f(t)}\right)^n.$$

Proof We simply verify that the matrix D is the inverse of $(1, f(t))$. Let us consider the generic element $v_{n,k}$ of the product $D * (1, f(t))$ and prove that it is $\delta_{n,k}$ (the Kronecker delta). We have

$$v_{n,k} = \sum_{j=0}^{k} d_{n,j}[y^j]f(y)^k = \sum_{j=0}^{k} \frac{j}{n}[t^{n-j}]\left(\frac{t}{f(t)}\right)^n [y^j]f(y)^k.$$

By the rule for differentiation of the coefficient of operator, we find

$$j[y^j]f(y)^k = [y^{j-1}]\frac{d}{dy}f(y)^k = k[y^j]yf'(y)f(y)^{k-1}.$$

Therefore

$$v_{n,k} = \frac{k}{n}\sum_{j=0}^{\infty}[t^{n-j}]\left(\frac{t}{f(t)}\right)^n [y^j]yf'(y)f(y)^{k-1} = \frac{k}{n}[t^n]\left(\frac{t}{f(t)}\right)^n tf'(t)f(t)^{k-1},$$

where we applied the convolution rule. We now distinguish between the case $k = n$ and $k \neq n$. When $k = n$ we have

$$v_{n,n} = \frac{n}{n}[t^n]t^n tf(t)^{-1}f'(t) = [t^0]f'(t)\left(\frac{t}{f(t)}\right) = 1;$$

in fact, $f'(t) = f_1 + 2f_2t + \cdots$ and $(f(t)/t)^{-1} = f_1^{-1} + \cdots$, so that the constant term in $f'(t)(t/f(t))$ is $f_1/f_1 = 1$. When $k \neq n$:

$$v_{n,k} = \frac{k}{n}[t^n]t^n tf(t)^{k-n-1}f'(t) = \frac{k}{n}[t^{-1}]\frac{1}{k-n}\cdot\frac{d}{dt}(f(t)^{k-n}) = 0$$

because $f(t)^{k-n}$ is a f.L.s. and, as we observed, the residue of its derivative is 0 (see Sect. 2.2). □

Since $d_{n,k} = [t^n]\overline{f}(t)^k$, we obtain a proof of the *Lagrange Inversion Formula* (LIF), already illustrated in Sect. 2.3:

Corollary 3.1 *If $f(t) \in \mathcal{F}_1$ and $\overline{f}(t)$ is its compositional inverse, then*

$$[t^n]\overline{f}(t) = \overline{f}_n = \frac{1}{n}[t^{n-1}]\left(\frac{t}{f(t)}\right)^n$$

and, more in general:

$$[t^n]\overline{f}(t)^k = \frac{k}{n}[t^{n-k}]\left(\frac{t}{f(t)}\right)^n.$$

Proof In $(1, \overline{f}(t))$ column 1 is composed of the coefficients of $\overline{f}(t)$, so that the first result is the case $k = 1$ of the previous theorem. The other columns are the powers of $\overline{f}(t)$ and therefore the second result follows. □

Example 3.7 Let us find the inverse P^{-1} of the Pascal triangle. We have $h(t) = t/(1-t)$ and so we should have: $h(\overline{h}(t)) = \overline{h}(h(t)) = t$. By solving this equation in $\overline{h}(t)$ we find $\overline{h}(t) = t/(1+t)$. Besides, we have

$$\overline{d}(t) = \frac{1}{d(\overline{h}(t))} = 1 - \overline{h}(t) = 1 - \frac{t}{1+t} = \frac{1}{1+t}.$$

Therefore

$$P^{-1} = \left(\frac{1}{1+t}, \frac{t}{1+t}\right).$$

Table 3.4 The inverse of the Catalan type triangle

	0	1	2	3	4	5
0	1					
1	−2	1				
2	0	−3	1			
3	0	2	−4	1		
4	0	0	5	−5	1	
5	0	0	−2	9	−6	1

and its generic element is

$$\overline{p}_{n,k} = [t^n]\frac{1}{1+t}\left(\frac{t}{1+t}\right)^k = [t^{n-k}]\frac{1}{(1+t)^{k+1}} = \binom{-k-1}{n-k} = (-1)^{n-k}\binom{n}{k},$$

a well-known result.

Example 3.8 For the Catalan type triangle C, we find

$$h(\overline{h}(t)) = \frac{1-\sqrt{1-4\overline{h}(t)}}{2} = t \qquad \text{or} \qquad \overline{h}(t) = t(1-t).$$

From this, we compute

$$\overline{d}(t) = \left(\frac{1}{\sqrt{1-4\overline{h}(t)}}\right)^{-1} = \sqrt{1-4t+4t^2} = 1-2t.$$

Therefore, we find

$$C^{-1} = (1-2t,\ t(1-t))$$

and the generic element is

$$\overline{c}_{n,k} = [t^n](1-2t)(t(1-t))^k = [t^{n-k}](1-t)^{k+1} - [t^{n-k-1}](1-t)^k =$$

$$= \binom{k+1}{n-k}(-1)^{n-k} - \binom{k}{n-k-1}(-1)^{n-k-1} = \frac{n+1}{k+1}\binom{k+1}{n-k}(-1)^{n-k}.$$

In Table 3.4, we give the initial part of this p.R.a.

Example 3.9 For the triangle of the Stirling numbers of the second kind, we have

$$h(\overline{h}(t)) = e^{\overline{h}(t)} - 1 = t \qquad \text{or} \qquad \overline{h}(t) = \ln(1+t).$$

Therefore we have $\overline{S} = (1,\ \ln(1+t))$, the generic element of which is

$$\overline{s}_{n,k} = \frac{k!}{n!} \begin{bmatrix} n \\ k \end{bmatrix} (-1)^{n-k},$$

as we have seen before (see Table 3.3).

We conclude this section by giving a characterization of p.R.a.'s by means of bivariate generating functions:

Theorem 3.4 *Let* $\delta(t, w)$ *be a bivariate f.p.s. and* $\left(\delta_{n,k}\right)_{n,k\in\mathbb{N}}$ *be defined as* $\delta_{n,k} = [t^n][w^k]\delta(t, w)$. *Then* $\left(\delta_{n,k}\right)_{n,k\in\mathbb{N}}$ *is a p.R.a. if and only if two f.p.s.* $d(t) \in \mathcal{F}_0,\ h(t) \in \mathcal{F}_1$ *exist such that*

$$\delta(t, w) = \frac{d(t)}{1 - wh(t)}. \tag{3.1.4}$$

Proof Let us suppose that $\delta(t, w)$ be as in (3.1.4), then we have

$$[t^n w^k]\delta(t, w) = [t^n w^k]d(t) \sum_{k=0}^{\infty} w^k h(t)^k = [t^n]d(t)h(t)^k$$

and this shows that $\left(\delta_{n,k}\right)_{n,k\in\mathbb{N}}$ is a p.R.a. On the other hand, if $(d(t), h(t))$ is a p.R.a., then the generating function of the kth column is $d(t)h(t)^k$ and the bivariate generating function relative to the whole array is

$$\sum_{k=0}^{\infty} d(t)h(t)^k w^k = d(t) \cdot \frac{1}{1 - wh(t)}$$

as desired. □

Remark 3.2 A proper Riordan array is characterized by a bivariate generating function $\delta(t, w) = \dfrac{d(t)}{1 - wh(t)}$, where $d(t) \in \mathcal{F}_0,\ h(t) \in \mathcal{F}_1$; $d(t)$ is the generating function of column 0; and $h(t)$ is the ratio between any two consecutive columns.

It is now useful to have a formula to perform the row-by-column product of two p.R.a.'s when we have them as bivariate generating functions:

Theorem 3.5 *If* $\delta(t, w)$ *and* $\epsilon(t, w)$ *are two p.R.a.'s bivariate generating functions, then we have*

$$\delta(t, w) * \epsilon(t, w) = \delta(t, 0) \left[\epsilon(y, w) \;\middle|\; y = 1 - \frac{\delta(t, 0)}{\delta(t, 1)} \right].$$

Proof Let $\delta(t, w) = d(t)/(1 - wh(t))$ for some $d(t) \in \mathcal{F}_0$ and $h(t) \in \mathcal{F}_1$. We have

$$\delta(t, 0) = d(t) \qquad \text{and} \qquad \frac{1}{\delta(t, 1)} = \frac{1 - h(t)}{d(t)}$$

so that $1 - \delta(t, 0)/\delta(t, 1) = h(t)$. If we also have $\epsilon(t, w) = a(t)/(1 - wb(t))$, then the substitution in the assertion is equivalent to

$$\frac{d(t)a(h(t))}{1 - wb(h(t))}$$

which is the bivariate generating function of the row-by-column product according to Eq. (3.1.2). □

An important property of Riordan arrays concerns the computation of combinatorial sums. In particular, we have the following result (see, e.g., [8, 11, 16]):

$$\sum_{k=0}^{n} d_{n,k} f_k = [t^n]d(t) f(h(t)), \tag{3.1.5}$$

that is, every combinatorial sum involving a Riordan array can be computed by extracting the coefficient of t^n from the generating function $d(t) f(h(t))$ where $f(t) = \mathcal{G}(f_k) = \sum_{k\geq 0} f_k t^k$ is the generating function of the sequence $(f_k)_{k\in\mathbb{N}}$. Due to its importance, relation (3.1.5) is often called the *fundamental rule of Riordan arrays*. The proof consists in a straightforward computation:

$$\sum_{k=0}^{n} d_{n,k} f_k = \sum_{k=0}^{\infty} d_{n,k} f_k = \sum_{k=0}^{\infty} [t^n]d(t)h(t)^k f_k =$$

$$= [t^n]d(t) \sum_{k=0}^{\infty} f_k h(t)^k = [t^n]d(t) f(h(t)).$$

Note, however, that this relation is equivalent to Theorem 3.1.

Remark 3.3 The *fundamental rule* of Riordan arrays states that every combinatorial sum involving a Riordan array can be computed as follows:

$$\sum_{k=0}^{n} d_{n,k} f_k = [t^n]d(t) f(h(t)), \text{ where } f(t) = \mathcal{G}(f_k).$$

In Chap. 5, many applications of this rule will be illustrated.

3.2 Some Special Subgroups

There are several interesting subgroups of the Riordan group $\mathcal{R}$. In this section, we introduce the special subgroups and briefly describe some of their properties [7]. The detailed proof is left as an exercise to the reader.

Appell subgroup A Riordan array of the form $(g(t), t)$ is called an *Appell matrix* and the set

$$\mathcal{A} = \{(g(t), t) \in \mathcal{R} \mid g(t) \in \mathcal{F}_0\}$$

forms a subgroup of the Riordan group. We call $\mathcal{A}$ the *Appell subgroup*. If $g(t) = \sum_{n\geq 0} g_n t^n$ then the Appell matrix $(g(t), t) = (a_{ij})$ is of the Toeplitz form in which $a_{ij} = g_{i-j}$ if $i \geq j$ and $a_{ij} = 0$ otherwise:

$$(g(t), t) = \begin{bmatrix} g_0 & 0 & 0 & 0 & \cdots \\ g_1 & g_0 & 0 & 0 & \cdots \\ g_2 & g_1 & g_0 & 0 & \cdots \\ g_3 & g_2 & g_1 & g_0 & \cdots \\ \vdots & \vdots & \vdots & \vdots & \ddots \end{bmatrix}.$$

The product of two Appell matrices is also an Appell matrix; by (3.1.2) we have

$$(g(t), t) * (f(t), t) = (g(t)f(t), t)$$

and so this product corresponds to the product of two f.p.s.'s. Therefore, if $g(t)^{-1}$ is the f.p.s. inverse of $g(t)$, then the Appell matrix $(g(t), t)$ has as inverse $(g(t)^{-1}, t)$. An important property of the Appell subgroup is

Theorem 3.6 *The Appell subgroup is a normal subgroup of the Riordan group.*

Proof Let $D = (d(t),\ h(t)) \in \mathcal{R}$ and $A = (g(t),\ t) \in \mathcal{A}$; we must prove that $DAD^{-1} \in \mathcal{A}$. Actually, we have

$$DA = (d(t),\ h(t))(g(t),\ t) = (d(t)g(h(t)),\ h(t));$$

we now use the inverse of D:

$$(d(t)g(h(t)),\ h(t))\left(\frac{1}{d(\bar{h}(t))},\ \bar{h}(t)\right) = \left(\frac{d(t)g(h(t))}{d(\bar{h}(h(t)))},\ \bar{h}(h(t))\right) = (g(h(t)),\ t)$$

which is an Appell matrix. □

The Appell subgroup is isomorphic to the group $\mathcal{F}_0$ of invertible formal power series under multiplication.

Table 3.5 A Lagrange matrix

$$(1, f(t)) = \begin{bmatrix} 1 & 0 & 0 & 0 & 0 & \cdots \\ 0 & f_1 & 0 & 0 & 0 & \cdots \\ 0 & f_2 & f_1^2 & 0 & 0 & \cdots \\ 0 & f_3 & 2f_1f_2 & f_1^3 & 0 & \cdots \\ 0 & f_4 & 2f_1f_3 + f_2^2 & 3f_1^2f_2 & f_1^4 & \cdots \\ \vdots & \vdots & \vdots & \vdots & \vdots & \ddots \end{bmatrix}.$$

Lagrange subgroup A Riordan array of the form $(1, f(t))$ is called a *Lagrange matrix* and the set

$$\mathcal{L} = \{(1, f(t)) \in \mathcal{R} \mid f(t) \in \mathcal{F}_1\}$$

forms a subgroup of the Riordan group. We call $\mathcal{L}$ the *Lagrange subgroup* and it is sometimes called the *associated subgroup*. Clearly, $\mathcal{L}$ is isomorphic to the group $\mathcal{F}_1$ of formal power series under composition. If $D \in \mathcal{L}$ and $D = (1, f(t))$, then its generic element is

$$d_{n,k} = [t^n]f(t)^k$$

showing that the columns of D are generated by the successive powers of $f(t)$ (see Table 3.5). This is very important and we can see how the Riordan array product acts on these matrices:

$$(1, f(t))(1, g(t)) = (1, g(f(t)))$$

so that the product of two Lagrange matrices is also a Lagrange matrix. Obviously, the identity is $(1, t)$ and the inverse of $(1, f(t))$ is $(1, \overline{f}(t))$. Therefore, the set of Lagrange matrices is a group, isomorphic to $(\mathcal{F}_1, \circ)$, if $\circ$ denotes composition.

In this way, we have solved, at least from a theoretical point of view, the problem of inverting a Lagrange matrix. However, as we are now going to see, the result can be extended to all p.R.a.'s:

Theorem 3.7 *Every p.R.a. can be seen as the product of an Appell matrix by a Lagrange matrix.*

Proof Let $(d(t), f(t))$ be any p.R.a., then we clearly have

$$(d(t), f(t)) = (d(t) \cdot 1, f(t)) = (d(t), t)(1, f(t))$$

as desired. □

As a consequence, we have a method to invert any p.R.a.:

Theorem 3.8 *Let $(d(t), f(t))$ be any p.R.a., then the inverse p.R.a. is*

$$(d(t), f(t))^{-1} = \left(\frac{1}{d(\overline{f}(t))}, \overline{f}(t)\right).$$

Proof As we have seen, $(d(t), f(t)) = (d(t), t)(1, f(t))$ and so

$$(d(t), f(t))^{-1} = (1, f(t))^{-1}(d(t), t)^{-1} = (1, \overline{f}(t))(d(t)^{-1}, t) = \left(\frac{1}{d(\overline{f}(t))}, \overline{f}(t)\right).$$

This concludes the proof. □

Bell subgroup A Riordan array of the form $(g(t), tg(t))$, or $(f(t)/t, f(t))$ is called a *Bell matrix* and the set

$$\mathcal{B} = \{(g(t), tg(t)) \in \mathcal{R} \mid g(t) \in \mathcal{F}_0\} \text{ or } \{(f(t)/t, f(t)) \in \mathcal{R} \mid f(t) \in \mathcal{F}_1\}$$

forms a subgroup of the Riordan group. We call $\mathcal{B}$ the *Bell subgroup*. Bell matrices frequently arise in combinatorics, e.g., the Pascal matrix $(1/(1-t), t/(1-t))$; the Catalan matrix $(C(t), tC(t))$; and the Motzkin matrix $(m(t), tm(t))$. Every proper Riordan array can be also factorized by the Appell matrix and the Bell matrix as follows:

$$(g(t), f(t)) = (tg(t)/f(t), t)(f(t)/t, f(t)).$$

Power-Bell subgroup For a fixed real number r, a Riordan array of the form $(g(t)tg(t)^r)$ is called a *power-Bell matrix* or an *r-Bell matrix*. The set

$$\mathcal{B}_r = \{\left(g(t), tg(t)^r\right) \in \mathcal{R} \mid g(t) \in \mathcal{F}_0\}$$

forms a subgroup of the Riordan group. We call $\mathcal{B}_r$ the *power-Bell subgroup* or the *r-Bell subgroup*. Clearly, $\mathcal{B}_0$ is the Appell subgroup and $\mathcal{B}_1$ is the Bell subgroup. For $r = 2$, we give an example with $F(t) = \frac{1}{1-t-t^2}$, the generating function for the Fibonacci numbers. Let $g(t) = F(t)^{\frac{1}{2}}$. Then

$$(F(t)^{\frac{1}{2}}, tF(t)) = \begin{bmatrix} 1 & 0 & 0 & 0 & 0 & \dots \\ 1/2 & 1 & 0 & 0 & 0 & \dots \\ 7/8 & 3/2 & 1 & 0 & 0 & \dots \\ 17/16 & 27/8 & 5/2 & 1 & 0 & \dots \\ 203/128 & 95/16 & 55/8 & 7/2 & 1 & \dots \\ \vdots & \vdots & \vdots & \vdots & \vdots & \ddots \end{bmatrix}.$$

Note that if we take $g(t) = 1/\sqrt{1-4t}$, then we get an element of the 2-Bell subgroup with integral entries. Another element of the power-Bell subgroup with integral entries can also be obtained by using the Fine numbers (sequence A000957). The Fine numbers count Dyck paths with no hills.

Checkerboard subgroup For an even function $g_e(t)$ and an odd function $g_o(t)$, a Riordan array of the form $(g_e(t),\ f_o(t))$ is called a *checkerboard matrix* and the set

$$C = \{(g_e(t),\ f_o(t)) \in \mathcal{R} \mid g_e(t) \in \mathcal{F}_0;\ f_o(t) \in \mathcal{F}_1\}$$

forms a subgroup of the Riordan group. We call C the *checkerboard subgroup*. An important property is that the centralizer of the element $M = (1, -t)$ of the Riordan group is the checkerboard subgroup (see Exercise 3.2).

Multi-checkerboard subgroup For a fixed positive integer m, a Riordan array of the form $(g(t), tf(t^m))$ is called a *multi-checkerboard matrix* or *m-checkerboard matrix*, and the set

$$C_m = \{(g(t), tf(t^m)) \in \mathcal{R} \mid g(t),\ f(t) \in \mathcal{F}_0\}$$

forms a subgroup of the Riordan group. We call C_m the *multi-checkerboard subgroup* or *m-checkerboard subgroup*. We note that the checkerboard subgroup is a subgroup of the multi-checkerboard subgroup (see Exercise 3.3). The multi-checkerboard subgroup is sometimes called the *Cheon subgroup* [7].

Derivative subgroup A Riordan array of the form $(f'(t),\ f(t))$ is called a *derivative matrix* where $f'(t)$ denotes the first derivative of $f(t)$, and the set

$$\mathcal{D} = \{(f'(t),\ f(t)) \in \mathcal{R} \mid f(t) \in \mathcal{F}_1\}$$

forms a subgroup of the Riordan group. We call $\mathcal{D}$ the *derivative subgroup*. For example,

$$(1+2t, t(1+t)) = \begin{bmatrix} 1\,0\,0\,0\,0 \ldots \\ 2\,1\,0\,0\,0 \ldots \\ 0\,3\,1\,0\,0 \ldots \\ 0\,2\,4\,1\,0 \ldots \\ 0\,0\,5\,5\,1 \ldots \\ \vdots\ \vdots\ \vdots\ \vdots\ \vdots\ \ddots \end{bmatrix}.$$

Hitting time subgroup A Riordan array of the form $(tf'(t)/f(t),\ f(t))$ is called a *hitting time matrix*, and the set

$$\mathcal{H} = \{(tf'(t)/f(t),\ f(t)) \in \mathcal{R} \mid f(t) \in \mathcal{F}_1\}$$

forms a subgroup of the Riordan group. We call $\mathcal{H}$ the *hitting time subgroup*.

Let p be a prime number. It is well known that p divides $\binom{p}{k}$ for $k = 1, \ldots, p-1$. It is also known [12] that the entries of each matrix in the hitting time subgroup exhibit the divisibility property. Indeed, by observing that $\binom{n}{k}$ are entries of the Pascal matrix, the divisibility property can be generalized to a large set of Riordan arrays as follows (see [12]).

(Divisibility property) A subset of Riordan matrices is said to have the *divisibility property* if each matrix $M = (m_{nk})_{n,k\geq 0}$ of the subset satisfies the property that n divides $k \cdot m_{nk}$ whenever $0 < k < n$.

Stabilizer subgroup Eigenvalues and eigenvectors of a square matrix A that satisfy the matrix equation $Av = \lambda v$, $v \neq 0$ are often introduced in the context of linear algebra.

Let $A = (g(t), f(t))$ be a Riordan matrix with eigenvalue $\lambda = 1$. If there exists a nonzero vector $v = (v_0, v_1, \ldots)^T$ such that $Av = v$, we say that A *stabilizes* v. Applying the FTRA together with the generating function $v(t) = \sum_{n\geq 0} v_n t^n$, we obtain $g(t)v(f(t)) = v(t)$, i.e., $g(t) = v(t)/v(f(t))$. For a given nonzero vector $v = (v_0, v_1, \ldots)^T$, the set

$$\mathcal{S}_v = \{(g(t), f(t)) \in \mathcal{R} \mid (g(t), f(t))v = v\} \text{ or } \left\{\left(\frac{v(t)}{v(f(t))}, f(t)\right) \in \mathcal{R} \,\middle|\, f(t) \in \mathcal{F}_1\right\}$$

forms a subgroup of the Riordan group. We call $\mathcal{S}_v$ the *stabilizer subgroup* of v.

It might be interesting to find a vector v for which $\mathcal{S}_v$ has special algebraic or combinatorial properties. For example, $\mathcal{S}_{e_1}$ with $e_1 = (1, 0, \ldots)$ is the Lagrange subgroup. In particular, the stabilizer subgroup of $e = (1, 1, \ldots)^T$ is called the *stochastic subgroup* and its elements are called *stochastic matrices*. Since a stochastic matrix here is infinite lower triangular and some of its entries can be negative, it differs from the usual stochastic matrix which is non-negative.

3.3 Several Aspects of the Riordan Group

In this section, we observe the Riordan group from an algebraic and topological perspective. First, we observe the Riordan group in the semidirect product of $\mathcal{F}_0$ and $\mathcal{F}_1$, and then we present its representation as an inverse limit of the inverse system of finite Riordan groups. We will also briefly discuss the Riordan group as the set of Krylov matrices obtained from the coefficients of formal power series.

Semidirect product Let $\mathcal{R}$ be the Riordan group with the identity $I = (1, t)$. Since $\mathcal{A} \lhd \mathcal{R}$ and

$$\mathcal{R} = \{(g, f) = (g, t)(1, f) \mid g \in \mathcal{F}_0, f \in \mathcal{F}_1\} = \mathcal{A}\mathcal{L}, \quad \mathcal{A} \cap \mathcal{L} = \{I\},$$

we see that the Riordan group is the semidirect product of the Appell subgroup $\mathcal{A}$ and the Lagrange subgroup $\mathcal{L}$ written as

$$\mathcal{R} = \mathcal{A} \rtimes \mathcal{L}.$$

Now we observe the Riordan group as a semidirect product of $\mathcal{F}_0$ and $\mathcal{F}_1$ as follows. Recall that $\mathcal{A} \cong \mathcal{F}_0$ and $\mathcal{L} \cong \mathcal{F}_1$. We denote by $\mathrm{Aut}(\mathcal{F}_0)$ the group of all automorphisms of $\mathcal{F}_0$. For every $g = g_0 + \sum_{i=1}^{\infty} g_i t^i \in \mathcal{F}_0$ and $f \in \mathcal{F}_1$, the substitution $g \circ f = g_0 + \sum_{i=1}^{\infty} g_i f^i$ is also in $\mathcal{F}_0$. Let $\varphi : \mathcal{F}_1 \to \mathrm{Aut}(\mathcal{F}_0)$ be a homomorphism defined by $\varphi(f)(g) = \varphi_f(g)$ where $\varphi_f(g) = g \circ f$ for every $g \in \mathcal{F}_0$ and $f \in \mathcal{F}_1$. It is clear that φ is well defined and indeed it is a homomorphism. Thus, the semidirect product $\mathcal{F}_0 \rtimes_\varphi \mathcal{F}_1$ is a group with the following properties:

(i) the underlying set is the Cartesian product:

$$\mathcal{F}_0 \times \mathcal{F}_1 = \{(g, f) \mid g \in \mathcal{F}_0, f \in \mathcal{F}_1\};$$

(ii) the operation $\cdot$ is determined by the homomorphism φ:

$$(g_1, f_1) \cdot (g_2, f_2) = (g_1 \varphi_{f_1}(g_2), f_1 \circ f_2) = (g_1 g_2(f_1), f_1 \circ f_2),$$

for every $(g_1, f_1), (g_2, f_2) \in \mathcal{F}_0 \times \mathcal{F}_1$.

Clearly, $(1, z)$ is the identity of $\mathcal{F}_0 \rtimes_\varphi \mathcal{F}_1$, and for every $(g, f) \in \mathcal{F}_0 \rtimes_\varphi \mathcal{F}_1$, its inverse is

$$(g, f)^{-1} = (\varphi_{f^{-1}}(g^{-1}), f^{-1}) = (g^{-1}(\bar{f}), \bar{f}),$$

where $f(\bar{f}) = \bar{f}(f) = t$.

Thus, it is easily shown that the Riordan group $\mathcal{R}$ is isomorphic to the semidirect product of $\mathcal{F}_0$ and $\mathcal{F}_1$ with respect to φ defined by $\varphi(f)(g) = g \circ f$ for $g \in \mathcal{F}_0$ and $f \in \mathcal{F}_1$, i.e.,

$$\mathcal{R} \cong \mathcal{F}_0 \rtimes_\varphi \mathcal{F}_1.$$

Inverse limit group By using the inverse limit tool, the Riordan group can be obtained in its infinite and bi-infinite representations from a family of groups of finite Riordan matrices [1, 9]. An *inverse system* $(G_n, \varphi_n)_{n\in\mathbb{N}}$ of groups consists of a family $(G_n)_{n\in\mathbb{N}}$ of groups and a family of homomorphisms $(\varphi_n : G_{n+1} \to G_n)_{n\in\mathbb{N}}$ for every $n \in \mathbb{N}$.

As every Riordan matrix is a lower triangular matrix, for every $m \in \mathbb{N}$, we have a natural homomorphism

$$\pi_m : \mathcal{R}(\mathbb{K}) \to GL_{m+1}(\mathbb{K})$$

defined by

$$\pi_m((d_{i,j})_{i,j\in\mathbb{N}}) := (d_{i,j})_{i,j=0,\ldots,m}.$$

For every $m \in \mathbb{N}$, let $\mathcal{R}_m(\mathbb{K}) := \pi_m(\mathcal{R}(\mathbb{K}))$. Then the natural map

$$\varphi_m : \mathcal{R}_{m+1}(\mathbb{K}) \to \mathcal{R}_m(\mathbb{K})$$

given by

$$\varphi_m((d_{i,j})_{i,j=0,\ldots,m+1}) := (d_{i,j})_{i,j=0,\ldots,m}$$

is a homomorphism. Then the Riordan group is isomorphic to the *inverse limit group*

$$\varprojlim\{(\mathcal{R}_m(\mathbb{K}), \varphi_m)_{m\in\mathbb{N}})\}.$$

The algebraic and topological properties of the Riordan group $\mathcal{R}(\mathbb{K}) = \{(g, f)|g \in \mathcal{F}_0, f \in \mathcal{F}_1\}$ depend heavily on the structure of the ring $\mathbb{K}$. When $\mathbb{K}$ is $\mathbb{R}$ or $\mathbb{C}$, many properties of the Riordan group $\mathcal{R}(\mathbb{K})$ can be investigated by using its Lie group structures, see [1, 9]. Indeed, as we shall see in the next chapter, a proper Riordan matrix (g, f) can be also characterized by both $g = g_0 + g_1 t + g_2 t^2 + \cdots$ and its A-sequence $a_0, a_1, a_2, \ldots$. Thus, $\mathcal{R}(\mathbb{K})$ has a Lie group structure in the sense that there exists an isomorphism ϕ such that

$$\phi : \mathcal{U}_\infty \to \mathcal{R}(\mathbb{K}),$$

where $\mathcal{U}_\infty = \{u = (g_0, a_0, g_1, a_1, \ldots) \in \mathbb{K}^{\mathbb{N}}\}$.

Krylov matrix representation A *Krylov matrix* of a matrix $A \in M_n(\mathbb{R})$ by a column vector $\mathrm{b} \in \mathbb{R}^n$ is defined as

$$K(A, \mathrm{b}) = \left[\mathrm{b}, A\mathrm{b}, \ldots, A^{n-1}\mathrm{b}\right].$$

Krylov matrices play a fundamental role in matrix computations. It is known that they are key tools in understanding and developing numerical methods for solving eigenvalue problems and systems of linear equations. In particular, Krylov subspace methods are known as the most successful methods for finding a few eigenvalues of a large sparse matrix in numerical linear algebra.

If A is an infinite lower triangular matrix then the Krylov matrix $K(A, \mathrm{b})$ is infinite and it is well defined. Infinite lower triangular matrices appear often in representations of operators on the space of formal power series. Let $g = \sum_{n=0}^{\infty} g_n t^n$ and $f = \sum_{n=0}^{\infty} f_n t^n$. Associating with the formal power series f the Riordan array of the Appell form

$$(f,t)=\begin{bmatrix} f_0 & & & & \\ f_1 & f_0 & & O & \\ f_2 & f_1 & f_0 & & \\ f_3 & f_2 & f_1 & f_0 & \\ \vdots & \ddots & \ddots & \ddots & \ddots \end{bmatrix},$$

we obtain $(f,t)g = fg$ by FTRA. The matrix (f,t) is a regular representation of the linear map on the space of formal power series that sends g to fg. By a matrix computation of $(f,t)g$, every equation of the form $fg = h$, where $h = \sum_{n\geq 0} h_n t^n$, gives us convolution of f and g where $h_n = \sum_{k=0}^{n} f_{n-k} g_k$.

Further, we have

$$(f,t)^k g = (f^k, t)g = f^k g, \; k = 0, 1, \ldots.$$

Therefore if $f_0 = f(0) = 0$, by letting

$$A_f := (f,t) = \begin{bmatrix} 0 & 0 & 0 & 0 & \ldots \\ f_1 & 0 & 0 & 0 & \ldots \\ f_2 & f_1 & 0 & 0 & \ldots \\ f_3 & f_2 & f_1 & 0 & \ldots \\ \vdots & \vdots & \vdots & \vdots & \ddots \end{bmatrix}, \quad \mathrm{g} = \begin{bmatrix} g_0 \\ g_1 \\ g_2 \\ \vdots \end{bmatrix},$$

we obtain

$$(g, f) = \left[\mathrm{g}, A_f \mathrm{g}, A_f^2 \mathrm{g}, \cdots\right] = K(A_f, \mathrm{g}).$$

This shows that every Riordan array is a Krylov matrix (also see [2]).

Now let $\mathcal{K} = \{K(A_f, \mathrm{g}) \mid g_0 \neq 0, f_0 = 0, f_1 \neq 0\}$. Explicitly, the set $\mathcal{K}$ can be described as the group of Krylov matrices with the multiplication

$$K(A_f, \mathrm{g})K(A_\ell, \mathrm{h}) = K(A_{\ell(f)}, \mathrm{d}),$$

where $\mathrm{d} = K(A_f, \mathrm{g})\mathrm{h}$. The identity is $K(S, \mathrm{e}_1)$ where

$$S = (t,t) = \begin{bmatrix} 0 & & & & \\ 1 & 0 & & O & \\ 0 & 1 & 0 & & \\ 0 & 0 & 1 & 0 & \\ \vdots & \ddots & \ddots & \ddots & \ddots \end{bmatrix}$$

and $\mathrm{e}_1 = (1, 0, \ldots)^T$. The inverse of $K(A_f, \mathrm{g})$ is $K(A_{\overline{F}}, \overline{\mathrm{g}})$ where $A_{\overline{F}} = (\overline{f}, t)$ and $\overline{\mathrm{g}}$ is the column vector corresponding to the formal power series $1/g(\overline{f})$ for the compositional inverse $\overline{f}$ of f. We call the group $\mathcal{K}$ the *Krylov group*. The FTRA can be also interpreted as

$$K(A_f, \mathrm{g})\mathrm{h} = h(A_f)\mathrm{g}, \tag{3.3.1}$$

where $h(t)$ is the generating series of a vector h. Note that Eq. (3.3.1) is valid for any square matrix A of order n instead of A_f, not necessarily Riordan matrix. It might be very useful for studying combinatorial sums and solving linear equations.

Exercises

3.1 Prove that the Lagrange subgroup is isomorphic to the Bell subgroup.

3.2 Prove that the centralizer of the element $M = (1, -t)$ of the Riordan group is the checkerboard subgroup.

3.3 Prove that the checkerboard subgroup is a subgroup of the multi-checkerboard subgroup.

3.4 Prove that the set of all Riordan matrices of the form $(C(t)/C(f(t)), f(t))$ is a (stabilizer) subgroup of the Riordan group.

3.5 Prove that the derivative subgroup is isomorphic to the Bell subgroup.

3.6 Let $\widehat{\mathcal{R}} := \{(\hat{g}, \hat{f}) \in \mathcal{R} \mid [t^0]\hat{g} = 1, [t^1]\hat{f} = 1\}$ and $\mathcal{D} := \{(g_0, f_1 t) \in \mathcal{R} \mid g_0, f_1 \neq 0\}$. Prove that $\mathcal{R} = \widehat{\mathcal{R}} \rtimes \mathcal{D}$.

3.7 Prove that if $A = (g(t), f(t))$ is a pseudo-involution, i.e., $(AM)^2 = I$ with $M = (1, -t)$, then $A^{-1} = (g(-t), -f(-t))$.

3.8 Let $A, B \in \mathcal{R}$ such that $B = P^{-1}AP$ where P is a checkerboard matrix. Then A is a pseudo-involution if and only if B is a pseudo-involution.

3.9 Prove that the hitting time subgroup has the divisibility property. In particular, the Pascal matrix is an element of the hitting time subgroup.

References

1. G.-S. Cheon, A. Luzón, M. Morón, F. Prieto-Martinez, M. Song, Finite and infinite dimensional Lie group structures on Riordan groups. Adv. Math. **319**, 522–566 (2017)
2. G.-S. Cheon, M. Song, A new aspect of Riordan arrays via Krylov matrices. Linear Algebr. Appl. **554**, 329–341 (2016)
3. G.P. Egorychev, E.V. Zima, Decomposition and group theoretic characterization of pairs of inverse relations of the Riordan type. Acta Appl. Math. **85**(1–3), 93–109 (2005)

4. H.W. Gould, T.-X. He, Characterization of (c)-Riordan arrays, Gegenbauer-Humbert-type polynomial sequences, and (c)-Bell polynomials. J. Math. Res. Appl. **33**(5), 505–527 (2013)
5. T.-X. He, L.C. Hsu, P.J.-S. Shiue, The Sheffer group and the Riordan group. Discret. Appl. Math. **155**(15), 1895–1909 (2007)
6. T.-X. He, R. Sprugnoli, Sequence characterization of Riordan arrays. Discret. Math. **309**(12), 3962–3974 (2009)
7. C. Jean-Louis, A. Nkwanta, Some algebraic structure of the Riordan group. Linear Algebr. Appl. **438**, 2018–2035 (2013)
8. A. Luzón, D. Merlini, M.A. Morón, R. Sprugnoli, Identities induced by Riordan arrays. Linear Algebr. Appl. **436**(3), 631–647 (2012)
9. A. Luzón, D. Merlini, M. Morón, F. Prieto-Martinez, R. Sprugnoli, Some inverse limit approaches to the Riordan group. Linear Algebr. Appl. **491**, 239–262 (2016)
10. D. Merlini, D.G. Rogers, R. Sprugnoli, M.C. Verri, On some alternative characterizations of Riordan arrays. Can. J. Math. **49**, 301–320 (1997)
11. D. Merlini, R. Sprugnoli, M.C. Verri, Combinatorial sums and implicit Riordan arrays. Discret. Math. **309**(2), 475–486 (2009)
12. P. Peart, W. Woan, A divisibility property for a subgroup of Riordan matrices. Discret. Appl. Math. **98**, 255–263 (2000)
13. D.G. Rogers, Pascal triangles, Catalan numbers and renewal arrays. Discret. Math. **22**, 301–310 (1978)
14. L.W. Shapiro, Bijections and the Riordan group. Theor. Comput. Sci. **307**(2), 403–413 (2003)
15. L.W. Shapiro, S. Getu, W.-J. Woan, L. Woodson, The Riordan group. Discret. Appl. Math. **34**, 229–239 (1991)
16. R. Sprugnoli, Riordan arrays and combinatorial sums. Discret. Math. **132**(1–3), 267–290 (1994)
17. R. Sprugnoli, Riordan arrays and the Abel-Gould identity. Discret. Math. **142**(1–3), 213–233 (1995)
18. W. Wang, T. Wang, Generalized Riordan arrays. Discret. Math. **308**(24), 6466–6500 (2008)
19. X. Zhao, S. Ding, T. Wang, Some summation rules related to the Riordan arrays. Discret. Math. **281**(1–3), 295–307 (2004)

Chapter 4
Characterization of Riordan Arrays by Special Sequences

Abstract As already described in the previous chapters, the concept of a Riordan array was introduced in 1991 by Shapiro, Getu, Woan and Woodson [21], with the aim of defining a class of infinite lower triangular arrays with properties analogous to those of the Pascal triangle. This concept was subsequently studied by Sprugnoli [22] in the context of the computation of combinatorial sums. In these papers, Riordan arrays correspond to matrices $(d_{n,k})_{n,k\in\mathbb{N}}$ where each element is described by a linear combination of the elements in the previous row, starting from the previous column; this characterization was already observed by Rogers [18]. The coefficients of this linear combination are independent of n and k, with $k \neq 0$, and constitute a specific sequence called the A-sequence of the Riordan array; when $k = 0$ another sequence called the Z- sequence is involved, as illustrated in Sect. 4.1. Later, several new characterizations of Riordan arrays were given in Merlini, Rogers, Sprugnoli and Verri [11]: the main result in that paper shows that a lower triangular array is of Riordan type whenever its generic element $d_{n+1,k+1}$ linearly depends on the elements $d_{r,s}$ lying in a well-defined zone of the array. The coefficients of this dependence constitute the so-called A-matrix and are illustrated in Sect. 4.2. There is no difference between Riordan arrays defined in either way: the A-sequence is a particular case of A-matrix and, given a Riordan array defined by an A-matrix, this corresponds to a well-defined A-sequence. However, there are some examples in which a Riordan array can be easily studied by means of the A-matrix while the A-sequence is very complex. From a combinatorial point of view, this means that it is very challenging to find a construction allowing us to obtain objects of size $n + 1$ from objects of size n. Instead, the existence of a simple A-matrix corresponds to a possible construction from objects of different sizes less than $n + 1$. Some combinatorial problems studied in terms of A-matrix can be found in [9–12]. Algebraic characterizations of Riordan arrays in terms of A- and Z- sequences are illustrated in [8]; many combinatorial examples where the A- and Z- sequences are strictly related are examined in [4]. In Sect. 4.3 it is shown that the A- and Z- sequences are related to the concept of production matrix and it is highlighted how the A-sequence, A-matrix and production matrix concepts can be used to prove the Riordan array nature of a problem.

© The Author(s), under exclusive license to Springer Nature Switzerland AG 2022

L. Shapiro et al., *The Riordan Group and Applications*, Springer Monographs in Mathematics, https://doi.org/10.1007/978-3-030-94151-2_4

4.1 The A- and Z- Sequences

For a proper Riordan array we have another important characterization, due to Rogers [18]: every element $d_{n+1,k+1}$, $n, k \in \mathbb{N}$, can be expressed as a linear combination of the elements in the preceding row, i.e.:

$$d_{n+1,k+1} = a_0 d_{n,k} + a_1 d_{n,k+1} + a_2 d_{n,k+2} + \cdots = \sum_{j=0}^{\infty} a_j d_{n,k+j}. \tag{4.1.1}$$

The sum is actually finite and the sequence $A = (a_k)_{k \in \mathbb{N}}$ is fixed. More precisely, we can prove the following theorem:

Theorem 4.1 *An infinite lower triangular array $D = \left(d_{n,k}\right)_{n,k \in \mathbb{N}}$ is a p.R.a. if and only if a sequence $A = \{a_0 \neq 0, a_1, a_2, \ldots\}$ exists such that for every $n, k \in \mathbb{N}$ relation (4.1.1) holds.*

Proof Let us suppose that D is the Riordan array $\mathcal{R}(d(t), h(t))$ and let us consider the Riordan array $\mathcal{R}(d(t)h(t)/t, h(t))$; we define the Riordan array $\mathcal{R}(A(t), M(t))$ by the relation:

$$\mathcal{R}(A(t), M(t)) = \mathcal{R}(d(t), h(t))^{-1} * \mathcal{R}(d(t)h(t)/t, h(t))$$

or:

$$\mathcal{R}(d(t), h(t)) * \mathcal{R}(A(t), M(t)) = \mathcal{R}(d(t)h(t)/t, h(t)).$$

By performing the product we find:

$$d(t)A(h(t)) = d(t)h(t)/t \qquad \text{and} \qquad M(h(t)) = h(t).$$

The latter identity implies $M(t) = t$. Therefore we have $\mathcal{R}(d(t), h(t)) * \mathcal{R}(A(t), t) = \mathcal{R}(d(t)h(t)/t, h(t))$. The element $f_{n,k}$ of the left hand member is $\sum_{j=0}^{\infty} d_{n,j} a_{k-j} = \sum_{j=0}^{\infty} d_{n,k+j} a_j$, if as usual we interpret a_{k-j} as 0 when $k < j$. The same element in the right hand member is:

$$[t^n]\frac{d(t)h(t)}{t} h(t)^k = [t^{n+1}]d(t)h(t)^{k+1} = d_{n+1,k+1}.$$

By equating these two quantities, we have the identity (4.1.1). For the converse, let us observe that (4.1.1) uniquely defines the array D when the elements $(d_{0,0}, d_{1,0}, d_{2,0}, \ldots)$ of column 0 are given. Let $d(t)$ be the generating function of this column, $A(t)$ the generating function of the sequence A and define $h(t)$ as the solution of the functional equation $h(t) = tA(h(t))$, which is uniquely determined because of our hypothesis $a_0 \neq 0$. We can therefore consider the proper Riordan array $\widehat{D} = \mathcal{R}(d(t), h(t))$; by the first part of the theorem, $\widehat{D}$ satisfies relation (4.1.1), for every

$n, k \in \mathbb{N}$ and therefore, by our previous observation, it must coincide with D. This completes the proof. □

The sequence $A = (a_k)_{k\in\mathbb{N}}$ is called the *A-sequence* of the Riordan array $D = \mathcal{R}(d(t), h(t))$ and it only depends on $h(t)$. In fact, as we have shown during the proof of the theorem, we have:

$$h(t) = tA(h(t)) \tag{4.1.2}$$

and this uniquely determines A when $h(t)$ is given and, vice versa, $h(t)$ is uniquely determined when A is given, since we have:

$$A(t) = \left[\frac{h(y)}{y} \;\middle|\; t = h(y)\right]. \tag{4.1.3}$$

As a simple consequence of this theorem, we also prove that a generic element $d_{n,k}$ of a Riordan array can be expressed as a linear combination of the elements in the successive row $n + 1$, starting with the $(k + 1)$-th element. The coefficients (independent of n or k), are exactly the coefficients of $A(t)^{-1}$:

Theorem 4.2 *Let $d_{n,k}$ be an element of the p.R.a. $\mathcal{R}(d(t), h(t))$; then we have:*

$$d_{n,k} = b_0 d_{n+1,k+1} + b_1 d_{n+1,k+2} + b_2 d_{n+1,k+3} + \cdots = \sum_{j=0}^{\infty} b_j d_{n+1,k+j+1} \tag{4.1.4}$$

(the sum is actually finite), where $B(t) = \sum_{j=0}^{\infty} b_j t^j = A(t)^{-1}$.

Proof Let us consider the identity in the proof of the previous theorem:

$$\mathcal{R}(A(t), t) = \mathcal{R}(d(t), h(t))^{-1} * \mathcal{R}(d(t)h(t)/t, h(t)).$$

We invert this equation and write it as:

$$\mathcal{R}(d(t)h(t)/t, h(t)) * \mathcal{R}(A(t), t)^{-1} = \mathcal{R}(d(t), h(t)).$$

Now we set $B(t) = A(t)^{-1}$ and extract the element in position (n, k) from both sides; the right hand side gives $d_{n,k}$ while the left hand side gives:

$$\sum_{j=0}^{\infty} d^*_{n,j} b_{k-j} = \sum_{j=0}^{n} d^*_{n,k+j} b_j$$

where:

$$d^*_{n,k+j} = [t^n]\frac{d(t)h(t)}{t} h(t)^{k+j} = [t^{n+1}]d(t)h(t)^{k+j+1} = d_{n+1,k+j+1}$$

as desired. □

On some occasions, we call the sequence $(b_k)_{k\in\mathbb{N}}$ the B-sequence of the Riordan array, but here we will use this term only occasionally.

As will be apparent from the following examples, a simple relation exists between the A-sequence and the compositional inverse of the $h(t)$ function or, if one prefers, between the A-sequence and the $h(t)$ function of the inverse p.R.a.:

Theorem 4.3 *The following relation holds:*

$$A(t) = \frac{t}{\overline{h}(t)}.$$

Proof By setting $y = h(t)$ in $h(t) = tA(h(t))$, we have:

$$A(y) = \left[\frac{h(t)}{t} \;\middle|\; y = h(t)\right];$$

however, $t = \overline{h}(h(t))$ or $t = \overline{h}(y)$, and therefore this is equivalent to the formula in the assertion. □

Example 4.1 The A-sequence for the Pascal triangle is the solution $A(y)$ of the functional equation $1/(1-t) = tA(t/(1-t))$. The simple substitution $y = t/(1-t)$ gives $A(y) = 1+y$, corresponding to the well-known basic recurrence of the Pascal triangle: $\binom{n+1}{k+1} = \binom{n}{k} + \binom{n}{k+1}$. We also observe that $B(t) = A(t)^{-1} = 1/(1+t)$, so that $b_j = (-1)^j$, for all j. The Eq. 4.1.4 becomes:

$$\binom{n}{k} = \sum_{j=0}^{\infty}(-1)^j\binom{n+1}{k+j+1}$$

a well-known property of binomial coefficients. At this point, we realize that we could have started with this recurrence relation and directly found $A(y) = 1+y$. Now, $h(t)$ is defined by (4.1.2) as the solution of $h(t) = t(1+h(t))$, and this immediately gives $h(t) = t/(1-t)$. Furthermore, since column 0 is $(1, 1, 1, \ldots)$, we have proved that the Pascal triangle corresponds to the Riordan array $\mathcal{R}(1/(1-t), t/(1-t))$ as initially stated.

Example 4.2 For the Catalan type triangle C, we apply Eq. (4.1.2) with $h(t) = (1-\sqrt{1-4t})/2$. If we set $y = h(t)$, we find $t = y(1-y)$ and therefore:

$$A(y) = \frac{1-\sqrt{1-4y+4y^2}}{2y(1-y)} = \frac{1}{1-y}.$$

This means that every element in C not belonging to column 0 is the sum of all the elements in the previous row, starting with the element on its left:

$$\binom{2n+2-k}{n+1-k} = \sum_{j=0}^{\infty} \binom{2n-k-j}{n-k-j}.$$

Now, $B(t) = 1 - t$ and Eq. (4.1.4) becomes:

$$\binom{2n-k}{n-k} = \binom{2n-k+1}{n-k} - \binom{2n-k}{n-k-1}$$

another occurrence of the basic recurrence of binomial coefficients.

Example 4.3 For what concerns the Stirling numbers of the second kind, Eq. (4.1.2) gives:

$$h(t) = e^t - 1 = y \qquad \text{or} \qquad t = \ln(1+y).$$

The A-sequence is therefore:

$$A(y) = \frac{e^{\ln(1+y)} - 1}{\ln(1+y)} = \frac{y}{\ln(1+y)}.$$

This is the exponential generating function of the so-called *Cauchy numbers of the first kind* (see [3] and [13]); they can be defined by:

$$\frac{C_n}{n!} = \int_0^1 \binom{x}{n} dx$$

and their initial values are:

n	0	1	2	3	4	5	6	7
C_n	1	1/2	−1/6	1/4	−19/30	9/4	−863/84	1375/24

Therefore, the A-sequence rule implies:

$$\frac{(k+1)!}{(n+1)!} \left\{ {n+1 \atop k+1} \right\} = \sum_{j=0}^{n} \frac{(k+j)!}{n!} \left\{ {n \atop k+j} \right\} \frac{C_j}{j!}.$$

This expression can be simplified and we obtain a non-obvious combinatorial identity:

$$\frac{k+1}{n+1} \left\{ {n+1 \atop k+1} \right\} = \sum_{j=0}^{n} \left\{ {n \atop k+j} \right\} \binom{k+j}{j} C_j,$$

involving Stirling numbers, Cauchy numbers and binomial coefficients. The generating function of the B-sequence is $B(t) = \ln(1+t)/t$, so that $b_j = (-1)^j/(j+1)$. The Eq. 4.1.4 now gives:

$$\frac{k!}{n!}\left\{ {n \atop k} \right\} = \sum_{j=0}^{n} \frac{(-1)^j}{j+1} \frac{(k+j+1)!}{(n+1)!} \left\{ {n+1 \atop k+j+1} \right\};$$

again, we can simplify and get:

$$\left\{ {n \atop k} \right\} = \frac{1}{n+1} \sum_{j=0}^{n} (-1)^j j! \binom{k+j+1}{k} \left\{ {n+1 \atop k+j+1} \right\}.$$

Example 4.4 As a last example, let us consider the p.R.a. associated to the signed Stirling numbers of the first kind. Since $h(t) = \ln(1+t)$, we find $A(t) = t/(e^t - 1)$, which is the well-known exponential generating function of the Bernoulli numbers. Therefore, we also have $B(t) = (e^t - 1)/t$, which implies $b_n = 1/(n+1)!$. By using these quantities, we find the following combinatorial identities:

$$\frac{k+1}{n+1} \left[{n+1 \atop k+1} \right] = \sum_{j=0}^{n} \left[{n \atop k+j} \right] \binom{k+j}{k} (-1)^j B_j$$

$$\left[{n \atop k} \right] = \frac{1}{n+1} \sum_{j=0}^{n} (-1)^j \binom{k+j+1}{k} \left[{n+1 \atop k+j+1} \right].$$

Remark 4.1 A proper Riordan array is characterized by a pair of formal power series $d(t), A(t) \in \mathcal{F}_0$ where $d(t)$ is the generating function of column 0 and $A(t)$ is the generating function of the A-sequence.

Actually, we can go further:

Theorem 4.4 *Let $(d_{n,k})_{n,k \in \mathbb{N}}$ be any infinite, lower triangular array with $d_{n,n} \neq 0, \forall n \in \mathbb{N}$ (in particular, let it be a proper Riordan array); then a unique sequence $Z = (z_k)_{k \in \mathbb{N}}$ exists such that every element in column 0 can be expressed as a linear combination of all the elements in the preceding row, i.e.:*

$$d_{n+1,0} = z_0 d_{n,0} + z_1 d_{n,1} + z_2 d_{n,2} + \cdots = \sum_{j=0}^{\infty} z_j d_{n,j}. \tag{4.1.5}$$

Proof Let $z_0 = d_{1,0}/d_{0,0}$. Now we can uniquely determine the value of z_1 by expressing $d_{2,0}$ in terms of the elements in row 1, i.e.:

$$d_{2,0} = z_0 d_{1,0} + z_1 d_{1,1} \qquad \text{or} \qquad z_1 = \frac{d_{0,0} d_{2,0} - d_{1,0}^2}{d_{0,0} d_{1,1}}.$$

In the same way, we determine z_2 by expressing $d_{3,0}$ in terms of the elements in row 2, and by substituting the values just obtained for z_0 and z_1. By proceeding in the same way, we determine the sequence Z in a unique way. □

The sequence Z is called the *Z-sequence* for the (Riordan) array; it characterizes column 0, except for the element $d_{0,0}$.

Remark 4.2 A proper Riordan array is characterized by a number $d_{0,0}$ and a pair of formal power series $Z(t), A(t) \in \mathcal{F}_0$ where $d_{0,0}$ is the starting value of the array, $Z(t)$ is the generating function of the Z-sequence and $A(t)$ is the generating function of the A-sequence.

To see how the Z-sequence is obtained by starting with the usual definition of a Riordan array, let us prove the following:

Theorem 4.5 *Let $\mathcal{R}(d(t), h(t))$ be a proper Riordan array and let $Z(t)$ be the generating function of its Z-sequence. We have:*

$$d(t) = \frac{d_{0,0}}{1 - tZ(h(t))}.$$

Proof By the preceding theorem, the Z-sequence exists and is unique. Therefore, Eq. (4.1.5) is valid for every $n \in \mathbb{N}$, and we can pass to the generating functions. Since $d(t)h(t)^k$ is the generating function for column k, we have:

$$\begin{aligned}\frac{d(t) - d_{0,0}}{t} &= z_0 d(t) + z_1 d(t)h(t) + z_2 d(t)h(t)^2 + \cdots \\ &= d(t)(z_0 + z_1 h(t) + z_2 h(t)^2 + \cdots) = d(t)Z(h(t)).\end{aligned}$$

By solving this equation in $d(t)$, we immediately find the relation desired. □

The relation can be inverted and this gives us the formula for the Z-sequence:

$$Z(y) = \left[\frac{d(t) - d_{0,0}}{td(t)} \;\middle|\; y = h(t)\right].$$

The following theorem characterizes renewal arrays by means of the A- and Z-sequences:

Theorem 4.6 *Let $d(0) = h(0) \neq 0$. Then $d(t) = h(t)/t$ if and only if $A(y) = d(0) + yZ(y)$.*

Proof Let us assume that $A(y) = d(0) + yZ(y)$ or $Z(y) = (A(y) - d(0))/y$. By the previous theorem, we have:

$$d(t) = \frac{d(0)}{1 - tZ(h(t))} = \frac{d(0)}{1 - (tA(h(t)) - d(0))/h(t)} = \frac{d(0)h(t)}{d(0)t} = \frac{h(t)}{t},$$

because $tA(h(t)) = h(t)$ by formula (4.1.2). Vice versa, by the formula for $Z(y)$, we obtain from the hypothesis $td(t) = h(t)$:

$$d(0)+yZ(y)=\left[d(0)+y\left(\frac{1}{t}-\frac{d(0)}{h(t)}\right)\mid y=h(t)\right]=$$

$$=\left[d(0)+\frac{h(t)}{t}-\frac{d(0)h(t)}{h(t)}\mid y=h(t)\right]=$$

$$=\left[\frac{h(t)}{t}\mid y=h(t)\right]=A(y)$$

by Eq. 4.1.3. □

Example 4.5 Let us consider the following model of walks: a walk of length n is represented by a sequence $((0, y_0), (1, y_1), \cdots, (n, y_n))$ of $n+1$ points in $\mathbb{N}^2$, where $y_0 = 0$ and $y_k = y_{k-1} + s_k$ such that when $y_{k-1} = j$, the s_k are constrained to belong to a fixed set $\mathcal{S}_j$. In particular, we consider in this example $\mathcal{S}_j = \{-1^\gamma, 0^\beta, 1^\alpha\}$ for $j > 0$ and $\mathcal{S}_0 = \{0^\beta, 1^\alpha\}$, where α, β and γ represent the number of colours for each kind of step ($\alpha, \gamma \neq 0$). We can represent all the walks of length $\leq n$ as a tree of height n, where the root (at level 0 by convention) is labelled with 0 and where the label of each node at level n encodes a possible position of the walk. More precisely, a walk of length n corresponds to a branch of length $n+1$ in the generating tree where the root is labeled with 0 and a node with label k has children labeled according to the following rule:

$$\begin{cases} root : (0) \\ rule : (0) \rightarrow (0)^\beta(1)^\alpha \\ \qquad\quad (k) \rightarrow (k-1)^\gamma(k)^\beta(k+1)^\alpha \quad k > 0. \end{cases} \tag{4.1.6}$$

Let $M^{[\alpha,\beta,\gamma]} = (M_{n,k}^{[\alpha,\beta,\gamma]})_{n,k\in\mathbb{N}}$ be the matrix associated with the generating tree described by (4.1.6), that is, $M_{n,k}^{[\alpha,\beta,\gamma]}$ counts the number of nodes at level n having label k. The elements of this matrix satisfies the following recurrence relation:

$$M_{n+1,k+1}^{[\alpha,\beta,\gamma]} = \alpha M_{n,k}^{[\alpha,\beta,\gamma]} + \beta M_{n,k+1}^{[\alpha,\beta,\gamma]} + \gamma M_{n,k+2}^{[\alpha,\beta,\gamma]}$$

$$M_{n+1,0}^{[\alpha,\beta,\gamma]} = \beta M_{n,0}^{[\alpha,\beta,\gamma]} + \gamma M_{n,1}^{[\alpha,\beta,\gamma]}$$

with initial condition $M_{0,0}^{[\alpha,\beta,\gamma]} = 1$. In other words, $M_{n+1,k+1}^{[\alpha,\beta,\gamma]}$ is a Riordan array $(d^{[\alpha,\beta,\gamma]}(t), h^{[\alpha,\beta,\gamma]}(t))$ with A- and Z-sequences defined by the following generating functions:

$$A^{[\alpha,\beta,\gamma]}(t) = \alpha + \beta t + \gamma t^2, \quad Z^{[\alpha,\beta,\gamma]}(t) = \beta + \gamma t.$$

Hence, by using the results of this section we have:

$$d^{[\alpha,\beta,\gamma]}(t) = \frac{1-\beta t-\sqrt{1-2\beta t+(\beta^2-4\alpha\gamma)t^2}}{2\alpha\gamma t^2}, \quad h^{[\alpha,\beta,\gamma]}(t) = \alpha t d^{[\alpha,\beta,\gamma]}(t).$$

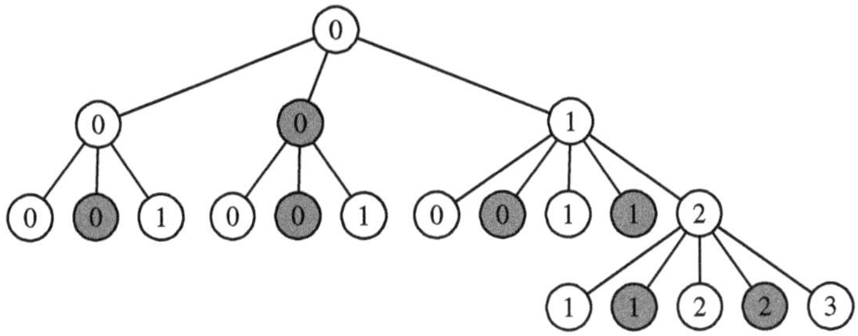

Fig. 4.1 A portion of the generating tree for the walks with $\beta = \gamma = 2$ and $\alpha = 1$

Table 4.1 The triangle $M^{[1,2,2]}$

n/k	0	1	2	3	4	5
0	1					
1	2	1				
2	6	4	1			
3	20	16	6	1		
4	72	64	30	8	1	
5	272	260	140	48	10	1

The case $[\alpha, \beta, \gamma] = [1, 1, 1]$ corresponds to the well-known Motzkin triangle while $[\alpha, \beta, \gamma] = [1, 0, 1]$ corresponds to the aerated Catalan triangle. Figure 4.1 illustrates a portion of the generating tree for the walks with $\beta = \gamma = 2$ and $\alpha = 1$; the branches with labels 0, 1, 2, 1, for example, denote the walk $((0, 0), (1, 1), (2, 2), (3, 1))$, where the last step can be of two different colours; in Table 4.1 we give the corresponding matrix.

The relation between the A- and Z- sequences of Riordan arrays and generating trees has been studied in [15]; the tree defined by rule (4.1.6) has been examined in [14] in the context of the average case analysis of algorithms related to the problem of ranking and generating uniformly at random the corresponding walks.

4.2 The A-matrix

As the concepts of A- and Z-sequences show, what seems essential in a Riordan array is the fact that the elements in a given row linearly depend on the elements of the row above it, starting from the element on the left. It is surprising that this dependence can be made much looser, as the following theorems show (see Merlini [9], Merlini, Rogers, Sprugnoli and Verri [11] and also the presentation of Shapiro

[19]); they greatly increase the applicability range of the Riordan array theory. We explicitly observe that the results presented in this section refer to Riordan arrays $\mathcal{R}(d(t), h(t))$ with $d(t) \in \mathcal{F}$ and $h(t) \in F_1$ (which is the currently used definition, see Chap. 3) and are slightly different from those in [9, 11] where an old equivalent definition $\mathcal{R}(d(t), th(t))$ with $d(t), h(t) \in \mathcal{F}$ was used. We also wish to point out that, by Theorem 4.4, the Z-sequence exists for every lower triangular array, and therefore we can implicitly assume its existence in all the subsequent theorems.

We are now going to prove that the generic element $d_{n+1,k+1}$ of a Riordan array can depend on the elements of an indefinite number of the previous rows, provided that it linearly depends on $d_{n,k}$ AND on the elements from the column k onwards. In other words, there exists an array $\left(\alpha_{n,k}\right)_{n,k\in\mathbb{N}}$, with $\alpha_{0,0} \neq 0$, such that every $d_{n+1,k+1}$ $(n, k \geq 0)$ can be expressed as:

$$d_{n+1,k+1} = \sum_{i\geq 0}\sum_{j\geq 0} \alpha_{i,j} d_{n-i,k+j}. \tag{4.2.1}$$

This matrix will be called the A-matrix of the Riordan array. However, while the A-sequence is unique for a given Riordan array, the A-matrix is not. For example, the following A-matrices, and many others, all define the Pascal triangle (the proof is quite obvious and relies on the basic recurrence for the binomial coefficients):

$$\begin{bmatrix} 1\,1\,0\,0 \ldots \\ 0\,0\,0\,0 \ldots \\ 0\,0\,0\,0 \ldots \\ 0\,0\,0\,0 \ldots \\ \vdots\,\vdots\,\vdots\,\vdots\,\ddots \end{bmatrix} \quad \begin{bmatrix} 1\,0\,0\,0 \ldots \\ 1\,1\,0\,0 \ldots \\ 0\,0\,0\,0 \ldots \\ 0\,0\,0\,0 \ldots \\ \vdots\,\vdots\,\vdots\,\vdots\,\ddots \end{bmatrix} \quad \begin{bmatrix} 1\,0\,0\,0 \ldots \\ 1\,0\,0\,0 \ldots \\ 1\,1\,0\,0 \ldots \\ 0\,0\,0\,0 \ldots \\ \vdots\,\vdots\,\vdots\,\vdots\,\ddots \end{bmatrix} \quad \begin{bmatrix} 1\,0\,0\,0 \ldots \\ 1\,0\,0\,0 \ldots \\ 1\,0\,0\,0 \ldots \\ 1\,1\,0\,0 \ldots \\ \vdots\,\vdots\,\vdots\,\vdots\,\ddots \end{bmatrix}.$$

These could equivalently be represented by

$$\begin{bmatrix} 1, 1 \end{bmatrix}, \quad \begin{bmatrix} 1, 0 \\ 1, 1 \end{bmatrix}, \quad \begin{bmatrix} 1, 0 \\ 1, 0 \\ 1, 1 \end{bmatrix}, \quad \text{and} \quad \begin{bmatrix} 1, 0 \\ 1, 0 \\ 1, 0 \\ 1, 1 \end{bmatrix}.$$

Instead of treating a global generating function for the A-matrix, let us examine a sequence of generating functions $P^{[0]}(t), P^{[1]}(t), P^{[2]}(t), \ldots$ corresponding to the rows $0, 1, 2, \ldots$ of the A-matrix, i.e.:

$$P^{[0]}(t) = \alpha_{0,0} + \alpha_{0,1}t + \alpha_{0,2}t^2 + \alpha_{0,3}t^3 + \cdots$$

$$P^{[1]}(t) = \alpha_{1,0} + \alpha_{1,1}t + \alpha_{1,2}t^2 + \alpha_{1,3}t^3 + \cdots$$

and so on. Our first basic result is:

Theorem 4.7 *A lower triangular array $D = \left(d_{n,k}\right)_{n,k\in\mathbb{N}}$ is Riordan if and only if there exists another array $\left(\alpha_{n,k}\right)_{n,k\in\mathbb{N}}$, with $\alpha_{0,0} \neq 0$, such that every $d_{n+1,k+1}$ $(n, k \geq 0)$ can be expressed as Eq. 4.2.1. Besides, if $D = \mathcal{R}(d(t),\ h(t))$ and $A(t)$ is its A-sequence, then $h(t)$ and $A(t)$ are implicitly defined by the two equations:*

$$\frac{h(t)}{t} = \sum_{i\geq 0} t^i P^{[i]}(h(t)) \qquad A(t) = \sum_{i\geq 0} t^i A(t)^{-i} P^{[i]}(t). \tag{4.2.2}$$

Proof If the array is Riordan, let $(a_k)_{k\in\mathbb{N}}$ be its A-sequence: the array defined as $\alpha_{0,j} = a_j$, $\forall j \in \mathbb{N}$, and $\alpha_{i,j} = 0$, $\forall i > 0$, $j \geq 0$, is exactly as we desired. The proof of the "if" part proceeds in the following way. Let us suppose that the A-matrix defines a Riordan array, say $\mathcal{R}(d(t),\ h(t))$. In that case we should have:

$$d_{n+1,k+1} = \sum_{i\geq 0}\sum_{j\geq 0} \alpha_{i,j} d_{k-i,k+j}$$

or, passing to column generating functions:

$$\frac{d_{k+1}(t)}{t} = \sum_{i\geq 0}\sum_{j\geq 0} \alpha_{i,j} d_{k+j}(t) \quad \text{or} \quad \frac{d(t)h(t)^{k+1}}{t} = \sum_{i\geq 0}\sum_{j\geq 0} \alpha_{i,j} t^i d(t)h(t)^{k+j}.$$

If we divide everything by $d(t)h(t)^k$ we get:

$$\frac{h(t)}{t} = \sum_{i\geq 0} t^i \sum_{j\geq 0} \alpha_{i,j} h(t)^j.$$

For fixed i, we have $\sum_{j\geq 0} \alpha_{i,j} h(t)^j = P^{[i]}(h(t))$ and so:

$$\frac{h(t)}{t} = \sum_{i\geq 0} t^i P^{[i]}(h(t)).$$

We now develop this equation and equate like coefficients, that is we extract the coefficient of t^n, for $n = 0, 1, 2, \ldots$:

$$[t^{n+1}]h(t) = \sum_{i\geq 0} [t^{n-i}] P^{[i]}(h(t)).$$

Therefore, h_{n+1} only depends on $h_0, h_1, \ldots, h_n$ and can be iteratively evaluated. Since $\alpha_{0,0} \neq 0$ the system has one and only one solution, so that the function $h(t)$ is unique. For example, the first instances are:

$$
\begin{aligned}
h_1 &= \alpha_{0,0} \\
h_2 &= \alpha_{0,1}h_1 + \alpha_{1,0} \\
h_3 &= \alpha_{0,1}h_2 + \alpha_{0,2}h_1^2 + \alpha_{1,1}h_1 + \alpha_{2,0} \\
h_4 &= \alpha_{0,1}h_3 + 2\alpha_{0,2}h_1h_2 + \alpha_{0,3}h_1^3 + \alpha_{1,1}h_2 + \alpha_{1,2}h_1^2 + \alpha_{2,1}h_1 + \alpha_{3,0}
\end{aligned}
$$

and the explicit values are:

$$
\begin{aligned}
h_1 &= \alpha_{0,0} \\
h_2 &= \alpha_{0,1}\alpha_{0,0} + \alpha_{1,0} \\
h_3 &= \alpha_{0,0}\alpha_{1,1} + \alpha_{0,1}\alpha_{1,0} + \alpha_{0,2}\alpha_{0,0}^2 + \alpha_{0,1}^2\alpha_{0,0} + \alpha_{2,0}
\end{aligned}
$$

and so on. Now, let us call $d(t)$ the generating function of column 0 in D and define $\widehat{D} = \mathcal{R}(d(t),\ h(t))$; clearly $D = \widehat{D}$ and so D is a Riordan array, as claimed. Finally, the formula for the A-sequence is obtained by using the relation in Theorem 4.3. □

This theorem shows that we can characterize a Riordan array by means of an A-matrix, rather than by a simple A-sequence.

Remark 4.3 A proper Riordan array is characterized by a formal power series $d(t) \in \mathcal{F}_0$ where $d(t)$ is the generating function of column 0 and an A-matrix $(\alpha_{n,k})_{n,k\in N}$ such that Eq. 4.2.1 holds, $\forall n, k \in \mathbb{N}$.

We can extend the linear dependence of the generic element $d_{n+1,k+1}$ to allow for elements in its own row, starting from $d_{n+1,k+2}$. In fact, we can prove the following characterization:

Theorem 4.8 *A lower triangular array $\left(d_{n,k}\right)_{n,k\in\mathbb{N}}$ is Riordan if and only if there exists another array $\left(\alpha_{n,k}\right)_{n,k\in\mathbb{N}}$, with $\alpha_{0,0} \neq 0$, and a sequence $(\rho_k)_{k\in\mathbb{N}}$ such that:*

$$
d_{n+1,k+1} = \sum_{i\geq 0}\sum_{j\geq 0} \alpha_{i,j} d_{n-i,k+j} + \sum_{j\geq 0} \rho_j d_{n+1,k+1+j}. \tag{4.2.3}
$$

Proof Here again, the “only if” part is obvious. As to the “if” part, we can eliminate $d_{n+1,k+2}$ from the recurrence, by applying relation (4.2.3) and by eventually changing the array $\left(\alpha_{n,k}\right)_{n,k\in\mathbb{N}}$ into $\left(\alpha'_{n,k}\right)_{n,k\in\mathbb{N}}$. In the same way, we can subsequently eliminate all the $d_{n+1,k+1+j}$'s. Since only a finite number of them actually appears in the evaluation of $d_{n+1,k+1}$, we can always reduce $d_{n+1,k+1}$ to depend on some array $\left(\bar{\alpha}_{n,k}\right)_{n,k\in\mathbb{N}}$, which is the left part of a limit array $\left(\alpha^*_{n,k}\right)_{n,k\in\mathbb{N}}$. Therefore, we can conclude that $\left(d_{n,k}\right)_{n,k\in\mathbb{N}}$ is a Riordan array. □

We can obtain the widest possible characterization of Riordan arrays (see Theorem 4.14 below):

Theorem 4.9 *A lower triangular array $\left(d_{n,k}\right)_{n,k\in\mathbb{N}}$ is Riordan if and only if there exists another array $\left(\alpha_{n,k}\right)_{n,k\in\mathbb{N}}$, with $\alpha_{0,0} \neq 0$, and s sequences $(\rho_k)_{k\in\mathbb{N}}$ $(i = 1, 2, \ldots, s)$ such that:*

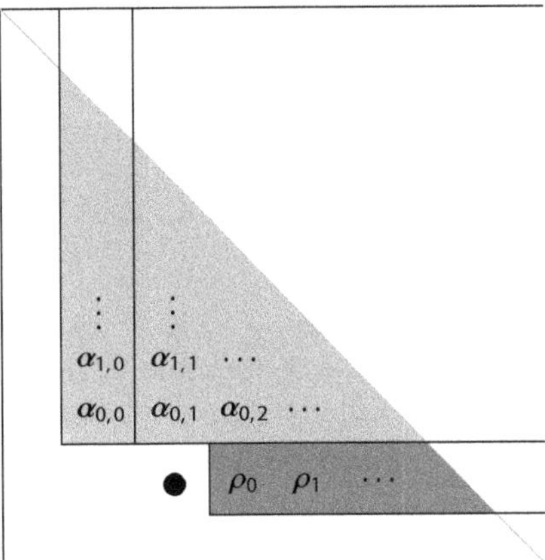

Fig. 4.2 The zones which $d_{n+1,k+1}$ can depend on

$$d_{n+1,k+1} = \sum_{i\geq 0}\sum_{j\geq 0} \alpha_{i,j} d_{n-i,k+j} + \sum_{i=1}^{S}\sum_{j\geq 0} \rho_j^{[i]} d_{n+i,k+i+j+1}. \tag{4.2.4}$$

Proof Repeated applications of the elimination technique used in the previous theorem's proof. □

In Fig. 4.2, we try to give a graphic representation of the zones which the generic element $d_{n+1,k+1}$ (denoted by a small disk or "bullet") is allowed to depend on, so that the array can be Riordan. The three zones correspond to Theorems 4.7, 4.8 and 4.9, and the only restrictions are that $\alpha_{0,0} \neq 0$ and that the number of rows *below* row n be finite.

Remark 4.4 A proper Riordan array is characterized by the generating function $d(t)$ of column 0 and the generating functions of the sequences corresponding to the dependence on any one or more of the three zones depicted in Fig. 4.2.

Up to now, we have assumed that $\alpha_{0,0} \neq 0$ because this condition assures that the resulting Riordan array is proper. However, if we change this hypothesis, but maintain that some $\alpha_{i,0} \neq 0$, for $i > 0$, then we obtain a non-proper Riordan array, i.e., a Riordan array with $h(0) = 0$. This happens under the following conditions:

Theorem 4.10 *Let $\left(d_{n,k}\right)_{n,k\in\mathbb{N}}$ be an array whose generic element $d_{n+1,k+1}$ is defined by a linear recurrence:*

$$d_{n+1,k+1} = \sum_{i\geq 0}\sum_{j\geq 0} \beta_{i,j} d_{v,k+j} \qquad v \leq n+S \quad \text{for some} \quad S \in \mathbb{N}.$$

Let γ be the minimum index for which $\beta_{\gamma,0} \neq 0$ and set $v' = v - \gamma j$. If $\forall \beta_{i,j} \neq 0$ we have $i = n - v'$ and, whenever $v' > n$, we also have $j \geq v' - n$, then $\left(d_{n,k}\right)_{n,k\in\mathbb{N}}$ is a non-proper Riordan array $(d(t), h(t))$ with $h(t) = t^\gamma v(t)$ and $v(0) \neq 0$.

Proof The theorem's conditions allow us to define a new array $\left(d'_{n,k}\right)_{n,k\in\mathbb{N}}$ whose generic element $d'_{n+1,k+1}$ is given by:

$$d'_{n+1,k+1} = \sum_{i\geq 0}\sum_{j\geq 0}\alpha_{i,j}d'_{n-i,k+j} + \sum_{i\geq 1}^{s}\sum_{j\geq 0}\rho_j^{[i]}d'_{n+i,k+i+j+1},$$

where $\alpha_{i,j} = \beta_{n-\nu',j}$ when $\nu' \leq n$, and $\rho_j^{[i]} = \beta_{\nu'-n,j}$ when $\nu' > n$. The number s exists thanks to the condition $\nu \leq n + S$, for some S. This is actually the definition of a proper Riordan array, in which $d'_{n,k} = d_{n+k\gamma,k}$ because the columns of $\left(d'_{n,k}\right)_{n,k\in\mathbb{N}}$ are the columns of $\left(d_{n,k}\right)_{n,k\in\mathbb{N}}$ moved γk positions up. If $\mathcal{R}(d(t), v(t))$ is the new proper Riordan array, then we should have $h(t) = t^{\gamma}v(t)$. □

As previously noted, the A-sequence and the function $h(t)$ of a Riordan array are strictly related to each other. This fact allows us to think that $h(t)$ can be deduced from the A-matrix $\left(\alpha_{n,k}\right)_{n,k\in\mathbb{N}}$ and the set of sequences $\left(\rho_k^{[i]}\right)_{k\in\mathbb{N}}$ for $i = 1, \cdots, s$. Then, after finding the function $h(t)$, we can also find the A-sequence by determining its generating function $A(t)$.

Let $Q^{[i]}(t)$ be the generating function for the sequence $\left(\rho_k^{[i]}\right)_{k\in\mathbb{N}}$. Thus we have:

Theorem 4.11 *If $\left(d_{n,k}\right)_{n,k\in\mathbb{N}}$ is a Riordan array whose generic element $d_{n+1,k+1}$ is defined by formula (4.2.4) through the A-matrix $\left(\alpha_{n,k}\right)_{n,k\in\mathbb{N}}$ and the set of sequences $\left(\rho_k^{[i]}\right)_{k\in\mathbb{N}}$, $i = 1, 2, \ldots, s$, then the functions $h(t)$ and $A(t)$ for $(d_{n,k})$ are given by the following implicit expressions:*

$$\frac{h(t)}{t} = \sum_{i\geq 0} t^i P^{[i]}(h(t)) + \sum_{i=1}^{s} t^{-i}h(t)^{i+1}Q^{[i]}(h(t)). \tag{4.2.5}$$

$$A(t) = \sum_{i\geq 0} t^i A(t)^{-i} P^{[i]}(t) + t\sum_{i=1}^{s} A(t)^i Q^{[i]}(t). \tag{4.2.6}$$

Proof Let $d_k(t) = d(t)(h(t))^k$ be the generating function of column k of the Riordan array; from (4.2.4) we deduce:

$$\frac{d_{k+1}(t)}{t} = \sum_{i\geq 0}\sum_{j\geq 0}\alpha_{i,j}t^i d_{k+j}(t) + \sum_{i=1}^{s}\sum_{j\geq 0}\rho_j^{[i]}t^{-i}d_{k+i+j+1}(t)$$

$$\frac{d(t)(h(t))^{k+1}}{t} = \sum_{i\geq 0}\sum_{j\geq 0}\alpha_{i,j}t^i d(t)(h(t))^{k+j} + \sum_{i=1}^{s}\sum_{j\geq 0}\rho_j^{[i]}t^{-i}d(t)(h(t))^{k+i+j+1}.$$

We can now divide everything by $d(t)(h(t))^k$:

$$\frac{h(t)}{t} = \sum_{i\geq 0} t^i \sum_{j\geq 0} \alpha_{i,j}(h(t))^j + \sum_{i=1}^{s} t^{-i}(h(t))^{i+1} \sum_{j\geq 0} \rho_j^{[i]}(h(t))^j.$$

We now go on to the generating functions $P^{[i]}(t)$ and $Q^{[i]}(t)$ and formula (4.2.5) immediately follows. Finally, by applying the formula in Theorem 4.3 we obtain the expression (4.2.6) for $A(t)$. □

As stated in the proof of Theorem 4.8, this theorem allows us to give some explicit formulas for the element a_n of the A-sequence. By extracting the coefficient of t^n, we find:

$$a_n = [t^n]A(t) = \sum_{i\geq 0}[t^{n-i}]B(t)^i P^{[i]}(t) + \sum_{i=1}^{s}[t^{n-1}]A(t)^i Q^{[i]}(t)$$

$$a_n = \sum_{i=0}^{n}\sum_{j=0}^{n-i} b_j^{(i)}\alpha_{i,n-i-j} + \sum_{i=1}^{s}\sum_{j=0}^{n-i} a_j^{(i)}\rho_{n-i-j}^{[i]},$$

where $a_j^{(i)}$ and $b_j^{(i)}$ denote the coefficients of t^j in the formal power series $A(t)^i$ and $B(t)^i = A(t)^{-i}$, respectively. As far as Theorem 4.8 is concerned, we have $s = 0$ and so:

$$a_n = \sum_{i=0}^{n}\sum_{j=0}^{n-i} b_j^{(i)}\alpha_{i,n-i-j},$$

which agrees with the values $a_0 = b_0^{(0)}\alpha_{0,0}$ and $a_1 = b_0^{(0)}\alpha_{0,1} + b_0^{(1)}\alpha_{1,0} = \alpha_{0,1} + b_0\alpha_{1,0} = \alpha_{0,1} + \alpha_{1,0}/\alpha_{0,0}$ (see the proof of Theorem 4.8 in [11]). As to Theorem 4.9, we have:

$$a_n = \sum_{i=0}^{n}\sum_{j=0}^{n-i} b_j^{(i)}\alpha_{i,n-i-j} + \sum_{j=0}^{n-i} a_j\rho_{n-i-j},$$

which only depends on the previously computed a_j values .

The generic element $d_{n+1,k+1}$ often only depends on the two previous rows and sometimes on the elements of its own row. In this case, the functional equation (4.2.6) reduces to a second degree equation in $A(t)$ and, as a result, we can give an explicit expression for the generating function of the A-sequence.

Theorem 4.12 *Let* $(d_{n,k})_{n,k\in\mathbb{N}}$ *be a Riordan array whose generic element* $d_{n+1,k+1}$ *only depends on the two previous rows and, possibly, on its own row. If* $P(t)$, $\bar{P}(t)$ *and* $Q(t)$ *are the generating functions for the coefficients of this dependence, i.e.,* $P(t) = P^{[0]}(t)$, $\bar{P}(t) = P^{[1]}(t)$ *and* $Q(t) = Q^{[1]}(t)$, *then we have:*

$$A(t) = \frac{P(t) + \sqrt{P(t)^2 + 4t\bar{P}(t)(1 - tQ(t))}}{2(1 - tQ(t))}. \tag{4.2.7}$$

Proof Formula (4.2.6) gives two solutions for $A(t)$ and the one having $A(0) = 0$ must be discarded because we always assume that $a_0 \neq 0$. □

It is worth noting that if $Q(t) = 0$, that is $d_{n+1,k+1}$ does not depend on the elements of its own row, then we have:

$$A(t) = \frac{P(t) + \sqrt{P(t)^2 + 4t\bar{P}(t)}}{2},$$

which is quite useful in several cases. When the dependence is more complicated, it is naturally more difficult to give an explicit expression for the A-sequence.

As shown in the previous section, $h(t)$ is related to $A(t)$ and $d(t)$ is related to $Z(t)$, the Z-sequence generating function. Since the Z-sequence exists for every lower triangular array (see Theorem 4.4), every recurrence defining $d_{n+1,0}$ in terms of the other elements in the array can be accepted as a good definition of column 0. Therefore, in analogy to (4.2.4), let us assume that we have the following linear relation:

$$d_{n+1,0} = \sum_{i\geq 0}\sum_{j\geq 0} \zeta_{i,j} d_{n-i,j} + \sum_{i=1}^{s}\sum_{j\geq 0} \sigma_j^{[i]} d_{n+i,i+j}. \tag{4.2.8}$$

In general, there is no connection between the $\zeta_{i,j}$'s and the $\alpha_{i,j}$'s or between the $\rho_j^{[i]}$'s and the $\sigma_j^{[i]}$'s and so we take the following generating functions into account:

$$R^{[0]}(t) = \zeta_{0,0} + \zeta_{0,1}t + \zeta_{0,2}t^2 + \zeta_{0,3}t^3 + \cdots$$

$$R^{[1]}(t) = \zeta_{1,0} + \zeta_{1,1}t + \zeta_{1,2}t^2 + \zeta_{1,3}t^3 + \cdots$$

(etc.) and $S^{[i]}(t) = \sum_{j\geq 0} \sigma_j^{[i]} t^j$. The coefficients defining $d_{n+1,k+1}$ and $d_{n+1,0}$ are sometimes the same ones, in the sense that:

$$\zeta_{i,j} = \alpha_{i,j+1} \quad \text{and} \quad \sigma_j^{[i]} = \rho_j^{[i]} \qquad \forall i, \forall j.$$

In this case, we say that *column 0 is unprivileged* and we obtain the following formulas for our generating functions:

$$R^{[i]}(t) = \frac{P^{[i]}(t) - \alpha_{i,0}}{t} \quad \text{and} \quad S^{[i]}(t) = Q^{[i]}(t)$$

for every i for which $R^{[i]}$, $P^{[i]}$, $S^{[i]}$ and $Q^{[i]}$ are well-defined.

At any rate, we can easily prove the following:

Theorem 4.13 *If $(d_{n,k})_{n,k\in\mathbb{N}}$ is a Riordan array whose elements in column 0 are defined by a relation (4.2.8), then the function $d(t)$ is given by the following formula:*

$$d(t) = \frac{d_{0,0}}{1 - \sum_{i\geq 0} t^{i+1} R^{[i]}(h(t)) - t\sum_{i=1}^{s} h(t)^i S^{[i]}(h(t))}. \tag{4.2.9}$$

Proof We go on to generating functions and find (4.2.9) by solving in $d(t)$. □

Example 4.6 An example of a Riordan array in which column 0 is *privileged* is the following: each element in column 0 is equal to the element of the previous row, found in column 1; every other element is given by the sum of the elements in the previous row, found in the same column and in the previous and next columns. This is equivalent to say that $P^{[0]}(t) = 1 + t + t^2$ and $R^{[0]}(t) = t$, with the generating functions for the other coefficients equal to 0. By Theorems 4.11 and 4.13 we have:

$$\frac{h(t)}{t} = P^{[0]}(h(t)) = 1 + h(t) + h(t)^2$$

$$d(t) = \frac{1}{1 - tR^{[0]}(h(t))} = \frac{1}{1 - th(t)}$$

which corresponds to the Riordan array:

$$\mathcal{R}\left(\frac{1+t-\sqrt{1-2t-3t^2}}{2t(1+t)}, \frac{1-t-\sqrt{1-2t-3t^2}}{2t}\right).$$

Note that in this case $P^{[0]}(t)$ and $R^{[0]}(t)$ correspond to the A- and Z- sequences.

When column 0 is unprivileged, the formula for $d(t)$ can be drastically simplified. For this reason, we state it as a separate theorem:

Corollary 4.1 *If $(d_{n,k})_{n,k\in\mathbb{N}}$ is a Riordan array whose column 0 is unprivileged, then $d(t)$ is given by the formula:*

$$d(t) = \frac{d_{0,0}h(t)}{t\sum_{i\geq 0}\alpha_{i,0}t^i}. \tag{4.2.10}$$

Proof We simply take the denominator in formula (4.2.9) and substitute $R^{[i]}(t)$ and $S^{[i]}(t)$ by their counterparts when column 0 is unprivileged; we then use the first result of Theorem 3.1. □

Besides being important for its own sake, this result also allows us to prove a very interesting characterization of "renewal arrays" or Bell matrices, i.e., Riordan arrays having $d(t) = h(t)/t$, when column 0 is unprivileged:

Corollary 4.2 *Let $(d_{n,k})_{n,k\in\mathbb{N}}$ be a Riordan array whose column 0 is unprivileged; then it is a renewal array if and only if the following two conditions are satisfied:*

i) $d_{n+1,k+1}$ *only depends on* $d_{n,k}$ *and not on any other element in column* k*; ii)* $\alpha_{0,0} = d_{0,0}$.

Proof If column 0 is unprivileged and $d(t) = h(t)$, then by (4.2.10) we have: $\sum_{i\geq 0} \alpha_{i,0} t^i = d_{0,0}$; therefore $\alpha_{i,0} = 0, \forall i \geq 1$ and this is equivalent to condition i). Only $\alpha_{0,0} = d_{0,0}$ is left and constitutes condition ii). Vice versa, if column 0 is unprivileged, then condition i) implies: $\sum_{i\geq 0} \alpha_{i,0} t^i = \alpha_{0,0}$, so $d(t) = d_{0,0} h(t)/\alpha_{0,0}$, and so condition ii) gives $d(t) = h(t)$. □

Example 4.7 Let us now consider another example of a Riordan array in which column 0 is privileged: each element in column 0 is equal to the sum of the elements of the previous row, found in columns 0 and 1; every other element is given by the sum of the elements in the previous row, found in the same column and in the previous and next columns, plus the next element in the same row. This is equivalent to say that $P^{[0]}(t) = 1 + t + t^2$, $Q^{[1]}(t) = 1$ and $R^{[0]}(t) = 1 + t$, with the generating functions for the other coefficients equal to 0. By Theorems 4.11 and 4.13 we have:

$$\frac{h(t)}{t} = P^{[0]}(h(t)) + t^{-1} h(t)^2 Q^{[1]}(h(t)) = 1 + h(t) + h(t)^2 + \frac{h(t)^2}{t}$$

$$d(t) = \frac{1}{1 - tR^{[0]}(h(t))} = \frac{1}{1 - t(1 + h(t))},$$

which corresponds to the Riordan array:

$$\mathcal{R}\left(\frac{2 - t - t^2 - t\sqrt{1 - 6t - 3t^2}}{2(1 - 2t + t^2 + t^3)}, \frac{1 - t - \sqrt{1 - 6t - 3t^2}}{2(1 + t)}\right).$$

The first rows of corresponding Riordan array $(d_{n,k})_{n,k\in\mathbb{N}}$ are given below (note, for example, $5 = 2 + 3$ and $117 = 45 + 24 + 7 + 41$):

n/k	0	1	2	3	4	5
0	1					
1	1	1				
2	**2**	**3**	1			
3	<u>5</u>	11	5	1		
4	16	**45**	**24**	**7**	1	
5	61	202	<u>**117**</u>	**41**	9	1

We wish to conclude this section by introducing an important result concerning the characterizations proven in this section. By means of generating functions, we can show that Theorem 4.10 gives the largest possible characterization of Riordan arrays. In other words, we can show that if $d_{n+1,k+1}$ depends on elements not contained in the grey zones of Fig. 4.2, then $\left(d_{n,k}\right)_{n,k\in\mathbb{N}}$ is not a Riordan array. It is worth noting that if $d_{n+1,k+1}$ depends on some elements $d_{\nu,\kappa}$ with $\nu > n$ and $\kappa < k + 1 + \nu - n$, then the recurrence is not well-defined, the computation of $d_{n+1,k+1}$ enters an infinite loop and its indexes keep growing, and, as a result, $(d_{n,k})_{n,k\in\mathbb{N}}$ is actually not defined. We must therefore show that $(d_{n,k})_{n,k\in\mathbb{N}}$ is not a Riordan array when $d_{n+1,k+1}$ depends on some element $d_{\nu,\kappa}$ with $\nu \leq n$ and $\kappa < k$. The following theorem shows this under the same conditions as Theorem 4.10 (no $\rho_j^{[i]}$ is involved) and with $\kappa \geq k-1$. Actually, this is sufficient for our purposes because the presence of some $\rho_j^{[i]}$'s does not change the proof. Moreover, the method is virtually the same when $\kappa < k-1$ (only a few technical aspects are slightly modified).

Theorem 4.14 *If the generic element $d_{n+1,k+1}$ in an array $(d_{n,k})_{n,k\in\mathbb{N}}$ is defined by the recurrence:*

$$d_{n+1,k+1} = \sum_{i\geq 0}\sum_{j\geq -1} \alpha_{i,j}^* d_{n-i,k+j} \qquad (d_{n,-1} = 0, \forall n \in N)$$

with some $\alpha_{i,-1}^* \neq 0$, then $(d_{n,k})_{n,k\in\mathbb{N}}$ is not a Riordan array.

Proof By assuming that the array is Riordan and by going on to generating functions, we obtain the contradiction that all the $\alpha_{i,-1}^*$ are zero. This proves the theorem. □

Example 4.8 The concept of the A-matrix has been used to study a model of under-diagonal paths in the integer lattice. In particular, in [11] it is shown that the paths from (0, 0) to $(n, n-k)$, under this model, correspond to a Riordan array and, in particular, there is a correspondence between some types of steps that can originate from a given point in $\mathbb{Z}^2$ and particular positions in the A-matrix, so that the enumeration of the paths can be easily carried out by using the results illustrated in this chapter. This model include classical Dyck, Motzkin and Scröder paths and many others. Moreover, the model allows privileged access to the main diagonal, that is, different steps can be used to reach the main diagonal and the function $d(t)$ of the Riordan array, counting the paths arriving on the main diagonal, can be computed by Theorem 4.13 in order to consider the different relation for the elements of the array in column 0. As an example without privileged access to the main diagonal, let us consider steps: E, N and EN^2 (where E and N correspond to east and north, respectively); the correspondence between steps and positions in the A-matrix is illustrated by the following figure

	k	$k+1$	$k+2$
n	①		①
$n+1$		●	①

which translates into the following recurrence relations

$$d_{n+1,0} = d_{n,1} + d_{n+1,1},$$

$$d_{n+1,k+1} = d_{n,k} + d_{n,k+2} + d_{n+1,k+2};$$

in fact, $d_{n,k}$ counts paths from $(0, 0)$ to $(n, n-k)$ and position $(n+1, n-k)$ can be reached from positions $(n, n-k)$, $(n, n-k+2)$ and $(n+1, n+k-1)$ with steps E, EN^2 and N, respectively; moreover, the main diagonal cannot be reached with an east step. The first rows of corresponding Riordan array $(d_{n,k})_{n,k\in\mathbb{N}}$ are given below (note, for example, $46 = 24 + 4 + 18$):

n/k	0	1	2	3	4	5
0	1					
1	1	1				
2	3	2	1			
3	9	7	3	1		
4	31	**24**	12	**4**	1	
5	113	89	**46**	**18**	5	1

If steps EN^2 occurs into two colours than we have

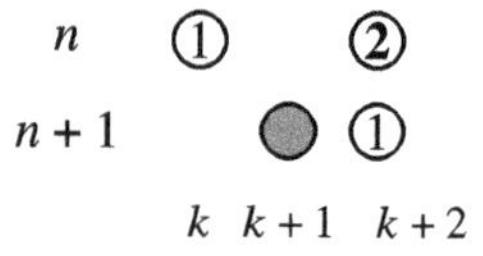

	k	$k+1$	$k+2$
n	①		②
$n+1$		●	①

and the recurrence relations

$$d_{n+1,0} = 2d_{n,1} + d_{n+1,1},$$

$$d_{n+1,k+1} = d_{n,k} + 2d_{n,k+2} + d_{n+1,k+2},$$

whose first values are given below (note, for example, $64 = 34 + 8 + 22$)

n/k	0	1	2	3	4	5
0	1					
1	1	1				
2	4	2	1			
3	13	9	3	1		
4	52	**34**	15	**4**	1	
5	214	146	**64**	**22**	5	1

Previous examples correspond to Exercises 4.4 and 4.6.

Example 4.9 Another problem in which the A-matrix simplifies the computation can be found in [10, 12] and concerns the enumeration of binary words with no occurrence of a pattern $\mathfrak{p} = p_0 \cdots p_{h-1}$. The fundamental notion is that of the *auto-correlation vector* of bits $c = (c_0, \ldots, c_{h-1})$ associated to a given $\mathfrak{p}$, that is, the vector of bits defined in terms of Iverson's bracket notation (for a predicate P, the expression $[\![P]\!]$ has value 1 if P is true and 0 otherwise) as follows:

$$c_i = [\![p_0 p_1 \cdots p_{h-1-i} = p_i p_{i+1} \cdots p_{h-1}]\!].$$

In other words, the bit c_i is determined by shifting $\mathfrak{p}$ right by i positions and setting $c_i = 1$ if and only if the remaining letters match the original. For example, when $\mathfrak{p} = 10101$ the autocorrelation vector is $c = (1, 0, 1, 0, 1)$, as illustrated below

1	0	1	0	1	Tails				
1	0	1	0	1					1
	1	0	1	0	1				0
		1	0	1	0	1			1
			1	0	1	0	1		0
				1	0	1	0	1	1

It can be proved that when $\mathfrak{p}$ satisfies particular conditions then the number $R_{n,k}^{[\mathfrak{p}]}$ of words avoiding $\mathfrak{p}$ and having n bits one and $n-k$ bits zero is a Riordan array which is related to the following A-matrix:

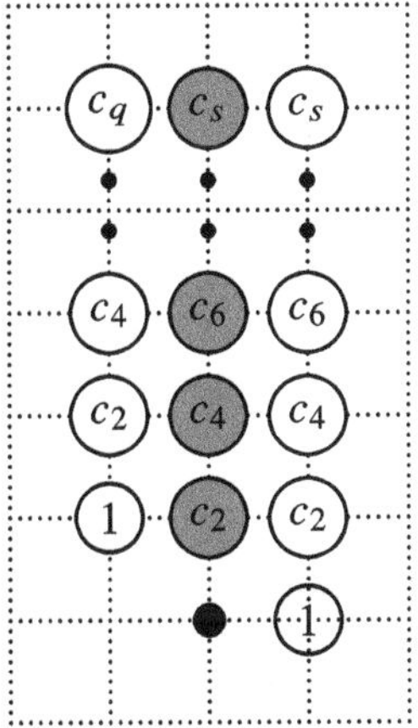

where the coefficients in the gray circles are negative and $s = q + 2$. The case $\mathfrak{p} = 10101$ corresponds in particular to an A-matrix with the following shape

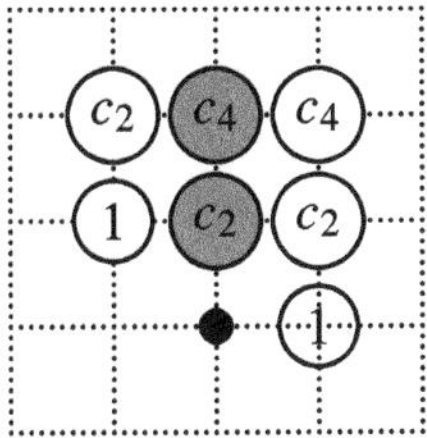

and is examined in Exercise 4.7.

We point out that in Example 4.9, as well as in Example 4.8 concerning a lattice path model, the A-matrix is simple while the A-sequence is rather complicated.

4.3 Is It a Riordan Array?

The A-sequence and A-matrix are important concepts that often allows us to identify the Riordan array nature of a certain problem, typically an enumeration problem in combinatorics. In fact, if the problem can be described by a recurrence relation such as those in Theorems 4.1, 4.7, 4.8 and 4.9 then we can infer that it is related to a Riordan array and use the corresponding theory to thoroughly study it. We have illustrated in the previous section very general applications concerning the enumeration of lattice paths. We can now see some other specific examples.

Example 4.10 Let us consider underdiagonal lattice paths starting from the origin $(0, 0)$ and composed by east and north steps and denote by $d_{n,k}$ the number of those that arrive at $(n, n-k)$. We easily find the recurrence relation:

$$d_{n+1,k+1} = d_{n,k} + d_{n+1,k+2}, \quad d_{n+1,0} = d_{n+1,1},$$

since position $(n+1, n-k)$ can be reached from position $(n, n-k)$ by an east step or from position $(n+1, n-k-1)$ by a north step; moreover, the main diagonal can be reached only by a north step. By using the results of this chapter we find $A(t) = \frac{1}{1-t}$ and the problem is easily brought back to the $\mathcal{R}(C(t), tC(t))$, with Dyck paths, represented by underdiagonal paths arriving at (n, n), counted on the first column by $C(t) = \frac{1-\sqrt{1-4t}}{2t}$.

Example 4.11 As another example, let us consider single source *directed animals*, a well-known class of combinatorial objects: a directed animal $\mathcal{A}$ with n nodes is a subset of n points in $\mathbb{N} \times \mathbb{N}$ such that the origin $(0, 0)$ is in $\mathcal{A}$ and is called the source and for every $P \in \mathcal{A}$ there is a lattice path from $(0, 0)$ to P with steps $(1, 0)$ and $(0, 1)$ all of whose vertices lie in $\mathcal{A}$.

Let $d_{n,k}$ be the number of directed animals with n nodes, k of which belong to the x-axis. The following relation can be found in [1, 20]:

$$d_{n+1,k+1} = d_{n,k} + d_{n-1,k} + d_{n-1,k+1} + d_{n-1,k+2} + \cdots$$

This proof is based on a nice combinatorial argument, which consists, given a configuration of size $n+1$, in removing the two right-most points on the x-axis and moving one step down, diagonally to the left, the set of those points which can no longer be reached from $(0, 0)$. This allows to obtain configurations with $n+1$ points from those with $n-1$ points. The recurrence relation translates into a Riordan array with A-matrix defined by:

$$P^{[0]}(t) = 1, \qquad P^{[1]}(t) = \frac{1}{1-t}$$

which correspond to the following A-sequence:

$$A(t) = \frac{1}{2}\left(1 + \sqrt{\frac{1+3t}{1-t}}\right) = 1 + t + t^3 - t^4 + 3t^5 - 6t^6 + 15t^7 - 36t^8 + \cdots$$

or

$$h(t) = \frac{1 + t - \sqrt{1 - 2t - 3t^2}}{2}.$$

The matrix begins as follows, by considering that for $k = 0$ only the term with $n = 0$ is different from zero:

n/k	0	1	2	3	4	5
0	1					
1	0	1				
2	0	1	1			
3	0	1	2	1		
4	0	2	3	3	1	
5	0	4	6	6	4	1

In many situations we can find ourselves faced with a triangle of numerical values that correspond to the enumeration of combinatorial structures for small dimensions, without initially having a general vision of the problem and therefore without knowing a recurrence relationship that describes it. By observing the values of the triangle one is sometimes able to recognize a pattern corresponding to simple A-sequences or simple A-matrices. For example, by looking for the number $d_{n,k}$ of binary words with n bits 1 and $n - k$ bits 0, avoiding the pattern $p = 001$, we find the triangle:

n/k	0	1	2	3	4	5
0	1					
1	2	1				
2	4	3	1			
3	8	7	4	1		
4	16	15	11	5	1	
5	32	31	26	16	6	1

In this case, it is very easy to recognize the A and Z-sequences defined by the generating functions $A(t) = 1 + t$ and $Z(t) = 2$, that is, the Riordan array:

$$\mathcal{R}\left(\frac{1}{1-2t}, \frac{t}{1-t}\right);$$

this result then should be proved by a combinatorial reasoning.

However, when the structure of the problem is complex, the simple observation of the triangle is not enough. Let us consider the following example, where $d_{n,k}$ counts the number of binary words with n bits 1 and $n - k$ bits 0, avoiding the pattern $p = 11100$ (as before, it can be obtained by generating the words):

n/k	0	1	2	3	4	5
0	1					
1	2	1				
2	6	3	1			
3	18	9	4	1		
4	58	29	13	5	1	
5	192	96	44	18	6	1

Is it a Riordan array? A simple inspection of the matrix does not allow to answer the question, but, in this case we are lucky because we can use the results in [12], where the Riordan array nature of the triangle is proved by means of the A-matrix concept. In fact, we have

$$d_{n+1,k+1} = d_{n,k} + d_{n+1,k+2} - d_{n-2,k},$$

which corresponds to

$$Q(t) = 1,\ P_0(t) = 1, \quad P_2(t) = -1,$$

and to an A-sequence which is the solution of the third degree equation

$$(1-t)A(t)^3 - A(t)^2 + t^2 = 0,$$

$$A(t) = 1 + t + 2t^3 - t^4 + 7t^5 - 12t^6 + 38t^7 - 99t^8 + \cdots ;$$

it is also proved that

$$Z(t) = 2 + 2t + 4t^3 - 2t^4 + 14t^5 - 24t^6 + 76t^7 - 198t^8 + \cdots$$

which yields the Riordan array

$$R = \mathcal{R}\left(\frac{1}{\sqrt{1-4t+4t^3}}, \frac{1-\sqrt{1-4t+4t^3}}{2}\right).$$

However, without the result in [12], it would not have been easy to understand the nature of the problem.

Now let $\overline{R}$ be the array R less its first row, and R^{-1} be the inverse of R and look at the 8×8 truncation of the product $R^{-1} \cdot \overline{R}$:

$$R^{-1}\cdot\overline{R}=\begin{bmatrix}1&0&0&0&0&0&0&0\\2&1&0&0&0&0&0&0\\6&3&1&0&0&0&0&0\\18&9&4&1&0&0&0&0\\58&29&13&5&1&0&0&0\\192&96&44&18&6&1&0&0\\650&325&151&64&24&7&1&0\\2232&1116&524&228&90&31&8&1\end{bmatrix}^{-1}\begin{bmatrix}2&1&0&0&0&0&0&0\\6&3&1&0&0&0&0&0\\18&9&4&1&0&0&0&0\\58&29&13&5&1&0&0&0\\192&96&44&18&6&1&0&0\\650&325&151&64&24&7&0&0\\2232&1116&524&228&90&31&8&1\\7746&3873&1833&813&333&123&39&9\end{bmatrix}$$

$$=\begin{bmatrix}1&0&0&0&0&0&0&0\\-2&1&0&0&0&0&0&0\\0&-3&1&0&0&0&0&0\\0&3&-4&1&0&0&0&0\\0&-5&7&.5&1&0&0&0\\0&12&-14&12&-6&1&0&0\\0&-28&35&-28&18&-7&1&0\\0&74&-88&74&-48&25&-8&1\end{bmatrix}\begin{bmatrix}2&1&0&0&0&0&0&0\\6&3&1&0&0&0&0&0\\18&9&4&1&0&0&0&0\\58&29&13&5&1&0&0&0\\192&96&44&18&6&1&0&0\\650&325&151&64&24&7&0&0\\2232&1116&524&228&90&31&8&1\\7746&3873&1833&813&333&123&39&9\end{bmatrix}$$

$$=\begin{bmatrix}2&1&0&0&0&0&0&0\\2&1&1&0&0&0&0&0\\0&0&1&1&0&0&0&0\\4&2&0&1&1&0&0&0\\-2&-1&2&0&1&1&0&0\\14&7&-1&2&0&1&1&0\\-24&-12&7&-1&2&0&1&1\\76&38&-12&7&-1&2&0&1\end{bmatrix}=P.$$

In the resulting matrix P, we can observe the Z-sequence in the leftmost column, the A-sequence in the 1-column and the shifted versions of the A-sequence in the subsequent columns. This is not an accident and in fact, if we observe that $\overline{R}$ can be obtained by computing the product $\overline{R}=U\cdot R$, where U denotes the upper shift matrix:

$$U=(\delta_{i+1,j})_{i,j\geq 0}=\begin{bmatrix}0&1&0&0&0&\cdots\\0&0&1&0&0&\cdots\\0&0&0&1&0&\cdots\\0&0&0&0&1&\cdots\\\vdots&\vdots&\vdots&\vdots&\vdots&\ddots\end{bmatrix},\tag{4.3.1}$$

we have the following theorem:

Theorem 4.15 *Let R be a Riordan array with A- and Z- sequences $A=(a_0,a_1,a_2,\cdots)$ and $Z=(z_0,z_1,z_2,\cdots)$ and U the upper shift matrix. Then the product $P=R^{-1}UR$ is given by*

$$P=\begin{bmatrix}z_0&a_0&0&0&0&\cdots\\z_1&a_1&a_0&0&0&\cdots\\z_2&a_2&a_1&a_0&0&\cdots\\z_3&a_3&a_2&a_1&a_0&\cdots\\\vdots&\vdots&\vdots&\vdots&\vdots&\ddots\end{bmatrix};\tag{4.3.2}$$

matrix P *is called the* production matrix *of* R.

In reality, a stronger result holds true and will be proved in Theorem 6.29: a lower triangle matrix R is a Riordan array if and only if $U \cdot R = R \cdot P$, where U and P are defined as in (4.3.1) and (4.3.2). These results are proved in [7] by using generating functions, as an extension of the concept of production matrices with non-negative integer entries introduced in [5, 6] in connection to the ECO method, and will be presented in detail in Chap. 6. We wish to observe that in [6] the concept of production matrix is studied in connection to succession rules and Riordan arrays, similarly to the study in [15], where the relation among A- and Z- sequences and generating trees is introduced. Many examples of production matrices will be illustrated in Chap. 9 in connection with orthogonal polynomials.

As already observed in [2], the production matrix gives us an alternative test to see if a candidate lower triangular array might or might not be a Riordan array. In particular, in some cases it allows us to quickly decide that an array is not a Riordan array. On the converse, since the test is generally done on a small portion of the triangle, finding that the production matrix has the form (4.3.2) does not prove that we are dealing with a Riordan array and in these cases it is necessary to deepen the study, for example from a combinatorial point of view, but still it can be an important suggestion.

Before closing this section, let us try to understand why the matrix P is called a production matrix, by coming back to the example concerning the enumeration of binary words avoiding the pattern $p = 11100$. If we compute P^2 we find a matrix having in the first row the elements of row $n = 2$ in R :

$$P^2 = \begin{bmatrix} 2 & 1 & 0 & 0 & 0 & 0 & 0 & 0 \\ 2 & 1 & 1 & 0 & 0 & 0 & 0 & 0 \\ 0 & 0 & 1 & 1 & 0 & 0 & 0 & 0 \\ 4 & 2 & 0 & 1 & 1 & 0 & 0 & 0 \\ -2 & -1 & 2 & 0 & 1 & 1 & 0 & 0 \\ 14 & 7 & -1 & 2 & 0 & 1 & 1 & 0 \\ -24 & -12 & 7 & -1 & 2 & 0 & 1 & 1 \\ 76 & 38 & -12 & 7 & -1 & 2 & 0 & 1 \end{bmatrix} \begin{bmatrix} 2 & 1 & 0 & 0 & 0 & 0 & 0 & 0 \\ 2 & 1 & 1 & 0 & 0 & 0 & 0 & 0 \\ 0 & 0 & 1 & 1 & 0 & 0 & 0 & 0 \\ 4 & 2 & 0 & 1 & 1 & 0 & 0 & 0 \\ -2 & -1 & 2 & 0 & 1 & 1 & 0 & 0 \\ 14 & 7 & -1 & 2 & 0 & 1 & 1 & 0 \\ -24 & -12 & 7 & -1 & 2 & 0 & 1 & 1 \\ 76 & 38 & -12 & 7 & -1 & 2 & 0 & 1 \end{bmatrix}$$

$$= \begin{bmatrix} \mathbf{6} & \mathbf{3} & \mathbf{1} & 0 & 0 & 0 & 0 & 0 \\ 6 & 3 & 2 & 1 & 0 & 0 & 0 & 0 \\ 4 & 2 & 1 & 2 & 1 & 0 & 0 & 0 \\ 14 & 7 & 4 & 1 & 2 & 1 & 0 & 0 \\ 6 & 3 & 2 & 4 & 1 & 2 & 1 & 0 \\ 40 & 29 & 12 & 2 & 4 & 1 & 2 & 1 \\ -28 & -14 & -6 & 12 & 2 & 4 & 1 & 2 \\ 164 & 82 & 48 & -6 & 12 & 2 & 4 & 1 \end{bmatrix},$$

similarly, if we compute P^3 we find a matrix having in the first row the elements of row $n = 3$ in R and, in general, if we compute P^n we find a matrix having at the top row the elements of row n, thus reproducing the rows of R. This concept of production matrix P defined by $U \cdot R = R \cdot P$ has been examined for quite general matrices R in [16, 17] and, in a sense, it generalizes the concepts of A- and Z-sequences for matrices that are not Riordan arrays.

Example 4.12 Let us consider the triangle R of Stirling numbers of second kind $\left\{ {n \atop k} \right\}$ and look at the 8×8 truncation of the product $R^{-1} \cdot \overline{R}$:

$$R^{-1} \cdot \overline{R} = \begin{bmatrix} 1 & 0 & 0 & 0 & 0 & 0 & 0 & 0 \\ 1 & 1 & 0 & 0 & 0 & 0 & 0 & 0 \\ 1 & 3 & 1 & 0 & 0 & 0 & 0 & 0 \\ 1 & 7 & 6 & 1 & 0 & 0 & 0 & 0 \\ 1 & 15 & 25 & 10 & 1 & 0 & 0 & 0 \\ 1 & 31 & 90 & 65 & 15 & 1 & 0 & 0 \\ 1 & 63 & 301 & 350 & 140 & 21 & 1 & 0 \\ 1 & 127 & 966 & 1701 & 1050 & 266 & 28 & 1 \end{bmatrix}^{-1} \begin{bmatrix} 1 & 1 & 0 & 0 & 0 & 0 & 0 & 0 \\ 1 & 3 & 1 & 0 & 0 & 0 & 0 & 0 \\ 1 & 7 & 6 & 1 & 0 & 0 & 0 & 0 \\ 1 & 15 & 25 & 10 & 1 & 0 & 0 & 0 \\ 1 & 31 & 90 & 65 & 15 & 1 & 0 & 0 \\ 1 & 63 & 301 & 350 & 140 & 21 & 1 & 0 \\ 1 & 127 & 966 & 1701 & 1050 & 266 & 28 & 1 \\ 1 & 255 & 3025 & 7770 & 6951 & 2646 & 462 & 36 \end{bmatrix}$$

$$= \begin{bmatrix} 1 & 1 & 0 & 0 & 0 & 0 & 0 & 0 \\ 0 & 2 & 1 & 0 & 0 & 0 & 0 & 0 \\ 0 & 0 & 3 & 1 & 0 & 0 & 0 & 0 \\ 0 & 0 & 0 & 4 & 1 & 0 & 0 & 0 \\ 0 & 0 & 0 & 0 & 5 & 1 & 0 & 0 \\ 0 & 0 & 0 & 0 & 0 & 6 & 1 & 0 \\ 0 & 0 & 0 & 0 & 0 & 0 & 7 & 1 \\ 0 & 0 & 0 & 0 & 0 & 0 & 0 & 8 \end{bmatrix} = P.$$

Since P has not the form (4.3.2) we can conclude that R is not a Riordan array. In fact, we already observed in Chap. 3 that in order to have a Riordan array we have to consider the triangle $\frac{k!}{n!} \left\{ {n \atop k} \right\}$.

However, we can observe that the rows of Stirling triangle can be reproduced by the production matrix P, as illustrated in the products P^2 and P^3 :

$$P^2 = \begin{bmatrix} 1 & 1 & 0 & 0 & 0 & 0 & 0 & 0 \\ 0 & 2 & 1 & 0 & 0 & 0 & 0 & 0 \\ 0 & 0 & 3 & 1 & 0 & 0 & 0 & 0 \\ 0 & 0 & 0 & 4 & 1 & 0 & 0 & 0 \\ 0 & 0 & 0 & 0 & 5 & 1 & 0 & 0 \\ 0 & 0 & 0 & 0 & 0 & 6 & 1 & 0 \\ 0 & 0 & 0 & 0 & 0 & 0 & 7 & 1 \\ 0 & 0 & 0 & 0 & 0 & 0 & 0 & 8 \end{bmatrix} \begin{bmatrix} 1 & 1 & 0 & 0 & 0 & 0 & 0 & 0 \\ 0 & 2 & 1 & 0 & 0 & 0 & 0 & 0 \\ 0 & 0 & 3 & 1 & 0 & 0 & 0 & 0 \\ 0 & 0 & 0 & 4 & 1 & 0 & 0 & 0 \\ 0 & 0 & 0 & 0 & 5 & 1 & 0 & 0 \\ 0 & 0 & 0 & 0 & 0 & 6 & 1 & 0 \\ 0 & 0 & 0 & 0 & 0 & 0 & 7 & 1 \\ 0 & 0 & 0 & 0 & 0 & 0 & 0 & 8 \end{bmatrix} = \begin{bmatrix} \mathbf{1} & \mathbf{3} & \mathbf{1} & 0 & 0 & 0 & 0 & 0 \\ 0 & 4 & 5 & 1 & 0 & 0 & 0 & 0 \\ 0 & 0 & 9 & 7 & 1 & 0 & 0 & 0 \\ 0 & 0 & 0 & 16 & 9 & 1 & 0 & 0 \\ 0 & 0 & 0 & 0 & 25 & 11 & 1 & 0 \\ 0 & 0 & 0 & 0 & 0 & 36 & 13 & 1 \\ 0 & 0 & 0 & 0 & 0 & 0 & 49 & 15 \\ 0 & 0 & 0 & 0 & 0 & 0 & 0 & 64 \end{bmatrix},$$

$$P^3 = \begin{bmatrix} 1 & 3 & 1 & 0 & 0 & 0 & 0 & 0 \\ 0 & 4 & 5 & 1 & 0 & 0 & 0 & 0 \\ 0 & 0 & 9 & 7 & 1 & 0 & 0 & 0 \\ 0 & 0 & 0 & 16 & 9 & 1 & 0 & 0 \\ 0 & 0 & 0 & 0 & 25 & 11 & 1 & 0 \\ 0 & 0 & 0 & 0 & 0 & 36 & 13 & 1 \\ 0 & 0 & 0 & 0 & 0 & 0 & 49 & 15 \\ 0 & 0 & 0 & 0 & 0 & 0 & 0 & 64 \end{bmatrix} \begin{bmatrix} 1 & 1 & 0 & 0 & 0 & 0 & 0 & 0 \\ 0 & 2 & 1 & 0 & 0 & 0 & 0 & 0 \\ 0 & 0 & 3 & 1 & 0 & 0 & 0 & 0 \\ 0 & 0 & 0 & 4 & 1 & 0 & 0 & 0 \\ 0 & 0 & 0 & 0 & 5 & 1 & 0 & 0 \\ 0 & 0 & 0 & 0 & 0 & 6 & 1 & 0 \\ 0 & 0 & 0 & 0 & 0 & 0 & 7 & 1 \\ 0 & 0 & 0 & 0 & 0 & 0 & 0 & 8 \end{bmatrix} = \begin{bmatrix} \mathbf{1} & \mathbf{7} & \mathbf{6} & \mathbf{1} & 0 & 0 & 0 & 0 \\ 0 & 8 & 19 & 9 & 1 & 0 & 0 & 0 \\ 0 & 0 & 27 & 37 & 12 & 1 & 0 & 0 \\ 0 & 0 & 0 & 64 & 61 & 15 & 1 & 0 \\ 0 & 0 & 0 & 0 & 125 & 91 & 18 & 1 \\ 0 & 0 & 0 & 0 & 0 & 216 & 127 & 21 \\ 0 & 0 & 0 & 0 & 0 & 0 & 343 & 169 \\ 0 & 0 & 0 & 0 & 0 & 0 & 0 & 512 \end{bmatrix}.$$

This example will be studied in terms of *exponential Riordan arrays* in Sect. 6.1, where many examples of production matrices for this kind of matrices will be shown.

We conclude with an example that has been treated by exponential Riordan arrays in [17].

Example 4.13 Let us consider the triangle R counting the number of forests of rooted trees on n labeled vertices having k trees; when $k \neq 0$ we have the explicit value $\binom{n}{k}kn^{n-k-1}$. By proceeding as before we have:

$$R = \begin{bmatrix} 1 & 0 & 0 & 0 & 0 & 0 & 0 & 0 \\ 0 & 1 & 0 & 0 & 0 & 0 & 0 & 0 \\ 0 & 2 & 1 & 0 & 0 & 0 & 0 & 0 \\ 0 & 9 & 6 & 1 & 0 & 0 & 0 & 0 \\ 0 & 64 & 48 & 12 & 1 & 0 & 0 & 0 \\ 0 & 625 & 500 & 150 & 20 & 1 & 0 & 0 \\ 0 & 7776 & 6480 & 2160 & 360 & 30 & 1 & 0 \\ 0 & 117649 & 100842 & 36015 & 6860 & 735 & 42 & 1 \end{bmatrix},$$

$$R^{-1} = \begin{bmatrix} 1 & 0 & 0 & 0 & 0 & 0 & 0 & 0 \\ 0 & 1 & 0 & 0 & 0 & 0 & 0 & 0 \\ 0 & -2 & 1 & 0 & 0 & 0 & 0 & 0 \\ 0 & 3 & -6 & 1 & 0 & 0 & 0 & 0 \\ 0 & -4 & 24 & -12 & 1 & 0 & 0 & 0 \\ 0 & 5 & -80 & 90 & -20 & 1 & 0 & 0 \\ 0 & -6 & 240 & -540 & 240 & -30 & 1 & 0 \\ 0 & 7 & -672 & 2835 & -2240 & 525 & -42 & 1 \end{bmatrix}$$

and

$$P = \begin{bmatrix} 0 & 1 & 0 & 0 & 0 & 0 & 0 & 0 \\ 0 & 2 & 1 & 0 & 0 & 0 & 0 & 0 \\ 0 & 5 & 4 & 1 & 0 & 0 & 0 & 0 \\ 0 & 16 & 15 & 6 & 1 & 0 & 0 & 0 \\ 0 & 65 & 64 & 30 & 8 & 1 & 0 & 0 \\ 0 & 326 & 325 & 160 & 50 & 10 & 1 & 0 \\ 0 & 1957 & 1956 & 975 & 320 & 75 & 12 & 1 \\ 0 & 13700 & 13699 & 6846 & 2275 & 560 & 105 & 14 \end{bmatrix}.$$

Since P has not the form (4.3.2) we can conclude that R is not a Riordan array, however, the top row of P^n corresponds to row n in R. For example, for $n = 4$ we find:

$$P^4 = \begin{bmatrix} 0 & \mathbf{64} & \mathbf{48} & \mathbf{12} & \mathbf{1} & 0 & 0 & 0 \\ 0 & 625 & 500 & 150 & 20 & 1 & 0 & 0 \\ 0 & 6526 & 5480 & 1860 & 320 & 28 & 1 & 0 \\ 0 & 72868 & 63462 & 23505 & 4760 & 558 & 36 & 1 \\ 0 & 869488 & 778424 & 307188 & 69600 & 9816 & 864 & 44 \\ \vdots & \vdots & \vdots & \vdots & \vdots & \vdots & \vdots & \vdots \end{bmatrix}.$$

As in the previous example, if we multiply every element of R by $\frac{k!}{n!}$ we find a new matrix R which is a Riordan array. In particular, the transformation gives:

$$R = \begin{bmatrix} 1 & 0 & 0 & 0 & 0 & 0 & 0 & 0 \\ 0 & 1 & 0 & 0 & 0 & 0 & 0 & 0 \\ 0 & 1 & 1 & 0 & 0 & 0 & 0 & 0 \\ 0 & \frac{3}{2} & 2 & 1 & 0 & 0 & 0 & 0 \\ 0 & \frac{8}{3} & 4 & 3 & 1 & 0 & 0 & 0 \\ 0 & \frac{125}{24} & \frac{25}{3} & \frac{15}{2} & 4 & 1 & 0 & 0 \\ 0 & \frac{54}{5} & 18 & 18 & 12 & 5 & 1 & 0 \\ 0 & \frac{16807}{720} & \frac{2401}{60} & \frac{343}{8} & \frac{98}{3} & \frac{35}{2} & 6 & 1 \end{bmatrix}$$

and the production matrix:

$$P = \begin{bmatrix} 0 & 1 & 0 & 0 & 0 & 0 & 0 & 0 \\ 0 & 1 & 1 & 0 & 0 & 0 & 0 & 0 \\ 0 & \frac{1}{2} & 1 & 1 & 0 & 0 & 0 & 0 \\ 0 & \frac{1}{6} & \frac{1}{2} & 1 & 1 & 0 & 0 & 0 \\ 0 & \frac{1}{24} & \frac{1}{6} & \frac{1}{2} & 1 & 1 & 0 & 0 \\ 0 & \frac{1}{120} & \frac{1}{24} & \frac{1}{6} & \frac{1}{2} & 1 & 1 & 0 \\ 0 & \frac{1}{720} & \frac{1}{120} & \frac{1}{24} & \frac{1}{6} & \frac{1}{2} & 1 & 1 \\ 0 & \frac{1}{5040} & \frac{1}{720} & \frac{1}{120} & \frac{1}{24} & \frac{1}{6} & \frac{1}{2} & 1 \end{bmatrix},$$

corresponding to an A-sequence with generating function $A(t) = e^t$.

Exercises

4.1 Find the A- and Z- sequences of the Riordan array $\mathcal{R}(\frac{1}{1-t}, \frac{t(1+rt)}{1-t})$ for $r \in \mathbb{R}$.

4.2 Find the A- and Z- sequences of the Riordan array $\mathcal{R}(\frac{1}{1-t-t^2}, t + t^2)$.

4.3 Find the A- and Z- sequences of the Riordan array $\mathcal{R}(d(t), td(t))$ where

$$d(t) = \frac{1 - t - \sqrt{1 - 2t - 3t^2}}{2t^2}$$

is the generating function of Motzkin numbers.

4.4 Find the Bell matrix corresponding to the relation:

$$d_{n+1,k+1} = d_{n,k} + d_{n,k+2} + d_{n+1,k+2}.$$

4.5 Find the matrix inverse of the matrix defined in Exercise 4.4.

4.6 Find the Bell matrix corresponding to the relation:

$$d_{n+1,k+1} = d_{n,k} + 2d_{n,k+2} + d_{n+1,k+2}.$$

4.7 Find the A-sequence corresponding to the A-matrix defined by:

$$Q(t) = 1, \quad P^{[0]}(t) = P^{[1]}(t) = 1 - t + t^2.$$

4.8 Find the A-sequence corresponding to the A-matrix defined by:

$$Q(t) = 1, \quad P^{[0]}(t) = 1 - t + t^2, P^{[1]}(t) = 1 - t, P^{[2]}(t) = 1.$$

4.9 Prove that the matrix R counting unsigned Stirling numbers of first kind $\left[{n \atop k}\right]$ is not a Riordan array. Here is the 8×8 truncation of the matrix:

$$R = \begin{bmatrix} 1 & 0 & 0 & 0 & 0 & 0 & 0 & 0 \\ 1 & 1 & 0 & 0 & 0 & 0 & 0 & 0 \\ 2 & 3 & 1 & 0 & 0 & 0 & 0 & 0 \\ 6 & 11 & 6 & 1 & 0 & 0 & 0 & 0 \\ 24 & 50 & 35 & 10 & 1 & 0 & 0 & 0 \\ 120 & 274 & 225 & 85 & 15 & 1 & 0 & 0 \\ 720 & 1764 & 1624 & 735 & 175 & 21 & 1 & 0 \\ 5040 & 13068 & 13132 & 6769 & 1960 & 322 & 28 & 1 \end{bmatrix}.$$

References

1. M. Aigner, Motzkin numbers. Europ. J. Comb. **19**, 663–675 (1998)
2. P. Barry, *Riordan Arrays: A Primer* (Logic Press, 2017)
3. L. Comtet, *Advanced Combinatorics* (Reidel, Dordrecht, 1974)
4. G.-S. Cheon, H. Kim, L.W. Shapiro, Combinatorics of Riordan arrays with identical A- and Z- sequences. Discrete Math. **312**(12–13), 2040–2049 (2012)
5. E. Deutsch, L. Ferrari, S. Rinaldi, Production matrices. Adv. Appl. Math. **34**(1), 101–122 (2005)

6. E. Deutsch, L. Ferrari, S. Rinaldi, Production matrices and Riordan arrays. Ann. Combin. **13**, 65–85 (2009)
7. T.X. He, Matrix characterizations of Riordan arrays. Linear Algebra Appl. **465**, 15–42 (2015)
8. T.-X. He, R. Sprugnoli, Sequence characterization of Riordan arrays. Discrete Math. **309**(12), 3962–3974 (2009)
9. D. Merlini, I Riordan array nell'analisi degli algoritmi. Tesi di Dottorato di Ricerca in Ingegneria Informatica e dell'Automazione, VIII ciclo, Università di Firenze (1996)
10. D. Merlini, M. Nocentini, Algebraic generating functions for languages avoiding Riordan patterns. J. Integer Seq. **21**(1), 1–25 (2018)
11. D. Merlini, D.G. Rogers, R. Sprugnoli, M.C. Verri, On some alternative characterizations of Riordan arrays. Canadian J. Math. **49**, 301–320 (1997)
12. D. Merlini, R. Sprugnoli, Algebraic aspects of some Riordan arrays related to binary words avoiding a pattern. Theoret. Comput. Sci. **412**(27), 2988–3001 (2011)
13. D. Merlini, R. Sprugnoli, M.C. Verri, The Cauchy numbers. Discrete Math. **306**(16), 1906–1920 (2006)
14. D. Merlini, R. Sprugnoli, The relevant prefixes of coloured Motzkin walks: an average case analysis. Theoret. Comput. Sci. **411**(1), 148–163 (2010)
15. D. Merlini, M.C. Verri, Generating trees and proper Riordan Arrays. Discrete Math. **218**(16), 167–183 (2000)
16. M. Pétréolle, A.D. Sokal, Lattice paths and branched continued fractions II. Multivariate Lah polynomials and Lah symmetric functions. Preprint. https://arxiv.org/abs/1907.02645 (2019)
17. A.D. Sokal, Total positivity of some polynomial matrices that enumerate labeled trees and forests I. Forests of rooted labeled trees. Preprint. https://arxiv.org/abs/2105.05583 (2021)
18. D.G. Rogers, Pascal triangles, Catalan numbers and renewal arrays. Discrete Math. **22**, 301–310 (1978)
19. L.W. Shapiro, A survey of the Riordan Group. Talk at a meeting of the American Mathematical Society, Richmond, Virginia (1994)
20. L.W. Shapiro, Bijections and the Riordan group. Theoret. Comput. Sci. **307**, 403–413 (2003)
21. L.W. Shapiro, S. Getu, W.-J. Woan, L. Woodson, The Riordan group. Discrete Appl. Math. **34**, 229–239 (1991)
22. R. Sprugnoli, Riordan arrays and combinatorial sums. Discrete Math. **132**(1–3), 267–290 (1994)

Chapter 5
Combinatorial Sums and Inversions

Abstract Traditional methods used for solving combinatorial sums (see, e.g., J. Riordan [18] or L. Comtet [1]) are used in R. L. Graham, D. E. Knuth, and O. Patashnik [8], where it is shown how to use the rules of binomial coefficients, Stirling numbers, and so on, for computing combinatorial sums. G. P. Egorychev [3] developed the method known as *integral representation of sums*. Gosper's method [7] and the Petkovšek-Wilf-Zeilberger approach [17, 22, 23] are other well-known methods which nowadays are embodied in every system for computer algebra. An interesting method for evaluating combinatorial sums has emerged in the Riordan arrays concept, with significant early contributions due to R. Sprugnoli [19, 20]. As a matter of fact, Riordan arrays correspond to a special application of the method of coefficients that allows one to compute a vast number of combinatorial sums and inversions in a uniform and often very simple way. In particular, R. Sprugnoli showed how to prove with the Riordan array approach almost all identities in H. W. Gould's book [5]. In this chapter the computation of combinatorial sums and inversions with Riordan arrays is presented in detail. Many other applications of the method can be found in [9, 11–16], while other characteristic combinatorial identities in several parameters have been recently studied with a Riordan array approach by A. Luzón, D. Merlini, M. A. Morón and R. Sprugnoli [10].

5.1 Combinatorial Sums

For the point of view of this chapter, the most important property is the fact that the sums involving the rows of a Riordan array can be performed by operating a suitable transformation on a generating function and then by extracting a coefficient from the resulting function, that is, the *fundamental rule of Riordan arrays* already illustrated in Chap. 3, where we also noticed the equivalence with Theorem 3.1:

$$\sum_{k=0}^{n} d_{n,k} f_k = [t^n] d(t) f(h(t)). \tag{5.1.1}$$

The first important contribution in this direction is due to Sprugnoli [19].

© The Author(s), under exclusive license to Springer Nature Switzerland AG 2022

L. Shapiro et al., *The Riordan Group and Applications*, Springer Monographs in Mathematics, https://doi.org/10.1007/978-3-030-94151-2_5

In the case of Pascal triangle we obtain the Euler transformation:

$$\sum_{k=0}^{n}\binom{n}{k}f_k=[t^n]\frac{1}{1-t}f\left(\frac{t}{1-t}\right). \tag{5.1.2}$$

By considering the generating functions $\mathcal{G}(1)=1/(1-t), \mathcal{G}\left((-1)^k\right)=1/(1+t)$, and $\mathcal{G}(k)=t/(1-t)^2$, from (5.1.1) we have

$$\begin{aligned}
\text{row sums } &\sum_k d_{n,k}=[t^n]\frac{d(t)}{1-h(t)}\\
\text{alternating row sums } &\sum_k(-1)^k d_{n,k}=[t^n]\frac{d(t)}{1+h(t)}\\
\text{weighted row sums } &\sum_k kd_{n,k}=[t^n]\frac{td(t)h(t)}{(1-h(t))^2}.
\end{aligned}$$

Moreover, by observing that $\mathcal{R}(d(t), th(t))$ is a Riordan array whose rows are the diagonals of $\mathcal{R}(d(t), h(t))$, we have

$$\text{diagonal sums} \qquad \sum_k d_{n-k,k}=[t^n]\frac{d(t)}{1-th(t)}.$$

Obviously, this observation can be generalized to find the generating function of any sum $\sum_k d_{n-sk,k}$ for every $s\geq 1$. We obtain the well-known results of the Pascal triangle; for example, diagonal sums give

$$\sum_k\binom{n-k}{k}=[t^n]\frac{1}{1-t}\frac{1}{1-t^2(1-t)^{-1}}=[t^n]\frac{1}{1-t-t^2}=F_{n+1}$$

connecting binomial coefficients and Fibonacci numbers.

Another general result can be obtained by means of two sequences $(f_k)_{k\in\mathbb{N}}$ and $(g_k)_{k\in\mathbb{N}}$ and their generating functions $f(t), g(t)$. For $p=1,2,\ldots$, the generic element of the Riordan array $(f(t), t^p)$ is

$$d_{n,k}=[t^n]f(t)(t^p)^k=[t^{n-pk}]f(t)=f_{n-pk}.$$

Therefore, by formula (5.1.1), we have

$$\sum_{k=0}^{\lfloor n/p\rfloor}f_{n-pk}g_k=[t^n]f(t)\big[g(y)\mid y=t^p\big]=[t^n]f(t)g(t^p).$$

This can be called the *rule of generalized convolution* since it reduces to the usual convolution rule for $p=1$. Suppose, for example, that we wish to sum one out of every three powers of 2, starting with 2^n and going down to the lowest integer exponent ≥ 0; we have

$$S_n = \sum_{k=0}^{\lfloor n/3 \rfloor} 2^{n-3k} = [t^n]\frac{1}{1-2t} \cdot \frac{1}{1-t^3}.$$

By a well-known result in asymptotics, an approximate value for this sum can be obtained by extracting the coefficient of the first factor and then by multiplying it by the second factor, in which t is substituted with the singularity of least modulus 1/2. This gives $S_n \approx 2^{n+3}/7$, and in fact we have the exact value $S_n = \lfloor 2^{n+3}/7 \rfloor$. This example wants to highlight that formula (5.1.1) can be used even when an exact extraction of the coefficient of $d(t)f(h(t))$ is not feasible; in such cases, in fact, an asymptotic approximation can be computed with standard techniques (see, for example, [4]).

In a sense, the theorem on the sums involving Riordan arrays is a characterization for them; in fact, we can prove a sort of inverse property:

Theorem 5.1 *Let $\left(d_{n,k}\right)_{n,k\in\mathbb{N}}$ be an infinite triangle such that for every sequence $(f_k)_{k\in\mathbb{N}}$ we have*

$$\sum_k d_{n,k} f_k = [t^n]d(t)f(th(t)),$$

where $f(t)$ is the generating function of the sequence and $d(t), h(t)$ are two f.p.s. not depending on $f(t)$. Then the triangle defined by the Riordan array $\mathcal{R}(d(t), h(t))$ coincides with $\left(d_{n,k}\right)_{n,k\in\mathbb{N}}$.

Proof For every $k \in \mathbb{N}$ take the sequence which is 0 everywhere except in the kth element $f_k = 1$. The corresponding generating function is $f(t) = t^k$ and we have $\sum_{i=0}^{\infty} d_{n,i} f_i = d_{n,k}$. Hence, according to the theorem's hypotheses, we find $\mathcal{G}_t\left(d_{n,k}\right)_{n\in\mathbb{N}} = d_k(t) = d(t)h(t)^k$, and this corresponds to the initial definition of column generating functions for a Riordan array, for every $k = 1, 2, \ldots$. □

Let us consider *simple binomial coefficients*, i.e., binomial coefficients of the form $\binom{n+ak}{m+bk}$, where a, b are two parameters and k is a non-negative integer variable. If we consider n a variable and m a parameter, or vice versa, we have two different infinite arrays $(d_{n,k})$ or $(\widehat{d}_{m,k})$, whose elements depend on the parameters a, b, m or a, b, n, respectively. In either case, if some conditions on a, b hold, we have Riordan arrays and therefore we can apply formula (5.1.1) to find the value of many sums.

Theorem 5.2 *Let $d_{n,k}$ and $\widehat{d}_{m,k}$ be as above. If $b > a$ and $b - a$ is an integer, then $D = (d_{n,k})$ is a Riordan array. If $b < 0$ is an integer, then $\widehat{D} = (\widehat{d}_{m,k})$ is a Riordan array. We have*

$$D = \left(\frac{t^m}{(1-t)^{m+1}}, \frac{t^{b-a}}{(1-t)^b}\right) \qquad \widehat{D} = \left((1+t)^n, \frac{t^{-b}}{(1+t)^{-a}}\right).$$

Proof By using well-known properties of binomial coefficients, we find

$$d_{n,k} = \binom{n+ak}{m+bk} = \binom{n+ak}{n-m+ak-bk} =$$

$$= \binom{-n-ak+n-m+ak-bk-1}{n-m+ak-bk} \times (-1)^{n-m+ak-bk} =$$

$$= \binom{-m-bk-1}{(n-m)+(a-b)k}(-1)^{n-m+ak-bk} = [t^{n-m+ak-bk}]\frac{1}{(1-t)^{m+1+bk}} =$$

$$= [t^n]\frac{t^m}{(1-t)^{m+1}}\left(\frac{t^{b-a}}{(1-t)^b}\right)^k;$$

and

$$\widehat{d}_{m,k} = [t^{m+bk}](1+t)^{n+ak} = [t^m](1+t)^n(t^{-b}(1+t)^a)^k.$$

The theorem now directly follows from Newton's rule. □

The corresponding sums take on two specific forms which are worth being stated explicitly:

$$\sum_k \binom{n+ak}{m+bk} f_k = [t^n]\frac{t^m}{(1-t)^{m+1}} f\left(\frac{t^{b-a}}{(1-t)^b}\right) \qquad b > a \tag{5.1.3}$$

$$\sum_k \binom{n+ak}{m+bk} f_k = [t^m](1+t)^n f(t^{-b}(1+t)^a) \qquad b < 0. \tag{5.1.4}$$

For $m = a = 0$ and $b = 1$ we find the Riordan array of the Pascal triangle and the Euler transformation, already introduced.

If m and n are independent of each other, these relations can also be stated as generating function identities. The binomial coefficient $\binom{n+ak}{m+bk}$ is so general that a large number of combinatorial sums can be solved by means of the two formulas (5.1.3) and (5.1.4). If n and m are not independent, we can apply the concept of implicit Riordan array to solve (i.e., to find a closed form) many identities.

Our first example requires some introductory considerations. By (2.4.2) we have $\mathcal{G}(k^p f_k) = \mathcal{S}_p(tD)\mathcal{G}(f_k)$, therefore it is a simple matter to show by mathematical induction that

$$\frac{d^r}{dt^r}\frac{1}{1-t} = \frac{r!}{(1-t)^{r+1}}$$

so we have (see also Graham, Knuth, and Patashnik [8, p. 337]):

$$\mathcal{G}\left(k^p\right) = \mathcal{G}\left(k^p \cdot 1\right) = \sum_{r=0}^{p} \left\{ {p \atop r} \right\} t^r \frac{d^r}{dt^r} \frac{1}{1-t} = \sum_{r=0}^{p} \left\{ {p \atop r} \right\} \frac{r!t^r}{(1-t)^{r+1}}$$

and this allows us to propose the following:

Example 5.1 We have, for $p \in \mathbb{N}$:

$$\sum_{k=0}^{n} (-1)^k \binom{n}{k} k^p = \left\{ {p \atop n} \right\} (-1)^n n!.$$

By using the generating function $\mathcal{G}\left(k^p\right)$ just found, the Riordan array approach reduces the proof to a straight-forward computation:

$$\sum_{k=0}^{n} \binom{n}{k} (-1)^k k^p = [t^n] \frac{1}{1-t} \left[\sum_{k=0}^{p} \left\{ {p \atop k} \right\} \frac{k!(-u)^k}{(1+u)^{k+1}} \;\middle|\; u = \frac{t}{1-t} \right] =$$

$$= [t^n] \sum_{k=0}^{p} \left\{ {p \atop k} \right\} k!(-1)^k t^k = \left\{ {p \atop n} \right\} n!(-1)^n$$

as desired.

We wish to point out that this result can be found in H. W. Gould's collection [5] in the case $p \le n$. Moreover, as suggested by one referee, it is related to a well-known fact about differences of polynomial that goes back to Euler (see, e.g., H. W. Gould [6]). In particular, if $P(k)$ is a polynomial of degree less than n then $\sum_{k=0}^{n} (-1)^k \binom{n}{k} P(k) = 0$ and if $P(k)$ is a polynomial of degree n with leading coefficient c then $\sum_{k=0}^{n} (-1)^k \binom{n}{k} P(k) = n!c$. However, among the various methods that can be used to prove the identity, we think that the approach with generating functions and Riordan arrays could be instructive.

Sometimes, some little trick is necessary to close a sum. The following example is relative to identities (1.41) and (1.42) in Gould's collection [5].

Example 5.2 We have

$$\sum_{k=0}^{n} (-1)^k \binom{n}{k} \frac{x}{x+k} = \binom{x+n}{n}^{-1} \quad \text{and} \quad \sum_{k=0}^{n} (-1)^k \binom{n}{k} \binom{x+k}{k}^{-1} = \frac{x}{x+n}.$$

The proof of the first identity can be done by multiplying and dividing by the same quantity; we use formula (5.1.4):

$$\sum_{k=0}^{n} (-1)^k \binom{n}{k} \binom{x+k}{k} \binom{x+k}{k}^{-1} \frac{x}{x+k} = \frac{n!x!}{(x+n)!} \sum_{k=0}^{n} (-1)^k \binom{x+n}{n-k} \binom{x+k-1}{k} =$$

$$= \frac{n!x!}{(x+n)!}[t^n](1+t)^{x+n}\left[\frac{1}{(1+u)^x} \;\Big|\; u=t\right] = \frac{n!x!}{(x+n)!}[t^n](1+t)^n = \binom{n+x}{n}^{-1}.$$

The other proof is more direct:

$$\sum_{k=0}^{n}(-1)^k\binom{n}{k}\binom{x+k}{k}^{-1} = \sum_{k=0}^{n}(-1)^k\frac{n!k!x!}{k!(n-k)!(x+k)!}$$

$$= \frac{n!x!}{(n+x)!}\sum_{k=0}^{n}\binom{n+x}{n-k}(-1)^k = \frac{n!x!}{(n+x)!}[t^n](1+t)^{n+x}\left[\frac{1}{1+u} \;\Big|\; u=t\right]$$

$$= \frac{n!x!}{(n+x)!}\binom{n+x-1}{n} = \frac{n!x!}{(n+x)!}\cdot\frac{(n+x-1)!}{n!(x-1)!} = \frac{x}{x+n}.$$

We have put together these two sums because they are an example of *inverse relations*. As we will see in the next section, the proof of any one of them implies the proof of the other one.

Identities between non-closed formulas can be proved by looking for the common generating function; this is identity (3.19) in Gould's collection [5], where we use the generating function $\mathcal{G}\left(\binom{p+k}{k}\right) = 1/(1-t)^{p+1}$:

Example 5.3 We have:

$$\sum_{k=0}^{n}\binom{n}{k}\binom{x}{k}2^k = \sum_{k=0}^{n}\binom{n}{k}\binom{x+k}{n}.$$

We apply an Euler transformation in the first part, while the second part reduces to a simple convolution:

$$(1)\quad \sum_{k=0}^{n}\binom{n}{k}\binom{x}{k}2^k = [t^n]\frac{1}{1-t}\left[(1+2u)^x \;\Big|\; u=\frac{t}{1-t}\right] = [t^n]\frac{(1+t)^x}{(1-t)^{x+1}}.$$

$$(2)\quad \sum_{k=0}^{n}\binom{x+k}{n}\binom{n}{k} = \sum_{k=0}^{n}\binom{x+k}{k}\binom{x}{n-k} = [t^n]\frac{(1+t)^x}{(1-t)^{x+1}}.$$

Let us consider an example with inverse binomial coefficient; here we can use the well-known integral representation:

$$\binom{n}{k}^{-1} = (n+1)\int_0^1 y^k(1-y)^{n-k}dy;$$

the Riordan array transformation is usually taken inside the integral.

Example 5.4 We have

$$\sum_{k=1}^{2n-1}(-1)^{k-1}k\binom{2n}{k}^{-1}=\frac{n}{n+1}.$$

Here we consider the generating function for the (weighted) partial sums of a geometric series:

$$\begin{aligned}
&\sum_{k=1}^{2n-1}(-1)^{k-1}k(2n+1)\int_0^1 y^k(1-y)^{2n-k}\mathrm{d}y\\
&=(2n+1)\int_0^1(1-y)^{2n}\left(\sum_{k=1}^{2n-1}(-1)^{k-1}k\left(\frac{y}{1-y}\right)^k\right)\mathrm{d}y\\
&=(2n+1)\int_0^1(1-y)^{2n}\left(\frac{\frac{y}{1-y}+\frac{(-y)^{2n+1}}{(1-y)^{2n+1}}}{\left(1+\frac{y}{1-y}\right)^2}-2n\frac{\left(\frac{-y}{1-y}\right)^{2n}}{\left(1+\frac{y}{1-y}\right)}\right)\mathrm{d}y\\
&=(2n+1)\int_0^1(1-y)^{2n+2}\left(\frac{y}{1-y}-\frac{y^{2n+1}}{(1-y)^{2n+1}}\right)\mathrm{d}y\\
&\quad+2n(2n+1)\int_0^1(1-y)^{2n+1}\frac{y^{2n}}{(1-y)^{2n}}\mathrm{d}y\\
&=(2n+1)\int_0^1 y(1-y)^{2n+1}\mathrm{d}y-(2n+1)\int_0^1 y^{2n+1}(1-y)\mathrm{d}y\\
&\quad+2n(2n+1)\int_0^1 y^{2n}(1-y)\mathrm{d}y\\
&=\frac{2n+1}{2n+3}\binom{2n+2}{2n+1}^{-1}-\frac{2n+1}{2n+3}\binom{2n+2}{2n+1}^{-1}+\frac{2n(2n+1)}{2n+2}\binom{2n+1}{2n}\\
&=\frac{n(2n+1)}{(n+1)(2n+1)}=\frac{n}{n+1}.
\end{aligned}$$

Often, the application of the Lagrange Inversion Theorem is required. Let us consider identity (3.26) in Gould's collection [5], where we use the bisection formulas.

Example 5.5 We have

$$\sum_{k=0}^{n}\binom{2x}{2k}\binom{x-k}{n-k}=\frac{x}{x+n}\binom{x+n}{2n}4^n=\frac{4^n}{(2n)!}\prod_{k=0}^{n-1}(x^2-k^2).$$

Here we use formula (5.1.4):

$$S = \sum_{k=0}^{n} \binom{2x}{2k}\binom{x-k}{n-k} = [t^n](1+t)^x \left[\frac{(1+\sqrt{u})^{2x} + (1-\sqrt{u})^{2x}}{2} \;\middle|\; u = \frac{t}{1+t} \right] =$$

$$= \frac{1}{2}[t^n]\left((1+2t+2\sqrt{t(1+t)})^x + (1-2t-2\sqrt{t(1+t)})^x \right).$$

By setting $1+2t+2\sqrt{t(1+t)} = 1+2\sqrt{y}$ we find $y = t(1+2\sqrt{y})$. The two expressions give the same value, so the value of the first one is our final result:

$$S = [t^n]\left[(1+2\sqrt{y})^x \;\middle|\; y = t(1+2\sqrt{y}) \right] = \frac{1}{n}[t^n]x\frac{(1+2\sqrt{t})^{x-1}}{\sqrt{t}}(1+2\sqrt{t})^n =$$

$$= \frac{x}{n}[w^{2n-1}](1+2w)^{x+n-1} = \frac{x}{n}2^{2n-1}\binom{x+n-1}{2n-1} = 4^n\frac{x}{x+n}\binom{x+n}{2n}.$$

The identity of K. von Szily [21] is an example of the use of implicit generating functions.

Example 5.6 We have

$$\sum_{k=-n}^{n} (-1)^k \binom{2n}{n-k}\binom{2r}{r-k} = \binom{2n}{n}\binom{2r}{r}\binom{n+r}{n}^{-1} = \frac{(2n)!(2r)!}{(n+r)!n!r!}.$$

We use again formula (5.1.4):

$$S = \sum_{k=-n}^{n} (-1)^k \binom{2n}{n-k}\binom{2r}{r-k}$$

$$= [t^n](1+t)^{2n}\left[\frac{(-1)^r(1-u)^{2r}}{u^r} \;\middle|\; u = t \right] = (-1)^r[t^n](1+t)^{2n}\frac{(1-t)^{2r}}{t^r}$$

$$= (-1)^r[t^n]\left[\frac{(1-w)^{2r}}{w^r}\cdot\frac{1+w}{1-w} \;\middle|\; w = t(1+w)^2 \right].$$

Now, we have to solve the second degree equation on the right; we find

$$w = \frac{1-2t \pm \sqrt{1-4t}}{2t}$$

and we have to choose the sign minus to let this be a f.p.s.. By substituting this value in the preceding expression, we have

$$\frac{(1-w)^2}{w} = \frac{1-4t}{t}; \qquad \frac{1+w}{1-w} = \frac{1}{\sqrt{1-4t}}.$$

Therefore, the final result is

$$S = (-1)^r [t^n](1-4t)^{r-1/2} = (-1)^r \binom{r-1/2}{n+r}(-4)^{n+r} = (-1)^n \binom{r-1/2}{n+r} 4^{n+r}.$$

In order to have a simpler expression, we can develop the binomial coefficient:

$$\binom{r-1/2}{n+r} = \frac{(r-1/2)(r-3/2)\cdots(-n+1/2)}{(n+r)!} =$$

$$= \frac{(2r-1)(2r-3)\cdots(-2n+1)}{2^{n+r}(n+r)!} = \frac{(2r)!(2n)!(-1)^n}{4^{n+r}n!r!(n+r)!}.$$

$$S = \frac{(2r)!(2n)!}{r!n!(n+r)!} = \binom{2r}{r}\binom{2n}{n}\binom{n+r}{n}^{-1}.$$

5.2 Combinatorial Inversions

In his book [18], Riordan considers the general problem of inverting combinatorial sums. Let an infinite set $a_n = \sum_{k=0}^{n} d_{n,k} b_k$ of identities be given, in which the sequence $(b_k)_{k\in N}$ is unknown, while the sequence $(a_k)_{k\in N}$ is known: how is it possible to determine the first sequence? In other words, is it possible to invert the sums and get $b_n = \sum_{k=0}^{n} d^*_{n,k} a_k$?. If we substitute the b_ks in the first set of sums, we immediately see that the two infinite arrays $(d_{n,k})_{n,k\in N}$ and $(d^*_{n,k})_{n,k\in N}$ should be inverse of each other, so the problem of inversion is equivalent to the inversion of the infinite array $(d_{n,k})_{n,k\in N}$. Besides, as we shall see, sometimes we can drop the hypothesis that D be a p.R.a. and assume that it is a general R.a..

When $(d_{n,k})_{n,k\in N}$ is a p.R.a., the problem can be attacked in a systematic way by the theory, as we are now going to show. Actually, all the cases considered by J. Riordan can be reduced to this case, that is to the inversion of explicit or implicit R.a.'s. Since p.R.a.'s have a group structure, in principle there is no theoretical problem: if $D = (d(t), h(t))$ then $D^{-1} = (d^*(t), h^*(t))$ exists and is unique. In practice, as the book of Riordan shows, things are quite different and the inversion of p.R.a.'s is not an obvious problem.

In general, as already observed in Chap. 3, if $D=(d(t), h(t))$ and $E=(f(t), g(t))$ are two p.R.a.'s, their product is defined by

$$\mathcal{R}(d(t), h(t)) * \mathcal{R}(f(t), g(t)) = \mathcal{R}(d(t)f(h(t)), g(h(t)))$$

and when $E = D^{-1}$ the result should be the identity $I = (1, t)$, remembering that

$$D^{-1} = \left(d^*(t), h^*(t)\right) = \left(\frac{1}{d(\overline{h}(t))}, \overline{h}(t)\right).$$

Surely, the simplest inversion is

$$a_n = \sum_{k=0}^{n} \binom{n}{k} b_k \qquad \Leftrightarrow \qquad b_n = \sum_{k=0}^{n} (-1)^{n-k} \binom{n}{k} a_k$$

corresponding to the Pascal triangle and its inverse.

Actually, when the starting sum can be expressed as a generating function relation, the method can be made more direct.

Theorem 5.3 *Let us suppose that* $(d_{n,k})_{n,k\in N}$ *is the p.R.a.* $\mathcal{R}(d(t), h(t))$ *and* $a(t)$, $b(t)$ *are the g.f.'s of the sequences* $(a_k)_{k\in N}$ *and* $(b_k)_{k\in N}$. *Then the inverse relation of* $a_n = \sum_{k=0}^{n} d_{n,k} b_k$ *is obtained by extracting the coefficient of* t^n *in the relation:*

$$b(t) = \left[\frac{a(w)}{d(w)} \;\middle|\; t = h(w) \right].$$

Proof By the summation formula for R.a.'s, the starting sum can be written as $a(t) = d(t)b(h(t))$ or $b(h(t)) = a(t)/d(t)$. Now the formula follows when we set $y = h(t)$ and change variables to adjust notation. □

In the Pascal case we have $b(t) = [a(w)(1-w) \mid w = t(1-w)]$ and by computing $w = t/(1+t)$ we find

$$b(t) = \frac{1}{1+t} a\left(\frac{t}{1+t}\right),$$

that is the inverse Euler transformation.

The direct computation of $d^*(t)$, $h^*(t)$ or the previous theorem can be sufficient to solve (at least in principle) the problem of explicit inversion. Before passing to implicit inversion, let us give a third approach to explicit inversion, which directly gives the form of $d^*_{n,k}$. In fact, we can prove:

Theorem 5.4 *Let* $D = \mathcal{R}(d(t), h(t))$ *be a p.R.a.; the generic element* $d^*_{n,k}$ *of the inverse p.R.a.* $D^{-1} = \mathcal{R}(d^*(t), h^*(t))$ *is given by*

$$d^*_{n,k} = [t^{n-k}] \frac{h'(t)}{d(t)} \left(\frac{t}{h(t)} \right)^{n+1}.$$

Proof As we have seen, the bivariate g.f. of D^{-1} is given by

$$\frac{d^*(t)}{1 - wh^*(t)} = \left[\frac{1}{d(y)} \frac{1}{1-wy} \;\middle|\; t = h(y) \right].$$

From this g.f. we extract the appropriate coefficient:

$$\begin{aligned}
d^*_{n,k} &= [t^n][w^k]\left[\frac{1}{d(y)}\frac{1}{1-wy} \;\middle|\; t=h(y)\right]\\
&= [t^n][w^k]\left[\frac{1}{d(y)}\sum_{k=0}^{\infty} w^k y^k \;\middle|\; t=h(y)\right]\\
&= [t^n]\left[\frac{y^k}{d(y)} \;\middle|\; t=h(y)\right] = [t^n]\frac{t^k}{d(t)}\frac{t^{n-1}}{h(t)^{n-1}}\left(\frac{t}{h(t)} - t\frac{h(t)-th'(t)}{h(t)^2}\right)\\
&= [t^{n-k}]\frac{t^{n+1}h'(t)}{d(t)h(t)^{n+1}}
\end{aligned}$$

as desired. □

This result can be used to express Theorem 5.3 in another form.

Theorem 5.5 *Let us suppose that the sum* $b_n = \sum_{k=0}^{n} d_{n,k}a_k$ *corresponds to the g.f. relation:* $b(t) = d(t)a(h(t))$, *that is* $(d_{n,k})$ *is the R.a.* $\mathcal{R}(d(t),\ h(t))$. *Then the inverse sum is*

$$a_n = \sum_{k=0}^{n}\left([t^{n-k}]\frac{h'(t)}{d(t)}\left(\frac{t}{h(t)}\right)^{n+1}\right) b_k.$$

Proof This is a simple consequence of Theorem 5.4 and of the fundamental property of sums containing a R.a. □

Example 5.7 Some inversions in the book of Riordan are immediate applications of this theorem. Let us consider inversion (2.2.2) in [18], that is,

$$a_n = \sum_{k=0}^{n}\left(\binom{p+qk-k}{n-k} + q\binom{p+qk-k}{n-k-1}\right) b_k \;\Leftrightarrow\; b_n = \sum_{k=0}^{n}(-1)^{n-k}\binom{p+qn-k}{n-k}a_k;$$

here the R.a. is found by developing the binomial coefficients:

$$[t^{n-k}](1+t)^{p+qk-k} + q[t^{n-k-1}](1+t)^{p+qk-k} = [t^n](1+qt)(1+t)^p(t(1+t)^{q-1})^k$$

$$D = \mathcal{R}((1+qt)(1+t)^p,\ t(1+t)^{q-1}).$$

We now apply Theorem 5.5, which gives

$$b_n = \sum_{k=0}^{n}\left([t^{n-k}]\frac{(1+t)^{q-1} + t(q-1)(1+t)^{q-2}}{(1+qt)(1+t)^p(1+t)^{(n+1)(q-1)}}\right) a_k =$$

$$= \sum_{k=0}^{n}\left(\frac{(1+qt)(1+t)^{q-2}}{(1+qt)(1+t)^p(1+t)^{nq-n+q-1}}\right) a_k =$$

$$= \sum_{k=0}^{n} \left([t^{n-k}](1+t)^{-p-nq+n-1}\right) a_k = \sum_{k=0}^{n} \binom{-p-nq+n-1}{n-k} a_k =$$

$$= \sum_{k=0}^{n} \binom{p+nq-n+1+n-k-1}{n-k} (-1)^{n-k} a_k = \sum_{k=0}^{n} (-1)^{n-k} \binom{p+qn-k}{n-k} a_k.$$

Inversion (2.6.3) in [18] can be proved in the same way.

Example 5.8 Inversions (3.1.3) and (3.1.3a) in [18] involve the use of exponential generating functions. Let us prove the last one, which is rather interesting:

$$a_n = \sum_{k=0}^{n} \binom{n}{k} (x+n)(x+k)^{n-k-1} b_k \quad \Leftrightarrow \quad b_n = \sum_{k=0}^{n} (-1)^{n-k} \binom{n}{k} (x+n)^{n-k} a_k.$$

By observing that $(x+n) = (x+k) + (n-k)$, the sum is so transformed:

$$\frac{a_n}{n!} = \sum_{k=0}^{n} \left(\frac{(x+k)^{n-k}}{(n-k)!} + \frac{(x+k)^{n-k-1}}{(n-k-1)!} \right) \frac{b_k}{k!}.$$

Let us denote by $\widehat{a}(t)$ the e.g.f. of a sequence (a_k), which is the same as the ordinary g.f. of the sequence $(a_k/k!)$. Here, the R.a. is found by observing that the two terms between parentheses are

$$[t^{n-k}]e^{(x+k)t} + [t^{n-k-1}]e^{(x+k)t} = [t^n](1+t)e^{xt}(te^t)^k.$$

Therefore we have $D = \mathcal{R}((1+t)e^{xt},\ te^t)$ and

$$\frac{b_n}{n!} = \sum_{k=0}^{n} \left([t^{n-k}] \frac{e^t + te^t}{(1+t)e^{xt}e^{(n+1)t}} \right) \frac{a_k}{k!} = \sum_{k=0}^{n} \left([t^{n-k}] e^{-xt} e^{-nt} \right) \frac{a_k}{k!} =$$

$$= \sum_{k=0}^{n} (-1)^{n-k} \frac{(x+n)^{n-k}}{(n-k)!} \frac{a_k}{k!}.$$

This is equivalent to the right part of the inversion, as desired. Inversion (3.1.3) in [18] is proved in a similar way.

When the expression to be inverted depends on n, a direct application of the Riordan array concepts is no longer feasible, and the rule of diagonalization should be applied. A typical example is given by the simple inversion:

$$a_n = \sum_{k=0}^{n} \binom{n}{k} b_k = \sum_{k=0}^{n} \binom{n}{n-k} b_k = [t^n](1+t)^n b(t).$$

Because of the $(1+t)^n$, this equality does not correspond to a g.f. relation of the type $a(t) = (1+t)^n b(t)$, and consequently we have to find a genuine g.f. equality. By the diagonalization rule we have

$$[t^n](1+t)^n b(t) = [t^n]\left[\frac{b(w)}{1-t} \;\middle|\; w = t(1+w)\right]$$
$$= [t^n]\left[(1+w)b(w) \;\middle|\; w = t(1+w)\right].$$

Now, the expression between the substitution brackets represents a g.f., because it does not depend on n; therefore, we have

$$a(t) = \left[(1+w)b(w) \;\middle|\; w = t(1+w)\right] \quad \text{or} \quad a\left(\frac{w}{1+w}\right) = (1+w)b(w)$$

and we have an expression only depending on w. By changing the indeterminate:

$$b(t) = \frac{1}{1+t}a\left(\frac{t}{1+t}\right) \quad \text{or} \quad b_n = \sum_{k=0}^{n}(-1)^{n-k}\binom{n}{k}a_k$$

according to the Pascal example. We can translate this derivation into a general result.

Theorem 5.6 *If the sum* $a_n = \sum_{k=0}^{n} d_{n,k}b_k$ *corresponds to the relation:* $a_n = [t^n]f(t)\phi(t)^n b(t)$, *where* $\phi(t) \in \mathbb{F}_0$, *then the inverse sum is*

$$b_n = \sum_{k=0}^{n}\left([t^{n-k}]\frac{\phi(t) - t\phi'(t)}{f(t)\phi(t)^{k+1}}\right)a_k.$$

Proof By applying the rule of diagonalization we have

$$a(t) = \left[\frac{f(w)b(w)}{1-t\phi'(w)} \;\middle|\; w = t\phi(w)\right].$$

We now substitute $w/\phi(w)$ to t in both members, and find

$$a\left(\frac{w}{\phi(w)}\right) = \frac{f(w)b(w)\phi(w)}{\phi(w) - w\phi'(w)} \quad \text{or} \quad b(w) = \frac{\phi(w) - w\phi'(w)}{f(w)\phi(w)}a\left(\frac{w}{\phi(w)}\right).$$

This formula proves that $b(w)$ is the transformation of $a(w)$ by the R.a.

$$\mathcal{R}\left(\frac{\phi(w) - w\phi'(w)}{f(w)\phi(w)}, \frac{w}{\phi(w)}\right)$$

whose generic element is

$$d^*_{n,k} = [w^n]\frac{\phi(w) - w\phi'(w)}{f(w)\phi(w)}\left(\frac{w}{\phi(w)}\right)^k = [w^{n-k}]\frac{\phi(w) - w\phi'(w)}{f(w)\phi(w)^{k+1}}.$$

The formula we are looking for now follows from the R.a. rule $b_n = \sum_{k=0}^{n} d^*_{n,k} a_k$. □

Example 5.9 As an example, let us consider inversion (2.4.1) in Riordan's book. We start from

$$a_n = \sum_k \binom{n}{k} b_{n-ck}$$

and observe that $b_{n-ck} = [t^n]b(t)(t^c)^k$; consequently, we consider b_{n-ck} as the generic element of the (not proper) R.a. $(b(t),\ t^c)$. The sum corresponds to

$$a_n = [t^n]b(t)[(1+y)^n \mid y = t^c] = [t^n]b(t)(1+t^c)^n.$$

Therefore, we can apply Theorem 5.6 with $f(t) = 1$ and $\phi(t) = 1 + t^c$:

$$b_n = \sum_{k=0}^{n} \left([t^{n-k}]\frac{(1+t^c) - ctt^{c-1}}{(1+t^c)^{k+1}}\right) a_k = \sum_{k=0}^{n} \left([t^{n-k}]\frac{1-(c-1)t^c}{(1+t^c)^{k+1}}\right) a_k =$$

$$= \sum_{k=0}^{n} \left(\binom{-k-1}{(n-k)/c} - (c-1)\binom{-k-1}{(n-k-c)/c}\right) a_k.$$

Let us now set $k = n - ch$ or $n - k = ch$:

$$b_n = \sum_{h=0}^{n/c} \left(\binom{n-ch+h}{h} + (c-1)\binom{n-ch+h-1}{h-1}\right) a_{n-ch}.$$

Example 5.10 In Egorychev [3, p. 91] we find the following problem: what is the value of f_k if we know that

$$\sum_{k=1}^{n} \binom{n-1}{k-1} n^{n-k}(k+p)!f_k = (2p)!n^{n+p}?$$

If we set $b_k = (k+p)!f_k$, we have to solve the inversion

$$a_n = \sum_{k=1}^{n} \binom{n-1}{k-1} n^{n-k} b_k$$

and eventually substitute to a_k its value $(2p)!k^{k+p}$. The sum can be rewritten:

$$a_n = \sum_{k=0}^{n} \binom{n}{k} n^{n-k-1} k b_k = \sum_{k=0}^{n} \binom{n}{k} n^{n-k-1}(n-(n-k))b_k.$$

By using exponential generating functions

$$\frac{a_n}{n!} = \sum_{k=0}^{n} \left(\frac{n^{n-k}}{(n-k)!} - \frac{n^{n-k-1}}{(n-k-1)!} \right) \frac{b_k}{k!} = [t^n]e^{nt}\widehat{b}(t) - [t^n]te^{nt}\widehat{b}(t)$$

$$\frac{a_n}{n!} = [t^n](1-t)e^{nt}\widehat{b}(t)$$

and we apply Theorem 5.6 with $f(t) = 1 - t$ and $\phi(t) = e^t$:

$$\frac{b_n}{n!} = \sum_{k=0}^{n} \left([t^{n-k}] \frac{e^k - te^t}{(1-t)e^{(k+1)t}} \right) \frac{a_k}{k!} = \sum_{k=0}^{n} (-1)^{n-k} \frac{k^{n-k}}{(n-k)!} \frac{a_k}{k!}$$

$$b_n = \sum_{k=0}^{n} (-1)^{n-k} \binom{n}{k} k^{n-k}.$$

We now substitute their values to a_k, b_n:

$$(n+p)! f_n = \sum_{k=0}^{n} (-1)^{n-k} \binom{n}{k} k^{n-k} (2p)! k^{k+p}$$

and derive Egorychev's solution:

$$f_n = \frac{(2p)!}{(n+p)!} \sum_{k=0}^{n} (-1)^{n-k} \binom{n}{k} k^{n+p}.$$

We can go a bit further and find a closed form for the last sum, by observing that it corresponds to the identity in Example 5.1 with $n + p$ instead of p. Therefore we have

$$\sum_{k=0}^{n} (-1)^{n-k} \binom{n}{k} k^{n+p} = \left\{ {n+p \atop n} \right\} n!,$$

and we conclude

$$f_n = \frac{(2p)! n!}{(n+p)!} \left\{ {n+p \atop n} \right\}.$$

We can now give a general result for the inversion of sums:

Theorem 5.7 *If the sum* $a_n = \sum_{k=0}^{n} d_{n,k} b_k$ *corresponds to the relation:* $a_n = [t^n] f(t)\phi(t)^n b(th(t))$, *where* $\phi(t) \in \mathbb{F}_0$, $h(t) \in \mathbb{F}_1$, *then the inverse sum is*

$$b_n = \left([t^{n-k}] \frac{(\phi(t) - t\phi'(t))(h(t) + th'(t))}{f(t)\phi(t)^{k+1}h(t)^{n+1}} \right) a_k.$$

Proof By setting $w = t\phi(w)$ we eliminate the dependence on n:

$$\begin{aligned} a_n &= [t^n] \left[\frac{f(w)b(wh(w))}{1 - w\phi'(w)/\phi(w)} \;\middle|\; w = t\phi(w) \right] \\ &= [t^n] \left[\frac{\phi(w)f(w)b(wh(w))}{\phi(w) - w\phi'(w)} \;\middle|\; w = t\phi(w) \right]. \end{aligned}$$

This is a g.f. identity and therefore we get

$$a\left(\frac{w}{\phi(w)}\right) = \frac{\phi(w)f(w)b(wh(w))}{\phi(w) - w\phi'(w)} \quad \text{or} \quad b(wh(w)) = \frac{\phi(w) - w\phi'(w)}{\phi(w)f(w)} a\left(\frac{w}{\phi(w)}\right).$$

We now set $y = wh(w)$ in order to obtain an expression for $b(y)$:

$$b(y) = \left[\frac{\phi(w) - w\phi'(w)}{\phi(w)f(w)} a\left(\frac{w}{\phi(w)}\right) \;\middle|\; w = \frac{y}{h(w)} \right].$$

An application of the Lagrange inversion formula gives us

$$\begin{aligned} b_n &= [t^n] \frac{\phi(t) - t\phi'(t)}{\phi(t)f(t)} a\left(\frac{t}{\phi(t)}\right) \frac{h(t) + th'(t)}{h(t)^{n+1}} = \\ &= \sum_{k=0}^{n} \left([t^{n-k}] \frac{\phi(t) - t\phi'(t))\,(h(t) + th'(t))}{f(t)\phi(t)^{k+1}h(t)^{n+1}} \right) a_k \end{aligned}$$

as desired. □

Example 5.11 Let us consider inversions of Abel's type in Riordan's book; the ingenious ways used by Riordan to prove them can be substituted by a more systematic approach. Let us consider the sums:

$$a_n = \sum_{k=0}^{n} \binom{n}{k} (x + pn + qk)^{n-k} b_k;$$

they are equivalent to

$$\frac{a_n}{n!} = \sum_{k=0}^{n} \frac{(x + pn + qk)^{n-k}}{(n-k)!} \frac{b_k}{k!}$$

and therefore we have

$$\frac{a_n}{n!} = [t^n] e^{xt} (e^{pt})^n \widehat{b}(te^{qt}).$$

The previous theorem allows us to invert:

$$\begin{aligned}
\frac{b_n}{n!} &= \sum_{k=0}^{n} \left([t^{n-k}] \frac{(1-pt)e^{pt}(1+qt)e^{qt}}{e^{xt}e^{(k+1)pt}e^{(n+1)qt}} \right) \frac{a_k}{k!} \\
&= \sum_{k=0}^{n} \left([t^{n-k}](1-pt)(1+qt)e^{-(x+pk+qn)t} \right) \frac{a_k}{k!} \\
&= \sum_{k=0}^{n} (-1)^{n-k} \left(1 + (p-q)\frac{n-k}{x+pk+qn} - pq\frac{(n-k)(n-k-1)}{(x+pk+qn)^2} \right) \\
&\quad \times \frac{(x+pk+qn)^{n-k}}{(n-k)!} \frac{a_k}{k!}.
\end{aligned}$$

The quantity between parentheses can be easily developed, and eventually we have

$$\begin{aligned}
\frac{b_n}{n!} &= \sum_{k=0}^{n} (-1)^{n-k} \left((p+q)^2 nk + x^2 + x(p+q)(n+k) + pq(n-k) \right) \\
&\quad \times \frac{(x+pk+qn)^{n-k-2}}{(n-k)!} \frac{a_k}{k!}
\end{aligned}$$

or

$$b_n = \sum_{k=0}^{n} (-1)^{n-k} \binom{n}{k} F(x,p,n,q,k)(x+pk+qn)^{n-k-2} a_k.$$

Inversions (3.1.2), (3.1.3), (3.1.3a), (3.1.4) and (3.1.5) in [18] are particular cases of this formula. In fact we have

$$\begin{array}{llll}
(3.1.2) & p=1 & q=-1 & F(x,p,n,q,k) = x^2 - n + k \\
(3.1.3) & p=0 & q=1 & F(x,p,n,q,k) = (x+k)(x+n) \\
(3.1.3a) & p=1 & q=0 & F(x,p,n,q,k) = (x+k)(x+n) \\
(3.1.4) & p=1 & q=1 & F(x,p,n,q,k) = (x+2n)(x+n+k) \\
(3.1.5) & x=0 & p=q=1 & F(x,p,n,q,k) = 4nk + n - k.
\end{array}$$

Obviously, other inversions are obtained by assigning different values to x, p, q. The reader may wish to try with some cases of interest.

We wish to conclude this chapter with a last example of a combinatorial inversion. Let us consider the relation

$$a_n = \sum_{k=0}^{\lfloor n/2 \rfloor} \binom{n}{2k} b_k, \tag{5.2.1}$$

which is related to the Riordan array $\mathcal{R}\left(1/(1-t), t^2/(1-t)^2\right)$, since

$$[t^n]\frac{1}{1-t}\left(\frac{t^2}{(1-t)^2}\right)^k = [t^{n-2k}]\frac{1}{(1-t)^{2k+1}} =$$

$$= \binom{-2k-1}{n-2k}(-1)^{n-2k} = \binom{n}{2k}.$$

Therefore, the generating function $a(t)$ of the sequence a_k is $a(t) = b(t^2/(1-t)^2)/(1-t)$, where $b(t)$ is the generating function of the sequence b_k. This relation can be inverted:

$$b\left(\frac{t^2}{(1-t)^2}\right) = (1-t)a(t), \qquad \text{or} \qquad b(y) = \left[(1-t)a(t)\,\middle|\, y = \frac{t^2}{(1-t)^2}\right].$$

The generic element b_n can be found by Lagrange inversion; in fact, we have

$$\begin{aligned}
b_n &= [y^n]\,b(y) = [y^n]\left[(1-t)a(t)\,\middle|\, t = y^{1/2}(1-t)\right] \\
&= [w^{2n}]\,[(1-t)a(t)\,|t = w(1-t)] = \frac{1}{2n}[t^{2n-1}]\,((1-t)a'(t) - a(t))(1-t)^{2n} \\
&= \frac{1}{2n}\left(\sum_{k=0}^{2n-1}(-1)^{2n-1-k}\binom{2n+1}{2n-k-1}(k+1)a_{k+1}\right. \\
&\qquad \left. - \sum_{k=0}^{2n-1}(-1)^{2n-1-k}\binom{2n}{2n-k-1}a_k\right) \\
&= \frac{1}{2n}\left(\sum_{k=1}^{2n}(-1)^k\binom{2n+1}{2n-k}ka_k + \sum_{k=0}^{2n-1}(-1)^k\binom{2n+1}{2n-k}\frac{2n-k}{2n+1}a_k\right) \\
&= \frac{1}{2n}\left(\sum_{k=0}^{2n}(-1)^k\binom{2n+1}{2n-k}\frac{2nk+k+2n-k}{2n+1}a_k\right).
\end{aligned}$$

After some simplification we find the inverse relation:

$$b_n = \sum_{k=0}^{2n}(-1)^k\binom{2n}{k}a_k, \tag{5.2.2}$$

which cannot be found in the book of Riordan [18]. In fact, we can observe that in (5.2.1) the involved Riordan array is not proper, that is, the elements on the main diagonal are zero with the exception of element $d_{0,0}$, and therefore the array cannot be inverted in the usual sense. However, the inversion is correct and if we define

$$P = \left\{\binom{n}{2k}\,\middle|\, n,k\in\mathbb{N}\right\}, \qquad \bar{P} = \left\{(-1)^k\binom{2n}{k}\,\middle|\, n,k\in\mathbb{N}\right\},$$

we can easily verify that $\bar{P}P = I$ and $P\bar{P}P = P$, that is, matrix $\bar{P}$ is a *left inverse* of P. The problem of left inversion with Riordan arrays has been studied in C. Corsani, D. Merlini, and R. Sprugnoli [2], where a series of significant examples can be found. The main result in that paper is the following, which generalizes Theorem 5.4, thus solving the inversion problem when the involved Riordan array is not proper:

Theorem 5.8 *Let $D = \mathcal{R}(d(t), h(t))$ be a non-proper Riordan array, $h(t) = t^s v(t)$, $v(0) \neq 0$. Then, the generic element $d^*_{n,k}$ of the left inverse $D^{-1} = \mathcal{R}(d^*(t), h^*(t))$ is given by the following formula:*

$$d^*_{n,k} = \frac{1}{sn}\left[t^{(sn-k)}\right]\left(k - t\frac{d'(t)}{d(t)}\right)\frac{1}{d(t)v(t)^n} \qquad n > 0, \qquad d^*_{0,k} = \delta_{k,0}/d_0.$$

Proof We have

$$d^*_{n,k} = \left[y^{sn}\right]d^*(y)\left(h^*(y)\right)^k = \left[y^{sn}\right]\left[\frac{t^k}{d(t)}\,\middle|\, y = (h(t))^{1/s}\right] =$$

$$= \left[y^{sn}\right]\left[\frac{t^k}{d(t)}\,\middle|\, y = tv(t)^{1/s}\right] = \frac{1}{sn}\left[t^{sn-1}\right]\frac{d}{dt}\left(\frac{t^k}{d(t)}\right)\frac{1}{v(t)^n} =$$

$$= \frac{1}{sn}\left[t^{sn-1}\right]\frac{kt^{k-1}d(t) - t^k d'(t)}{d(t)^2}\frac{1}{v(t)^n}$$

and then we can conclude

$$d^*_{n,k} = \frac{1}{sn}\left[t^{sn-k}\right]\left(k - t\frac{d'(t)}{d(t)}\right)\frac{1}{d(t)v(t)^n} \qquad n > 0.$$

When $n = 0$, we have

$$d^*_{0,k} = \left[y^0\right]\left[\frac{t^k}{d(t)}\,\middle|\, t = yv(t)^{-1/s}\right] = \left[t^0\right]\frac{t^k}{d(t)} - \left[t^{-1}\right]\frac{t^k v(t)^{1/s}\left(v(t)^{-1/s}\right)}{d(t)} =$$

$$= \left[t^{-k}\right]\frac{1}{d(t)} + \frac{1}{s}\left[t^{-1}\right]\frac{t^k v(t)^{-1}v'(t)}{d(t)} = \left[t^{-k}\right]d(t)^{-1}$$

or $d^*_{0,k} = \delta_{k,0}/d_0$. □

The previous example can be solved in a natural way using Theorem 5.8 with $s = 2$, $d(t) = (1-t)^{-1}$ and $v(t) = (1-t)^{-2}$:

$$d^*_{n,k} = \frac{1}{2n}[t^{2n-k}]\left(k - t\frac{1-t}{(1-t)^2}\right)(1-t)(1-t)^{2n} =$$

$$= \frac{k}{2n}[t^{2n-k}](1-t)^{2n+1} - \frac{1}{2n}[t^{2n-k-1}](1-t)^{2n} =$$

$$= \frac{k}{2n}\binom{2n+1}{2n-k}(-1)^k - \frac{1}{2n}\binom{2n}{2n-k-1}(-1)^{k-1} =$$

$$= (-1)^k\left(\frac{k}{2n}\binom{2n+1}{k+1} + \frac{1}{2n}\binom{2n}{k+1}\right) =$$

$$= \binom{2n}{k}(-1)^k\left(\frac{k(2n+1)}{2n(k+1)} + \frac{2n-k}{2n(k+1)}\right) = \binom{2n}{k}(-1)^k,$$

as expected.

Exercises

5.1 Compute the following combinatorial sum:

$$\sum_{k=1}^{n}\binom{n}{k}\frac{(-1)^{k-1}}{k}.$$

5.2 Prove the following combinatorial identity:

$$\sum_{k=0}^{n}\binom{n}{k}(k+1) = (n+2)2^{n-1}.$$

5.3 Prove the following combinatorial identity:

$$\sum_{k=0}^{n}(-1)^{n-k}\binom{n+k}{n-k}4^k = 2n+1.$$

5.4 Prove the following combinatorial identity:

$$\sum_{k=0}^{n}\binom{n+k}{m+2k}\binom{2k}{k}\frac{(-1)^k}{k+1} = \binom{n-1}{m-1}.$$

5.5 The Stirling numbers of first and second kind satisfy the following relations:

$$\mathcal{G}\left(\frac{k!}{n!}\left[{n \atop k}\right]\right) = \left(\ln\frac{1}{1-t}\right)^k, \quad \mathcal{G}\left(\frac{k!}{n!}\left\{{n \atop k}\right\}\right) = \left(e^t - 1\right)^k.$$

Prove that

$$\sum_{k=0}^{n}\begin{bmatrix}n\\k\end{bmatrix} = n! \quad \sum_{k=0}^{n}\begin{Bmatrix}n\\k\end{Bmatrix} = \mathcal{B}_n$$

where $\mathcal{B}_n$ is the nth Bell number, with exponential generating function $\sum_{n\geq 0}\mathcal{B}_n\frac{t^n}{n!} = \exp(e^t - 1)$.

5.6 Find the combinatorial identity inverse of the combinatorial identity corresponding to Exercise 5.3.

5.7 Prove that

$$a_n = \sum_{k=0}^{n}\frac{k+1}{n+1}\binom{2n-k}{n-k}b_k, \quad b_n = \sum_{k=0}^{n}(-1)^{n-k}\binom{k+1}{n-k}a_k$$

are inverse relations.

5.8 Invert the identity:

$$a_n = \sum_{k=0}^{n}\binom{n}{k}(n+k)^{n-k}b_k.$$

5.9 Invert the identity:

$$a_n = \sum_{k=0}^{\lfloor n/s\rfloor}\binom{n}{sk}b_k.$$

5.10 Invert the identity:

$$a_n = \sum_{k=0}^{\lfloor n/s\rfloor}\binom{n+p}{sk+p}b_k.$$

References

1. L. Comtet, *Advanced Combinatorics* (Reidel, Dordrecht, 1974)
2. C. Corsani, D. Merlini, R. Sprugnoli, Left-inversion of combinatorial sums. Discrete Math. **180**(1–3), 107–122 (1998)
3. G.P. Egorychev, *Integral Representation and the Computation of Combinatorial Sums* (trans. H. H. McFadden), vol. 59 (American Mathematical Society, Providence, 1984)
4. P. Flajolet, R. Sedgewick, *Analytic Combinatorics* (Cambridge University Press, 2009)
5. H.W. Gould, *Combinatorial Identities, A Standardized Set of Tables Listing 500 Binomial Coefficient Summations* (Morgantown W. Va., 1972)
6. H.W. Gould, Euler's formula for nth differences of powers. Am. Math. Mon. **85**(6), 450–467 (1978)
7. R. Gosper, Decision procedure for indefinite hypergeometric summation. Proc. Natl. Acad. Sci. USA **75**(1), 40–42 (1978)

8. R.L. Graham, D.E. Knuth, O. Patashnik, *Concrete Mathematics* (Addison-Wesley, New York, 1989)
9. T.-X. He, L.W. Shapiro, Row sums and alternating sums of Riordan arrays. Linear Algebra Appl. **507**, 77–95 (2016)
10. A. Luzón, D. Merlini, M.A. Morón, R. Sprugnoli, Identities induced by Riordan arrays. Linear Algebra Appl. **436**, 631–647 (2012)
11. D. Merlini, R. Sprugnoli, Arithmetic into geometric progressions through Riordan arrays. Discrete Math. **340**(2), 160–174 (2017)
12. D. Merlini, R. Sprugnoli, M.C. Verri, The Akiyama-Tanigawa transformation. Integers **5**(1), A5 12 pp. (2005)
13. D. Merlini, R. Sprugnoli, M.C. Verri, Human and constructive proof of combinatorial identities: an example from Romik, in *2005 International Conference on Analysis of Algorithms* (2005), pp. 383–391
14. D. Merlini, R. Sprugnoli, M.C. Verri, Lagrange inversion: when and how. Acta Appl. Math. **94**(3), 233–249 (2006)
15. D. Merlini, R. Sprugnoli, M.C. Verri, The Cauchy numbers. Discrete Math. **306**(16), 1906–1920 (2006)
16. D. Merlini, R. Sprugnoli, M.C. Verri, The method of coefficients. Am. Math. Mon. **114**(1), 40–57 (2007)
17. M. Petkovšek, H.S. Wilf, D. Zeilberger, *A=B* (AK Peters, Natick, MA, 1996)
18. J. Riordan, *Combinatorial Identities* (Wiley, New York, 1968)
19. R. Sprugnoli, Riordan arrays and combinatorial sums. Discrete Math. **132**, 267–290 (1994)
20. R. Sprugnoli, Riordan arrays and the Abel-Gould identity. Discrete Math. **142**(1–3), 213–233 (1995)
21. K. Von Szily, Über die quadratsummen der binomial coëfficienten. Math. Nat. Ber. Ung. **12**, 84–91 (1893)
22. H.S. Wilf, D. Zeilberger, Rational functions certify combinatorial identities. J. Am. Math. Soc. **3**, 147–158 (1990)
23. D. Zeilberger, A fast algorithm for proving terminating hypegeometric identites. Discrete Math. **80**, 207–211 (1990)

Chapter 6
Generalized Riordan Arrays

Abstract The previous chapters have shown that the theory of Riordan arrays is a powerful tool for studying combinatorial sums and special polynomial and number sequences. One of the well-known classes of polynomial sequences is the class of Sheffer sequences, including many important sequences such as Bernoulli polynomials, Euler polynomials, Abel polynomials, Hermite polynomials, Laguerre polynomials, etc. This class contains the subclasses of associated sequences and Appell sequences. In [57–59], Rota, Roman, et al. established a solid background for Sheffer sequences by using the theory of modern umbral calculus and finite operator calculus. In [56], Roman further developed the theory of umbral calculus and generalized the concept of Sheffer sequences so that more special polynomial sequences are included such as the sequences of Gegenbauer polynomials, Chebyshev polynomials, and Jacobi polynomials. Using Roman's notations, a generalized Sheffer sequence $(s_n(x))_{n\in\mathbb{N}}$ is defined by a generating function of the form

$$A(t)\varepsilon_x(B(t)) = \sum_{k=0}^{\infty} \frac{s_k(x)}{c_k} t^k, \tag{6.0.1}$$

where $\varepsilon_x(t) = \sum_{k=0}^{\infty} x^k t^k / c_k$ is a generalization of the exponential series, and $(c_k)_{k\geq 0}$ is a non-zero sequence with $c_0 = 1$. When $c_k = k!$, (6.0.1) defines the (exponential) Sheffer sequence and its generating function turns into $A(t)\mathrm{e}^{xB(t)}$. Note that there are several similar names presented in the literature, such as sequences of Sheffer A-type zero [63, 64] and generalized Appell sequences [9–12]. The connection between Riordan arrays and Sheffer sequences has already been pointed out by Shapiro et al. [62] and Sprugnoli [20, 65, 66]. In fact, the classical Riordan arrays are related to the 1-umbral calculus, and thus related to Sheffer sequences defined by (6.0.1) with $c_k = 1$, and the exponential Riordan arrays shown in Sect. 6.1 are related to (exponential) Sheffer sequences with $c_k = k!$. In other words, the exponential Riordan arrays are related to the $k!$-umbral calculus. In this chapter, we introduce the theory of exponential Riordan arrays and their production matrices in Sect. 6.1. Part of this section comes from Deutsch and Shapiro [26] and Barry [7, 8]. In Sects. 6.2 and 6.3,

© The Author(s), under exclusive license to Springer Nature Switzerland AG 2022
L. Shapiro et al., *The Riordan Group and Applications*, Springer Monographs in Mathematics, https://doi.org/10.1007/978-3-030-94151-2_6

we introduce the concept of generalized Riordan arrays, and give explicitly the relationships between the generalized Riordan arrays and generalized Sheffer sequences. Then, we present the important properties and applications of the generalized Riordan arrays. Furthermore, the determinantal definition for Sheffer sequences using the relations between Riordan arrays and Sheffer sequences is also given. In Sect. 6.4, we introduce some important special Riordan arrays and Sheffer sequences. We will also give the basic properties of these arrays and polynomial sequences by using the results obtained in the previous two sections. Finally, in Sect. 6.5, we present the double Riordan arrays and their related Sheffer polynomial sequence pairs as well as their applications in combinatorics and series summations. Readers can also refer to the further discussion on the (exponential) Sheffer sequence and the classical Riordan array in He et al. [40] and the discussion on the generalized Sheffer sequence and the generalized Riordan array in Gould et al. [29], He [31–33], Wang [71], and Wang et al. [73].

6.1 Exponential Riordan Arrays

The initial idea of the exponential Riordan arrays and related production matrices comes from Deutsch and Shapiro [26], and various further results on the exponential Riordan arrays can also be found in, for example, the works due to Barry [4–8], Cheon et al. [15, 16], Deutsch et al. [25], Gould and He [29], and Zhao and Wang [77].

Let $D = [d_{n,k}]_{n,k\in\mathbb{N}}$ be an infinite lower triangular matrix. If there exist exponential generating functions

$$g(t) = \sum_{k=0}^{\infty} g_k \frac{t^k}{k!} \quad \text{and} \quad f(t) = \sum_{k=1}^{\infty} f_k \frac{t^k}{k!},$$

with $g_0 \neq 0$ and $f_1 \neq 0$, such that the kth column of D has exponential generating function $g(t)(f(t))^k/k!$ for $k = 0, 1, 2, \ldots$; consequently,

$$d_{n,k} = \left[\frac{t^n}{n!}\right] g(t) \frac{(f(t))^k}{k!}, \tag{6.1.1}$$

then $D = [d_{n,k}]_{n,k\in\mathbb{N}}$ is called an *exponential Riordan array*, and denoted by $[g(t), f(t)]$.

Let $h(t) = \sum_{k=0}^{\infty} h_k t^k/k!$ be the exponential generating function of the sequence $(h_k)_{k\in\mathbb{N}}$. Then the fundamental theorem for the exponential Riordan arrays states that

$$\sum_{k=0}^{n} d_{n,k}h_k = \sum_{k=0}^{\infty}\left[\frac{t^n}{n!}\right]g(t)\frac{(f(t))^k}{k!}h_k = \left[\frac{t^n}{n!}\right]g(t)\sum_{k=0}^{\infty}h_k\frac{(f(t))^k}{k!}$$
$$= \left[\frac{t^n}{n!}\right]g(t)h(f(t)), \tag{6.1.2}$$

or symbolically,

$$[g(t), f(t)]h(t) = g(t)h(f(t)).$$

If $h_k = 1$, for $k = 0, 1, 2, \ldots$, then $h(t) = \mathrm{e}^t$, and we obtain the row sums of an exponential Riordan array:

$$\sum_{k=0}^{n} d_{n,k} = \left[\frac{t^n}{n!}\right]g(t)\,\mathrm{e}^{f(t)}.$$

Additionally, by fundamental theorem, the set of all exponential Riordan arrays is a group under matrix multiplication

$$[g(t), f(t)][h(t), l(t)] = [g(t)h(f(t)), l(f(t))],$$

and we denote this group by $\mathcal{R}_e$. The identity of $\mathcal{R}_e$ is the matrix $[1, t]$, and the inverse of the array $[g(t), f(t)]$ is $[1/g(\bar{f}(t)), \bar{f}(t)]$, where $\bar{f}(t)$ is the compositional inverse of $f(t)$.

Example 6.1 Based on the definition, the (n, k)th entry of the exponential Riordan array $D_0 = [\mathrm{e}^t, t]$ is

$$d_{n,k}^{(0)} = \left[\frac{t^n}{n!}\right]\mathrm{e}^t\frac{t^k}{k!} = \frac{n!}{k!}[t^{n-k}]\mathrm{e}^t = \binom{n}{k}.$$

Therefore, it is the Pascal triangle $(\frac{1}{1-t}, \frac{t}{1-t})$ by the notation of the classical Riordan arrays. Additionally, from [19, p. 209, Eq. (3c) and p. 215, Eq. (6e)], the (n, k)th entry of $D_1 = [\frac{1}{1-t}, \ln(\frac{1}{1-t})]$ is

$$d_{n,k}^{(1)} = \left[\frac{t^n}{n!}\right]\frac{1}{1-t}\frac{(-\ln(1-t))^k}{k!} = n!\sum_{j=k}^{n}\frac{1}{j!}\begin{bmatrix} j \\ k \end{bmatrix} = \begin{bmatrix} n+1 \\ k+1 \end{bmatrix},$$

and the (n, k)th entry of $D_2 = [\mathrm{e}^t, \mathrm{e}^t - 1]$ is

$$d_{n,k}^{(2)} = \left[\frac{t^n}{n!}\right]\mathrm{e}^t\frac{(\mathrm{e}^t-1)^k}{k!} = \sum_{j=k}^{n}\binom{n}{j}\begin{Bmatrix} j \\ k \end{Bmatrix} = \begin{Bmatrix} n+1 \\ k+1 \end{Bmatrix},$$

where $\left[{n \atop k}\right]$ and $\left\{{n \atop k}\right\}$ are the unsigned Stirling numbers of the first kind and the Stirling numbers of the second kind, respectively. See Table 3.2 in Chap. 3 for the related matrices (with the 0th row and column removed).

Next, let $D = [g(t), f(t)]$ be an exponential Riordan array, and $\bar{D}$ be the same matrix with the top row removed. Then the production matrix of D is given by

$$P = D^{-1}\bar{D} = D^{-1}UD\,,$$

where $U = (\delta_{i+1,j})_{i,j\geq 0}$, i.e.,

$$U = \begin{bmatrix} 0\ 1 & & & & & \\ 0\ 0\ 1 & & & & & \\ 0\ 0\ 0\ 1 & & & & & \\ 0\ 0\ 0\ 0\ \ 1 & & & & & \\ \vdots\ \vdots\ \vdots\ \vdots\ \ddots\ \ddots & & & & & \end{bmatrix}.$$

The (n, k)th entry of production matrix P is determined in the following theorem.

Theorem 6.1 *The (n, k)th entry of the production matrix $P = (p_{n,k})_{n,k\in\mathbb{N}}$ for the exponential Riordan array $D = [g(t), f(t)]$ satisfies*

$$p_{n,k} = \frac{n!}{k!}(c_{n-k} + kr_{n-k+1})\,, \tag{6.1.3}$$

where the sequences $(r_k)_{k\in\mathbb{N}}$ and $(c_k)_{k\in\mathbb{N}}$ are defined by the ordinary generating functions

$$r(t) = f'(\bar{f}(t)) = \sum_{k=0}^{\infty} r_k t^k \quad \text{and} \quad c(t) = \frac{g'(\bar{f}(t))}{g(\bar{f}(t))} = \sum_{k=0}^{\infty} r_k t^k\,, \tag{6.1.4}$$

respectively, and by convention, $r_k = c_k = 0$ for $k < 0$.

Proof The exponential generating function of the kth column of $\bar{D}$ is

$$\frac{\mathrm{d}}{\mathrm{d}t} g(t)\frac{(f(t))^k}{k!} = g'(t)\frac{(f(t))^k}{k!} + kg(t)\frac{(f(t))^{k-1}}{k!}f'(t)\,.$$

Therefore, the exponential generating function of the kth column of P is

$$\begin{aligned} p_k(t) &= \left[\frac{1}{g(\bar{f}(t))}, \bar{f}(t)\right]\frac{\mathrm{d}}{\mathrm{d}t} g(t)\frac{(f(t))^k}{k!} = \frac{g'(\bar{f}(t))}{g(\bar{f}(t))}\frac{t^k}{k!} + k\frac{t^{k-1}}{k!}f'(\bar{f}(t)) \\ &= \frac{1}{k!}(c(t)t^k + kt^{k-1}r(t))\,. \end{aligned}$$

Then $p_{n,k} = [t^n/n!]p_k(t)$, and the desired result follows by the extraction of the coefficients. □

By Theorem 6.1, it can be found that the production matrix P of the exponential Riordan array $D = [g(t), f(t)]$ has the general form

$$P = \begin{bmatrix} c_0 & r_0 \\ 1!c_1 & \frac{1!}{1!}(c_0 + r_1) & r_0 \\ 2!c_2 & \frac{2!}{1!}(c_1 + r_2) & \frac{2!}{2!}(c_0 + 2r_1) & r_0 \\ 3!c_3 & \frac{3!}{1!}(c_2 + r_3) & \frac{3!}{2!}(c_1 + 2r_2) & \frac{3!}{3!}(c_0 + 3r_1) & r_0 \\ 4!c_4 & \frac{4!}{1!}(c_3 + r_4) & \frac{4!}{2!}(c_2 + 2r_3) & \frac{4!}{3!}(c_1 + 3r_2) & \frac{4!}{4!}(c_0 + 4r_1) & r_0 \\ \vdots & \vdots & \vdots & \vdots & \vdots & \ddots & \ddots \end{bmatrix}.$$

Additionally, the bivariate generating function of the matrix P is

$$\begin{aligned} \phi_P(t,z) &= \sum_{n,k} p_{n,k} \frac{t^n}{n!} z^k = \sum_{k=0}^{\infty} \frac{1}{k!}(c(t)t^k + kt^{k-1}r(t))z^k \\ &= e^{zt}(c(t) + zr(t)). \end{aligned}$$

Example 6.2 The production matrix of the Pascal triangle $[e^t, t]$ is

$$P_0 = \begin{bmatrix} 1 & 1 \\ 0 & 1 & 1 \\ 0 & 0 & 1 & 1 \\ 0 & 0 & 0 & 1 & 1 \\ \vdots & \vdots & \vdots & \vdots & \vdots & \vdots & \ddots \end{bmatrix},$$

with $r(t) = c(t) = 1$. Similarly, the production matrices of the two Stirling matrices $[\frac{1}{1-t}, \ln(\frac{1}{1-t})]$ and $[e^t, e^t - 1]$ are

$$P_1 = \begin{bmatrix} 1 & 1 \\ 1 & 2 & 1 \\ 1 & 3 & 3 & 1 \\ 1 & 4 & 6 & 4 & 1 \\ \vdots & \vdots & \vdots & \vdots & \vdots & \vdots & \ddots \end{bmatrix}, \quad P_2 = \begin{bmatrix} 1 & 1 \\ 0 & 2 & 1 \\ 0 & 0 & 3 & 1 \\ 0 & 0 & 0 & 4 & 1 \\ \vdots & \vdots & \vdots & \vdots & \vdots & \vdots & \ddots \end{bmatrix},$$

respectively, with $r_1(t) = c_1(t) = e^t$, and $r_2(t) = 1 + t$, $c_2(t) = 1$.

If $D = [d_{n,k}]_{n,k\in\mathbb{N}} = [g(t), f(t)]$ is an exponential Riordan array, and $D^* = (d^*_{n,k})_{n,k\in\mathbb{N}} = (g(t), f(t))$ is the corresponding classical Riordan array, then by (6.1.1), $d_{n,k} = n!d^*_{n,k}/k!$, and the following matrix identity holds:

$$D = [g(t), f(t)] = F(g(t), f(t))F^{-1} = FD^*F^{-1}, \tag{6.1.5}$$

where $F = \text{diag}(0!, 1!, 2!, \ldots)$ is a diagonal matrix. Thus, the production matrix P of the exponential Riordan array $D = [g(t), f(t)]$ satisfies

$$P = D^{-1}\bar{D} = D^{-1}UD = (FD^*F^{-1})^{-1}U(FD^*F^{-1}) = F(D^*)^{-1}VD^*F^{-1},$$

where V is the matrix defined by

$$V = F^{-1}UF = \begin{bmatrix} 0 & 1 & & & & \\ 0 & 0 & 2 & & & \\ 0 & 0 & 0 & 3 & & \\ 0 & 0 & 0 & 0 & 4 & \\ \vdots & \vdots & \vdots & \vdots & \vdots & \ddots \end{bmatrix}.$$

In particular, according to (6.1.3), we have $k!p_{n,k}/n! = c_{n-k} + kr_{n-k+1}$, which shows that each diagonal of $F^{-1}PF = (D^*)^{-1}VD^*$ represents an arithmetic sequence.

Similar to the classical case, by using the production matrix P, the recurrences for the entries of the exponential Riordan array D can be established. Thus, from P and the top row of D, we can reveal all the rows of D.

Theorem 6.2 *Let $D = [d_{n,k}]_{n,k\in\mathbb{N}} = [g(t), f(t)]$ be an exponential Riordan array, and let P be the corresponding production matrix defined by (6.1.3). Then*

$$d_{n+1,0} = \sum_{i=0}^{n} i!c_i d_{n,i}, \tag{6.1.6}$$

$$d_{n+1,k} = r_0 d_{n,k-1} + \sum_{i=k}^{n} \frac{i!}{k!}(c_{i-k} + kr_{i-k+1})d_{n,i}, \quad \textit{for } k \geq 1. \tag{6.1.7}$$

Proof According to the definition of the production matrix, we have $\bar{D} = DP$, which, together with the convention $r_k = c_k = 0$ for $k < 0$, gives

$$d_{n+1,k} = \sum_{i=k-1}^{n} \frac{i!}{k!}(c_{i-k} + kr_{i-k+1})d_{n,i}, \tag{6.1.8}$$

for all $k \geq 0$. Then the desired recurrences hold. □

The readers are referred to Exercise 6.1 to see how to use the sequences $r(t)$ and $c(t)$ defined by (6.1.4) to characterize the corresponding exponential Riordan array $[g(t), f(t)]$ and its inverse $[1/g(\bar{f}(t)), \bar{f}(t)]$.

Finally, we present two more exponential Riordan arrays as examples.

Example 6.3 The classical Riordan array $(\frac{1}{1-t}, \frac{t}{1-t})$ is the Pascal triangle. Let us consider the exponential Riordan array $[\frac{1}{1-t}, \frac{t}{1-t}]$. According to the definitions, its

(n, k)th entry is

$$d_{n,k} = \left[\frac{t^n}{n!}\right] \frac{1}{1-t} \frac{1}{k!} \left(\frac{t}{1-t}\right)^k = \frac{n!}{k!}[t^{n-k}] \frac{1}{(1-t)^{k+1}} = \frac{n!}{k!}\binom{n}{k},$$

and the corresponding production matrix are determined by $r(t) = (1+t)^2$ and $c(t) = 1 + t$. By computation, this exponential Riordan array and its production matrix begin as follows:

$$\left[\frac{1}{1-t}, \frac{t}{1-t}\right] = \begin{bmatrix} 1 & & & & & \\ 1 & 1 & & & & \\ 2 & 4 & 1 & & & \\ 6 & 18 & 9 & 1 & & \\ 24 & 96 & 72 & 16 & 1 & \\ \vdots & \vdots & \vdots & \vdots & \vdots & \ddots \end{bmatrix}, \quad P = \begin{bmatrix} 1 & 1 & & & & & \\ 1 & 3 & 1 & & & & \\ 0 & 4 & 5 & 1 & & & \\ 0 & 0 & 9 & 7 & 1 & & \\ 0 & 0 & 0 & 16 & 9 & 1 & \\ \vdots & \vdots & \vdots & \vdots & \vdots & \vdots & \ddots \end{bmatrix}.$$

The readers are referred to Sect. 6.4 and He et al. [40] for the discussion of the more general exponential Riordan array $[(1-t)^{-\alpha-1}, t/(t-1)]$, which is the coefficient arrays of the well-known Laguerre polynomials $L_n^{(\alpha)}(x)$.

Example 6.4 The (n, k)th entry of the exponential Riordan array $[e^{t^2/2}, t] = [T(n, k)]_{n,k\in\mathbb{N}}$ is

$$T(n,k) = \left[\frac{t^n}{n!}\right] e^{\frac{t^2}{2}} \frac{t^k}{k!} = \frac{n!}{k!}[t^{n-k}]e^{\frac{t^2}{2}} = \begin{cases} 0, & \text{if } n-k \text{ is odd}, \\ \frac{n!}{k! 2^{\frac{n-k}{2}} (\frac{n-k}{2})!}, & \text{if } n-k \text{ is even}. \end{cases}$$

Therefore, $T(n, k)$ is the number of involutions having k fixed points (or equivalently, k cycles of length 1) in the symmetric group S_n on the set $\{1, 2, ..., n\}$. The corresponding production matrix is determined by $r(t) = 1$ and $c(t) = t$. Thus, the first few rows of $T(n, k)$ and its production matrix are

$$[e^{t^2/2}, t] = \begin{bmatrix} 1 & & & & & & \\ 0 & 1 & & & & & \\ 1 & 0 & 1 & & & & \\ 0 & 3 & 0 & 1 & & & \\ 3 & 0 & 6 & 0 & 1 & & \\ \vdots & \vdots & \vdots & \vdots & \vdots & \vdots & \ddots \end{bmatrix}, \quad P = \begin{bmatrix} 0 & 1 & & & & & & \\ 1 & 0 & 1 & & & & & \\ 0 & 2 & 0 & 1 & & & & \\ 0 & 0 & 3 & 0 & 1 & & & \\ 0 & 0 & 0 & 4 & 0 & 1 & & \\ \vdots & \vdots & \vdots & \vdots & \vdots & \vdots & \vdots & \ddots \end{bmatrix}.$$

By fundamental theorem, the total number of involutions $I(n)$ in the symmetric group S_n satisfies

$$I(n) = \sum_{k=0}^{n} T(n,k) = \left[\frac{t^n}{n!}\right] e^{\frac{t^2}{2}+t}.$$

Thus, there holds the recurrence

$$I(0) = I(1) = 1\,, \quad I(n) = I(n-1) + (n-1)I(n-2)\,, \quad \text{for } n \geq 2\,,$$

and the sequence $(I(n))_{n\in\mathbb{N}}$ begins as

$$1, 1, 2, 4, 10, 26, 76, \ldots .$$

The readers are referred to Sect. 6.4 for the connection with the Hermite polynomials, Cheon et al. [16] the Bessel numbers, and Knuth [45, Sect. 5.1.4] more discussions on the numbers $I(n)$.

6.2 Generalized Riordan Arrays and the Riordan Group

In this section, we introduce first the concepts of generalized Riordan arrays and the *generalized Riordan group*. Then, we give basic properties of the generalized Riordan arrays and the generalized Riordan group, including the *A-sequence* of a generalized Riordan array and three recurrence relations for the entries of a generalized Riordan array.

Definition and properties of generalized Riordan arrays

If $(f_n)_{n\in\mathbb{N}}$ is a sequence of real numbers, the formal power series $f(t) = \sum_{k=0}^{\infty} f_k t^k/c_k$ is called the *generating function* of the sequence, where $(c_n)_{n\in\mathbb{N}}$ is a fixed sequence of non-zero constants with $c_0 = 1$. In particular, $f(t)$ is an *ordinary generating function* if $c_n = 1$, and an *exponential generating function* if $c_n = n!$. A formal power series $f(t)$ is called an *invertible series* if $f(t) \in \mathcal{F}_0$, and a *delta series* if $f(t) \in \mathcal{F}_1$, where $\mathcal{F}_r$ is the set of formal power series $f \in \mathbb{R}[[t]]$ with $f^{(k)}(0) = 0$ for $k = 0, 1, \ldots, r-1$ and $f^{(r)}(0) \neq 0$.

As usual, the notation $[t^n]$ stands for the "coefficient" operator. Namely, $[t^n]f(t) = f_n$ if $f(t) = \sum_{k=0}^{\infty} f_k t^k$. Similarly, if $f(t) = \sum_{k=0}^{\infty} f_k t^k/c_k$, then $[t^n/c_n]f(t) = f_n$. Thus, $[t^n/c_n]f(t) = c_n[t^n]f(t)$.

Definition 6.1 A *generalized Riordan array* with respect to the sequence $(c_n)_{n\in\mathbb{N}}$ is a pair $(g(t), f(t))$ of formal power series, where $g(t) = \sum_{k=0}^{\infty} g_k t^k/c_k$ and $f(t) = \sum_{k=1}^{\infty} f_k t^k/c_k$ with $f_1 \neq 0$, i.e., $f(t)$ is a *delta series*. The Riordan array $(g(t), f(t))$ defines an infinite, lower triangular array $(d_{n,k})_{0\leq k\leq n<\infty}$ according to the rule

$$d_{n,k} = \left[\frac{t^n}{c_n}\right] g(t) \frac{(f(t))^k}{c_k}\,, \tag{6.2.1}$$

where the functions $g(t)(f(t))^k/c_k$ are called the *column generating functions* of the Riordan array.

By the definition, the *classical Riordan arrays* introduced and studied in the previous chapters correspond to the case of $c_n = 1$, and the *exponential Riordan arrays* presented in Sect. 6.1 correspond to $c_n = n!$. In Gould and He's paper [29], the generalized Riordan arrays are referred to as (c)-Riordan arrays.

One of the most important applications of the theory of Riordan arrays is to deal with summations of the form $\sum_{k=0}^{n} d_{n,k}h_k$. In the context of generalized Riordan arrays, we have the following theorem about the summation, which is referred to as the first fundamental theorem of generalized Riordan arrays.

Theorem 6.3 *Let $D = (g(t), f(t)) = (d_{n,k})_{n,k\in\mathbb{N}}$ be a Riordan array with respect to $(c_n)_{n\in\mathbb{N}}$ and let $h(t) = \sum_{k=0}^{\infty} h_k t^k/c_k$ be the generating function of the sequence $(h_n)_{n\in\mathbb{N}}$. Then we have*

$$\sum_{k=0}^{n} d_{n,k}h_k = \left[\frac{t^n}{c_n}\right] g(t)h(f(t)), \tag{6.2.2}$$

or equivalently, $(g(t), f(t))h(t) = g(t)h(f(t))$.

Proof Based on the definition, we have

$$\begin{aligned}\sum_{k=0}^{n} d_{n,k}h_k &= \sum_{k=0}^{\infty}\left[\frac{t^n}{c_n}\right] g(t)\frac{(f(t))^k}{c_k}h_k \\ &= \left[\frac{t^n}{c_n}\right] g(t)\sum_{k=0}^{\infty} h_k\frac{(f(t))^k}{c_k} = \left[\frac{t^n}{c_n}\right] g(t)h(f(t)).\end{aligned}$$

This completes the proof. □

Similar to the classical case [65, Theorem 1.2], we can prove the inverse of Theorem 6.3.

Theorem 6.4 *Let $D = (d_{n,k})_{0\le k\le n<\infty}$ be an infinite triangle such that for every sequence $(h_k)_{k\in\mathbb{N}}$, we have $\sum_{k=0}^{n} d_{n,k}h_k = [t^n/c_n]g(t)h(f(t))$, where $h(t) = \sum_{k=0}^{\infty} h_k t^k/c_k$ is the generating function of the sequence $(h_k)_{k\in\mathbb{N}}$, and $g(t)$, $f(t)$ are two formal power series not depending on $h(t)$. Then the triangle defined by the Riordan array $(g(t), f(t))$ coincides with $(d_{n,k})_{n,k\in\mathbb{N}}$.*

Proof It is the same as that of [65, Theorem 1.2]. For any $k \in \mathbb{N}$, take the sequence which is 0 everywhere except in the kth element $h_k = 1$. Then $h(t) = \sum_{i=0}^{\infty} h_i t^i/c_i = t^k/c_k$ and $\sum_{i=0}^{n} d_{n,i}h_i = d_{n,k} = [t^n/c_n]g(t)(f(t))^k/c_k$, which proves the assertion of the theorem. □

With Theorem 6.3, we can further compute the product of two Riordan arrays $(g(t), f(t))(h(t), l(t))$. In fact, the column generating function of $(h(t), l(t))$ is $h(t)(l(t))^k/c_k$. Thus, by matrix multiplication, the column generating function of the product $(g(t), f(t))(h(t), l(t))$ is

$$g(t)h(f(t))\frac{(l(f(t)))^k}{c_k},$$

which means that the product is also a Riordan array, i.e.,

$$(g(t), f(t))(h(t), l(t)) = (g(t)h(f(t)), l(f(t))) . \tag{6.2.3}$$

Similar to the classical case, the next theorem holds.

Theorem 6.5 *For any fixed sequence $(c_n)_{n\in\mathbb{N}}$, the set of all generalized Riordan arrays $(g(t), f(t))$ with invertible series $g(t)$ is a group under matrix multiplication. Moreover, the identity of this group is $(1, t)$ and the inverse of the array $(g(t), f(t))$ is $(1/g(\bar{f}(t)), \bar{f}(t))$, where $\bar{f}(t)$ is the compositional inverse of $f(t)$.*

Proof Denote the set by $\mathscr{R}$. Then $\mathscr{R}$ is closed under matrix multiplication according to (6.2.3), and the multiplication is associative. The array $(1, t)$ is the identity of $\mathscr{R}$. Additionally, for each array $(g(t), f(t)) \in \mathscr{R}$, there exists a unique array $(1/g(\bar{f}(t)), \bar{f}(t)) \in \mathscr{R}$ such that

$$(g(t), f(t))\left(\frac{1}{g(\bar{f}(t))}, \bar{f}(t)\right) = (1, t) = \left(\frac{1}{g(\bar{f}(t))}, \bar{f}(t)\right)(g(t), f(t)) .$$

Hence, $\mathscr{R}$ is a group and the proof is complete. □

The group introduced in Theorem 6.5 is called the *generalized Riordan group* with respect to $(c_n)_{n\in\mathbb{N}}$ (or the Riordan group with respect to $(c_n)_{n\in\mathbb{N}}$ for short). It should be noticed that, for any fixed sequence $(c_n)_{n\in\mathbb{N}}$, the identity $(1, t)$ of the Riordan group $\mathscr{R}$ is the usual infinite identity matrix I. Actually, by Eq. (6.2.1), the (n, k) entry of $(1, t)$ is

$$d_{n,k} = \left[\frac{t^n}{c_n}\right]\frac{t^k}{c_k} = \frac{c_n}{c_k}[t^{n-k}]1 = \delta_{n,k} ,$$

where $\delta_{n,k}$ is the *Kronecker delta* defined by $\delta_{n,n} = 1$ and $\delta_{n,k} = 0$ for $n \neq k$.

A large number of infinite lower triangular arrays are Riordan arrays, for example, the iteration matrices. With every formal power series $f(t) = \sum_{k=1}^{\infty} f_k t^k/c_k$, we associate the infinite lower *iteration matrix* with respect to $(c_n)_{n\in\mathbb{N}}$ [19, p. 145]:

$$B(f(t)) := \begin{bmatrix} 1 & 0 & 0 & 0 & 0 & \cdots \\ 0 & B_{1,1} & 0 & 0 & 0 & \cdots \\ 0 & B_{2,1} & B_{2,2} & 0 & 0 & \cdots \\ 0 & B_{3,1} & B_{3,2} & B_{3,3} & 0 & \cdots \\ 0 & B_{4,1} & B_{4,2} & B_{4,3} & B_{4,4} & \cdots \\ \vdots & \vdots & \vdots & \vdots & \vdots & \ddots \end{bmatrix},$$

where $B_{n,k} = B_{n,k}(f_1, f_2, \ldots)$ is the *Bell polynomial* with respect to $(c_n)_{n\in\mathbb{N}}$, defined as follows:

$$\frac{1}{c_k}(f(t))^k = \sum_{n=k}^{\infty} B_{n,k}\frac{t^n}{c_n}\,. \tag{6.2.4}$$

Therefore, $B_{n,k} = [t^n/c_n](f(t))^k/c_k$, which implies that the iteration matrix $B(f(t))$ is the Riordan array $(1, f(t))$. Now, the following important property of the iteration matrix

$$B(f(g(t))) = B(g(t))B(f(t))$$

is trivial in the context of the theory of Riordan arrays, i.e.,

$$(1, f(g(t))) = (1, g(t))(1, f(t))\,;$$

and the well-known Faà di Bruno formula is a specialization of the summation rule (6.2.2):

$$\sum_{k=1}^{n} B_{n,k}(g_1, g_2, \ldots, g_{n-k+1}) f_k = \left[\frac{t^n}{n!}\right] f(g(t))\,.$$

See [19, p. 145, Theorem A] and [19, p. 137, Theorem A]. Additionally, for any delta series $f(t)$, the inverse array $(1, \bar{f}(t)) = B(\bar{f}(t))$ is also an iteration matrix. Thus, the set of iteration matrices with respect to $(c_n)_{n\in\mathbb{N}}$, denoted by $\mathscr{B}$, is a nonempty subset of the Riordan group $\mathscr{R}$ with respect to $(c_n)_{n\in\mathbb{N}}$, closed under multiplication and having inverses in $\mathscr{R}$. These indicate that $\mathscr{B}$ is a subgroup of $\mathscr{R}$ and we call it the *associated subgroup*.

It can be shown that the set of Riordan arrays which have the form $(g(t), t)$, where $g(t)$ is an invertible series, is also a subgroup of $\mathscr{R}$. We call it the *Appell subgroup* and denote it by $\mathscr{A}$. Since

$$(g(t), f(t)) = (g(t), t)(1, f(t)) = (1, f(t))(g(\bar{f}(t)), t)\,,$$

then we have $\mathscr{A}\mathscr{B} = \mathscr{B}\mathscr{A} = \mathscr{R}$. The readers are referred to the paper [61] by Shapiro for more subgroups.

A-sequence of generalized Riordan arrays

Let us consider the A-sequence of a Riordan array with respect to $(c_n)_{n\in\mathbb{N}}$. For the classical Riordan array $(g(t), f(t))$, Rogers [55] found that every element $d_{n+1,k+1}$, $n, k \in \mathbb{N}$, can be expressed as a linear combination of the elements in the preceding row, i.e.,

$$d_{n+1,k+1} = a_0 d_{n,k} + a_1 d_{n,k+1} + a_2 d_{n,k+2} + \cdots = \sum_{j=0}^{\infty} a_j d_{n,k+j}\,.$$

The sequence $A = (a_k)_{k\in\mathbb{N}}$ is called the *A-sequence* of the Riordan array. Based on the theory of A-sequences for the classical Riordan arrays, illustrated in Chap. 4, we can further develop the corresponding theory for the generalized Riordan arrays.

Theorem 6.6 *The quantity $d_{n,k}$ is the (n, k) entry of the generalized Riordan array $(g(t), f(t))$ with respect to $(c_n)_{n\in\mathbb{N}}$ if and only if $c_k d_{n,k}/c_n$ is the (n, k) entry of the classical Riordan array $(g(t), f(t))$.*

Proof By Definition 6.1, we have

$$d_{n,k} = \left[\frac{t^n}{c_n}\right] g(t)\frac{(f(t))^k}{c_k} = \frac{c_n}{c_k}[t^n]g(t)(f(t))^k ,$$

which is equivalent to the fact that $c_k d_{n,k}/c_n = [t^n]g(t)(f(t))^k$. This completes the proof. □

Despite its simple proof, Theorem 6.6 is an important result, because it shows that the generalized Riordan arrays can always be reduced to the classical case. By Theorem 6.6, the next two theorems can be obtained without difficulty.

Theorem 6.7 *For any generalized Riordan array $(g(t), f(t)) = (d_{n,k})_{n,k\in\mathbb{N}}$, every element $d_{n+1,k+1}$, where $n, k \in \mathbb{N}$, can be expressed as follows:*

$$d_{n+1,k+1} = \sum_{j=0}^{\infty} \frac{c_{n+1}c_{k+j}}{c_{k+1}c_n} a_j d_{n,k+j} , \tag{6.2.5}$$

where the sum is actually finite and the sequence $A = (a_k)_{k\in\mathbb{N}}$ is fixed. It is called the A-sequence of the generalized Riordan array and it only depends on $f(t)$. Particularly, let $A(t) = \sum_{k=0}^{\infty} a_k t^k$, then

$$f(t) = tA(f(t)) \quad \text{and} \quad A(t) = \frac{t}{\bar{f}(t)} . \tag{6.2.6}$$

Proof According to the results of Rogers [55], for the classical Riordan array $(g(t), f(t)) = (d^*_{n,k})_{n,k\in\mathbb{N}}$, there exists a unique sequence $A = (a_k)_{k\in\mathbb{N}}$ that satisfies the statements of the theorem. Then

$$d^*_{n+1,k+1} = \sum_{j=0}^{\infty} a_j d^*_{n,k+j}$$

and $f(t) = tA(f(t))$. By Theorem 6.6, $d^*_{n,k} = c_k d_{n,k}/c_n$, so we have

$$\frac{c_{k+1}}{c_{n+1}} d_{n+1,k+1} = \sum_{j=0}^{\infty} a_j \frac{c_{k+j}}{c_n} d_{n,k+j} ,$$

which leads recurrence relation (6.2.5) at once. □

Theorem 6.8 *Let $(c_n)_{n\in\mathbb{N}}$ be a sequence of non-zero constants with $c_0 = 1$, and let $D := (d_{n,k})_{0\le k\le n<\infty}$ be an infinite triangle such that $d_{n,n} \neq 0$, $\forall n \in \mathbb{N}$, and for which the relation (6.2.5) holds for some sequence $A = (a_k)_{k\in\mathbb{N}}$, $a_0 \neq 0$. Then D is a generalized Riordan array $(g(t), f(t))$ with respect to c_n, where $g(t) = \sum_{k=0}^{\infty} d_{k,0}t^k/c_k$ and $f(t)$ is the unique solution of $f(t) = tA(f(t))$ with $A(t) = \sum_{k=0}^{\infty} a_k t^k$.*

Proof Define $d^*_{n,k} = c_k d_{n,k}/c_n$, then $d^*_{n,n} \neq 0$, $\forall n \in \mathbb{N}$ and $d^*_{n+1,k+1} = \sum_{j=0}^{\infty} a_j d^*_{n,k+j}$. In view of [55, 65], the infinite triangle $D^* = (d^*_{n,k})_{n,k\in\mathbb{N}}$ is the classical Riordan array $(g(t), f(t))$, where $g(t) = \sum_{k=0}^{\infty} d^*_{k,0}t^k$ and $f(t)$ is the unique solution of $f(t) = tA(f(t))$. Thus, $g(t) = \sum_{k=0}^{\infty} c_0 d_{k,0}t^k/c_k = \sum_{k=0}^{\infty} d_{k,0}t^k/c_k$, and by Theorem 6.6, $D = (d_{n,k})_{n,k\in\mathbb{N}}$ is the generalized Riordan array $(g(t), f(t))$ with respect to $(c_n)_{n\in\mathbb{N}}$. □

Three recurrences satisfied by entries of Riordan arrays

Next, we demonstrate another three recurrences related to the entries of the generalized Riordan arrays.

Theorem 6.9 *For any generalized Riordan array $(g(t), f(t)) = (d_{n,k})_{n,k\in\mathbb{N}}$, we have*

$$d_{n,k} - \frac{c_n}{nc_{n-1}}\tilde{d}_{n-1,k} = \sum_{l=k}^{n} \frac{c_n}{c_{l-1}c_{n-l+1}} \frac{n-l+1}{n} f_{n-l+1} \frac{kc_{k-1}}{c_k} d_{l-1,k-1}\,, \tag{6.2.7}$$

for $n, k \ge 1$, where $\tilde{d}_{n,k}$ is the (n,k) entry of the Riordan array $(g'(t), f(t))$, $g'(t)$ is the derivative of $g(t)$, and f_k are the coefficients of the delta series $f(t) = \sum_{k=1}^{\infty} f_k t^k/c_k$.

Proof The column generating function is

$$\sum_{n=k}^{\infty} d_{n,k}\frac{t^n}{c_n} = g(t)\frac{(f(t))^k}{c_k}\,. \tag{6.2.8}$$

Differentiating (6.2.8) with respect to t gives

$$\begin{aligned}
\sum_{n=k}^{\infty} d_{n,k}\frac{nt^{n-1}}{c_n} &= g'(t)\frac{(f(t))^k}{c_k} + \frac{kc_{k-1}}{c_k} g(t)f'(t)\frac{(f(t))^{k-1}}{c_{k-1}}\\
&= \sum_{n=k}^{\infty} \tilde{d}_{n,k}\frac{t^n}{c_n} + \frac{kc_{k-1}}{c_k}\sum_{i=k-1}^{\infty} d_{i,k-1}\frac{t^i}{c_i}\sum_{j=0}^{\infty}(j+1)f_{j+1}\frac{t^j}{c_{j+1}}\\
&= \sum_{n=k}^{\infty} \tilde{d}_{n,k}\frac{t^n}{c_n} + \frac{kc_{k-1}}{c_k}\sum_{n=k-1}^{\infty}\sum_{i=k-1}^{n} \frac{c_n}{c_i c_{n-i+1}}(n-i+1)f_{n-i+1}d_{i,k-1}\frac{t^n}{c_n}\,.
\end{aligned}$$

Identifying the coefficients of t^{n-1}/c_{n-1} in the equation above yields

$$\frac{nc_{n-1}}{c_n}d_{n,k} = \tilde{d}_{n-1,k} + \sum_{i=k-1}^{n-1} \frac{c_{n-1}}{c_i c_{n-i}}(n-i) f_{n-i} \frac{kc_{k-1}}{c_k} d_{i,k-1} ,$$

which, after some transformations, leads to (6.2.7) finally. □

For the iteration matrix with respect to c_n, we have $g'(t) = 0$. This fact indicates that $\tilde{d}_{n-1,k} = 0$. Thus, (6.2.7) reduces to

$$d_{n,k} = \sum_{l=k}^{n} \frac{c_n}{c_{l-1}c_{n-l+1}} \frac{n-l+1}{n} f_{n-l+1} \frac{kc_{k-1}}{c_k} d_{l-1,k-1} , \tag{6.2.9}$$

which has been given in [73, Lemma 3.1].

Theorem 6.10 *For any generalized Riordan array* $(g(t), f(t)) = (d_{n,k})_{n,k \in \mathbb{N}}$, *we have*

$$\frac{c_k}{c_{k-1}} d_{n,k} = \sum_{l=k}^{n} \frac{c_n}{c_{l-1}c_{n-l+1}} f_{n-l+1} d_{l-1,k-1} , \quad n, k \geq 1 , \tag{6.2.10}$$

where f_k *are the coefficients of the delta series* $f(t) = \sum_{k=1}^{\infty} f_k t^k / c_k$.

Proof From (6.2.8), we have

$$\begin{aligned} \sum_{n=k}^{\infty} \frac{c_k}{c_{k-1}} d_{n,k} \frac{t^n}{c_n} &= f(t)g(t)\frac{(f(t))^{k-1}}{c_{k-1}} = \sum_{i=1}^{\infty} f_i \frac{t^i}{c_i} \sum_{j=k-1}^{\infty} d_{j,k-1} \frac{t^j}{c_j} \\ &= \sum_{n=k}^{\infty} \sum_{j=k-1}^{n-1} f_{n-j} d_{j,k-1} \frac{c_n}{c_j c_{n-j}} \frac{t^n}{c_n} . \end{aligned}$$

By equating the coefficients of t^n / c_n, we obtain the desired result. □

Theorem 6.11 *For any generalized Riordan array* $(g(t), f(t)) = (d_{n,k})_{n,k \in \mathbb{N}}$, *we have*

$$\frac{c_{n+1}}{c_n} d_{n,k} = \sum_{l=k}^{n} \frac{c_{l+1}}{c_{l+1-k}c_k} \bar{f}_{l+1-k} d_{n+1,l+1} , \tag{6.2.11}$$

where $\bar{f}_j$ *are the coefficients of the delta series* $\bar{f}(t) = \sum_{j=1}^{\infty} \bar{f}_j t^j / c_j$, *and* $\bar{f}(t)$ *is the compositional inverse of* $f(t)$.

Proof Based on (6.2.1), we have

$$\sum_{n=k}^{\infty}\left(\sum_{l=k}^{n}\frac{c_{l+1}}{c_{l+1-k}c_k}\bar{f}_{l+1-k}\frac{d_{n+1,l+1}}{c_{n+1}}\right)t^n=\frac{1}{c_k}\sum_{l=k}^{\infty}\frac{\bar{f}_{l+1-k}c_{l+1}}{c_{l+1-k}}\sum_{n=l}^{\infty}d_{n+1,l+1}\frac{t^n}{c_{n+1}}$$
$$=\frac{g(t)(f(t))^k}{c_k t}\sum_{l=k}^{\infty}\frac{\bar{f}_{l+1-k}}{c_{l+1-k}}(f(t))^{l+1-k}$$
$$=\frac{g(t)(f(t))^k}{c_k t}\sum_{j=1}^{\infty}\frac{\bar{f}_j}{c_j}(f(t))^j=g(t)\frac{(f(t))^k}{c_k},$$

which is coincident with the generating function of $d_{n,k}/c_n$. Therefore, the recurrence (6.2.11) can be established. □

For convenience, we give the specializations of the recurrences (6.2.5), (6.2.7), (6.2.10), and (6.2.11) for the cases $c_n = 1$ and $c_n = n!$, respectively.

Corollary 6.1 *For any classical Riordan array $(g(t), f(t)) = (d_{n,k})_{n,k\in\mathbb{N}}$, we have*

$$d_{n+1,k+1}=\sum_{j=0}^{\infty}a_j d_{n,k+j}\,, \tag{6.2.12}$$

$$d_{n,k}=\sum_{l=k}^{n}\frac{k}{n}(n-l+1)f_{n-l+1}d_{l-1,k-1}+\frac{1}{n}\tilde{d}_{n-1,k}\,, \tag{6.2.13}$$

$$d_{n,k}=\sum_{l=k}^{n}f_{n-l+1}d_{l-1,k-1}\,, \tag{6.2.14}$$

$$d_{n,k}=\sum_{l=k}^{n}\bar{f}_{l+1-k}d_{n+1,l+1}\,. \tag{6.2.15}$$

Corollary 6.2 *For any exponential Riordan array $[g(t), f(t)] = [d_{n,k}]_{n,k\in\mathbb{N}}$, we have*

$$d_{n+1,k+1}=\sum_{j=0}^{\infty}\frac{n+1}{k+1}\binom{k+j}{j}j!a_j d_{n,k+j}\,, \tag{6.2.16}$$

$$d_{n,k}=\sum_{l=k}^{n}\binom{n-1}{l-1}f_{n-l+1}d_{l-1,k-1}+\tilde{d}_{n-1,k}\,, \tag{6.2.17}$$

$$kd_{n,k}=\sum_{l=k}^{n}\binom{n}{l-1}f_{n-l+1}d_{l-1,k-1}\,, \tag{6.2.18}$$

$$(n+1)d_{n,k}=\sum_{l=k}^{n}\binom{l+1}{k}\bar{f}_{l+1-k}d_{n+1,l+1}\,. \tag{6.2.19}$$

Note that, according to Theorem 6.6, the recurrences presented in Corollaries 6.1 and 6.2 are in fact equivalent to the general ones, i.e., recurrences (6.2.5), (6.2.7), (6.2.10), and (6.2.11).

6.3 Relations Between Riordan Arrays and Sheffer Sequences

This section is devoted to the relations between generalized Riordan arrays and generalized Sheffer sequences. We can see that the (generalized) Riordan group and the (generalized) Sheffer group are isomorphic, which is an extension of the classical result shown in [40] and is revealed in [29]. We also consider the inverse relations indicated by Riordan arrays, study the connection constants between two Sheffer sequences, and give the determinantal definition for Sheffer sequences. By using the determinantal definition in the theory of Riordan arrays, we give a Riordan approach to reveal some general properties of Sheffer sequences, including the conjugate representation, the generating function, the operator characterization, and the umbral composition. Furthermore, we consider the determinants consisting of the entries of Riordan arrays.

Definition of generalized Sheffer sequences and their composition

Let $\mathscr{P}$ be the algebra of polynomials in the single indeterminate x over $\mathbb{K}$ and $\mathscr{P}^*$ be the dual vector space of all linear functionals on $\mathscr{P}$. Let $(c_n)_{n\in\mathbb{N}}$ be a fixed sequence of non-zero constants. Then, following the notations in the umbral calculus introduced in [56], for each $f(t) = \sum_{k=0}^{\infty} a_k t^k$ in $\mathscr{F}$, we define a linear functional $f(t)$ in $\mathscr{P}^*$ by

$$\langle f(t)|x^n\rangle = c_n a_n\,, \tag{6.3.1}$$

and define a linear operator $f(t)$ on $\mathscr{P}$ by

$$f(t)x^n = \sum_{k=0}^{n} \frac{c_n}{c_{n-k}} a_k x^{n-k}\,. \tag{6.3.2}$$

Thus, an element of $\mathscr{F}$ plays three roles, but the notational difference between $\langle f(t)|p(x)\rangle$ and $f(t)p(x)$ will make the particular role of $f(t)$ clear. Using these notations, the definition of Sheffer sequences can be given (cf. [56, Sect. 5]).

Definition 6.2 Let $g(t)$ be an invertible series and $f(t)$ be a delta series; we say that the sequence $(s_n(x))_{n\in\mathbb{N}}$ is *Sheffer for the pair* $(g(t), f(t))$ if and only if

$$\langle g(t)f(t)^k|s_n(x)\rangle = c_n\delta_{n,k}$$

for all $n, k \geq 0$, where $\delta_{n,k}$ is the Kronecker delta function. In particular, the Sheffer sequence for $(1, f(t))$ is called the *associated sequence* for $f(t)$, and the Sheffer sequence for $(g(t), t)$ is called the *Appell sequence* for $g(t)$.

The Sheffer sequence can also be characterized by generating function.

Definition 6.3 Let $g(t)$ be an invertible series and $f(t)$ be a delta series; we say that the sequence $(s_n(x))_{n\in\mathbb{N}}$ is the Sheffer for the pair $(g(t), f(t))$ if and only if

$$\sum_{k=0}^{\infty} s_k(x)\frac{t^k}{c_k} = \frac{1}{g(\bar{f}(t))}\varepsilon_x(\bar{f}(t)), \tag{6.3.3}$$

where $\varepsilon_x(t) = \sum_{k=0}^{\infty} x^k t^k / c_k$ is the generalized exponential series.

Therefore, the sequence $(s_n(x))_{n\in\mathbb{N}}$ is associated with $f(t)$ if and only if

$$\sum_{n=0}^{\infty} s_n(x)\frac{t^n}{c_n} = \varepsilon_x(\bar{f}(t)),$$

and the sequence $(s_n(x))_{n\in\mathbb{N}}$ is Appell for $g(t)$ if and only if

$$\sum_{n=0}^{\infty} s_n(x)\frac{t^n}{c_n} = \frac{1}{g(t)}\varepsilon_x(t).$$

Note that $\varepsilon_x(t) = \mathrm{e}^{xt}$ for $c_n = n!$ and $\varepsilon_x(t) = 1/(1 - xt)$ for $c_n = 1$.

An alternative definition of the Sheffer sequences will be represented in next subsection. If $(p_n(x))_{n\in\mathbb{N}}$ and $(q_n(x))_{n\in\mathbb{N}}$ are two sequences of polynomials with $q_n(x) = \sum_{k=0}^{n} q_{n,k} x^k$, then the *umbral composition* of $q_n(x)$ with $p_n(x)$ is the sequence $(q_n(x) \circ p_n(x))_{n\in\mathbb{N}}$, where

$$q_n(x) \circ p_n(x) = q_n(\mathbf{p}(x)) = \sum_{k=0}^{n} q_{n,k} p_k(x),$$

in which $\mathbf{p}(x)$ stands for the sequence $(p_n(x))$ and $q_n(\mathbf{p}(x))$ is the abbreviation of $q_n(x) \circ p_n(x)$. In particular, for any fixed sequence $(c_n)_{n\in\mathbb{N}}$, if $p_n(x)$ is the Sheffer for $(g(t), f(t))$ and $q_n(x)$ is the Sheffer for $(h(t), l(t))$, then $q_n(\mathbf{p}(x))$ is the Sheffer for

$$(g(t)h(f(t)), l(f(t))),$$

and the set of all Sheffer sequences, denoted by $\mathscr{S}$, is a group under umbral composition, with the set of associated sequences and the set of Appell sequences its subgroups (cf. [56, Sect. 8], [57, Theorem 3.5.5] and [59, p. 708, Theorem 7]). We call $\mathscr{S}$ the *Sheffer group* with respect to $(c_n)_{n\in\mathbb{N}}$. For more details on the theory of umbral calculus and Sheffer sequences, the readers are referred to [30, 56–59].

Isomorphism theorem of the Riordan group and Sheffer group

The Riordan arrays determined by an invertible series and a delta series play a very important role in our study, and in this subsection, we will consider the relations between such Riordan arrays and Sheffer sequences. By Definitions 6.1 and 6.3, the following theorem can be established.

Theorem 6.12 *Let $s_n(x) = \sum_{k=0}^{n} d_{n,k}x^k$. If $d_{n,k}$ is the (n, k) entry of the Riordan array $(g(t), f(t))$, then the polynomial sequence $(s_n(x))_{n\in\mathbb{N}}$ is the Sheffer for $(1/g(\bar{f}(t)), \bar{f}(t))$. Conversely, if the sequence $(s_n(x))_{n\in\mathbb{N}}$ is the Sheffer for $(g(t), f(t))$, then the coefficient $d_{n,k}$ is the (n, k) entry of the Riordan array $(1/g(\bar{f}(t)), \bar{f}(t))$.*

Proof From the definition, we have $\sum_{n=k}^{\infty} d_{n,k}t^n/c_n = g(t)(f(t))^k/c_k$. Thus

$$\sum_{n=0}^{\infty} s_n(x)\frac{t^n}{c_n} = \sum_{n=0}^{\infty}\left(\sum_{k=0}^{n} d_{n,k}x^k\right)\frac{t^n}{c_n} = \sum_{k=0}^{\infty}\left(\sum_{n=k}^{\infty} d_{n,k}\frac{t^n}{c_n}\right)x^k$$
$$= \sum_{k=0}^{\infty} g(t)\frac{(f(t))^k}{c_k}x^k = g(t)\varepsilon_x(f(t)) .$$

Therefore, the first statement of the theorem can be proved by means of (6.3.3). To prove the second statement, we observe that

$$\sum_{n=0}^{\infty} s_n(x)\frac{t^n}{c_n} = \sum_{n=0}^{\infty}\left(\sum_{k=0}^{n} d_{n,k}x^k\right)\frac{t^n}{c_n}$$
$$= \sum_{k=0}^{\infty}\left(\sum_{n=k}^{\infty} d_{n,k}\frac{t^n}{c_n}\right)x^k = \frac{1}{g(\bar{f}(t))}\sum_{k=0}^{\infty}(\bar{f}(t))^k\frac{x^k}{c_k} .$$

By equating the coefficients of x^k in the last equation, we have

$$\sum_{n=k}^{\infty} d_{n,k}\frac{t^n}{c_n} = \frac{1}{g(\bar{f}(t))}\frac{(\bar{f}(t))^k}{c_k}$$

and then $d_{n,k}$ is the (n, k) entry of the Riordan array $(1/g(\bar{f}(t)), \bar{f}(t))$. □

From the proof of Theorem 6.12, we can see that a Sheffer sequence $p_n(x) = \sum_{k=0}^{n} d_{n,k}x^k$ has a generating function $g(t)\varepsilon_x(f(t))$, where $(d_{n,k})_{n,k\geq 0} = (g(t), f(t))$. In other words, a Sheffer sequence can be defined with $(g(t), f(t))$ instead of $(g(t), f(t))^{-1}$. There are several papers considering this modern view, such as [29, 40]. Readers can easily adopt this approach when researching and processing these literatures.

With some specializations, we obtain from Theorem 6.12 the relations between the iteration matrices and the associated sequences, which has already been indicated by Roman [57, Sect. 4.1.8] for the case $c_n = n!$.

Moreover, it can be easily seen that the product of two Riordan arrays and the umbral composition of two Sheffer sequences follow formally the same rule. Therefore, it is instructive to study the relation between the Riordan arrays and Sheffer sequences in the view of groups.

Theorem 6.13 *For any fixed sequence $(c_n)_{n\in\mathbb{N}}$, the Riordan group $\mathscr{R}$ and the Sheffer group $\mathscr{S}$ are isomorphic.*

Proof Define $\sigma : \mathscr{R} \to \mathscr{S}$ by $\sigma(g(t), f(t)) = s_n(x)$, where $s_n(x)$ is the Sheffer for $(1/g(\bar{f}(t)), \bar{f}(t))$. Because $\bar{f}(t)$ is uniquely determined by $f(t)$ and $s_n(x)$ is uniquely determined by the pair $(1/g(\bar{f}(t)), \bar{f}(t))$ [56, Theorem 5.1], the map σ is well defined. Now, let us prove that σ is an isomorphism.

In fact, if $s_n(x)$ is the Sheffer for $(h(t), l(t))$, then there exists a Riordan array $(1/h(\bar{l}(t)), \bar{l}(t))$ such that $\sigma(1/h(\bar{l}(t)), \bar{l}(t)) = s_n(x)$. This indicates that σ is surjective. Next, suppose

$$\sigma(g_1(t), f_1(t)) = \sigma(g_2(t), f_2(t)) = s_n(x),$$

where $s_n(x)$ is the Sheffer for $(h(t), l(t))$. Then $\bar{f_1}(t) = l(t) = \bar{f_2}(t)$ and $f_1(t) = \bar{l}(t) = f_2(t)$. Additionally, we have $g_1(\bar{f_1}(t)) = 1/h(t) = g_2(\bar{f_2}(t))$, which leads us to the fact that

$$g_1(\bar{f_1}(f_1(t))) = g_1(t) = \frac{1}{h(\bar{l}(t))} = g_2(\bar{f_2}(f_2(t))) = g_2(t).$$

Therefore, $(g_1(t), f_1(t)) = (g_2(t), f_2(t))$ and σ is injective.

We have shown that σ is a bijection, and it remains to check that σ preserves the group operation. To do this, suppose $\sigma(g(t), f(t)) = q_n(x)$ where $q_n(x)$ is the Sheffer for $(1/g(\bar{f}(t)), \bar{f}(t))$, and $\sigma(h(t), l(t)) = p_n(x)$ where $p_n(x)$ is the Sheffer for $(1/h(\bar{l}(t)), \bar{l}(t))$. By the umbral composition, we have

$$\sigma(g(t), f(t)) \circ \sigma(h(t), l(t)) = q_n(x) \circ p_n(x) = q_n(\mathbf{p}(x)),$$

where the sequence $q_n(\mathbf{p}(x))$ is the Sheffer for

$$\left(\frac{1}{h(\bar{l}(t))g(\bar{f}(\bar{l}(t)))}, \bar{f}(\bar{l}(t))\right).$$

On the other hand, by Eq. (6.2.3) we have

$$\sigma((g(t), f(t))(h(t), l(t))) = \sigma(g(t)h(f(t)), l(f(t))) = s_n(x),$$

where the sequence $s_n(x)$ is the Sheffer for

$$\left(\frac{1}{g(\bar{f}(\bar{l}(t)))h(\bar{l}(t))}, \bar{f}(\bar{l}(t))\right).$$

Thus, $q_n(\mathbf{p}(x)) = s_n(x)$, which shows that

$$\sigma((g(t), f(t))(h(t), l(t))) = \sigma(g(t), f(t)) \circ \sigma(h(t), l(t)).$$

Therefore, σ is indeed an isomorphism and $\mathscr{R} \cong \mathscr{S}$. □

Theorem 6.14 *The associated subgroup $\mathscr{A}$ and the group of associated sequences are isomorphic. The Appell subgroup $\mathscr{B}$ and the group of Appell sequences are isomorphic.*

More properties and applications of the isomorphism between the (generalized) Sheffer group and the (generalized) Riordan group can be found along the approach shown in [40] and [29]. For instance, since the group of Sheffer sequences coincides with the group of exponential Riordan arrays, [40] shows that some of these arrays corresponding to well-known orthogonal polynomial sequences can be treated uniformly. He et al. [40] also define exponential Riordan array pairs and generalized Stirling number pairs, where any pair is composed of an array and its inverse (in the group sense). These pairs allow one to solve some interesting inversion problems. Furthermore, the high-dimensional Riordan group and the high-dimensional Sheffer group are defined and discussed in [40]. Some extensions have been shown in [29], partials of which are presented in the next subsection.

Roman's result on inverse relations

Let $(g(t), f(t)) = (a_{n,k})_{n,k\in\mathbb{N}}$ be a generalized Riordan array, where $g(t)$ is an invertible series and $f(t)$ is a delta series. Then this array has an inverse $(1/g(\bar{f}(t)), \bar{f}(t)) = (b_{n,k})_{n,k\in\mathbb{N}}$, and we can further establish an inverse relation of the form:

$$y_n = \sum_{k=0}^{n} a_{n,k} x_k \quad \text{and} \quad x_n = \sum_{k=0}^{n} b_{n,k} y_k .$$

A particular instance is the *generalized Stirling number pair* introduced and studied by Hsu [43]. Let $f(t)$ be a delta series and let

$$\frac{1}{k!}(f(t))^k = \sum_{n=0}^{\infty} A_1(n,k)\frac{t^n}{n!} \quad \text{and} \quad \frac{1}{k!}(\bar{f}(t))^k = \sum_{n=0}^{\infty} A_2(n,k)\frac{t^n}{n!} .$$

Then $A_1(n,k)$ and $A_2(n,k)$ are called a *generalized Stirling number pair*. In the context of the theory of Riordan arrays, $A_1(n,k)$ and $A_2(n,k)$ are the (n,k) entries of the Riordan arrays $(1, f(t))$ and $(1, \bar{f}(t))$, respectively. So we have

$$y_n = \sum_{k=0}^{n} A_1(n,k)x_k \quad \text{and} \quad x_n = \sum_{k=0}^{n} A_2(n,k)y_k\,.$$

The interested readers may consult the relevant papers, e.g., [43, 70, 77], for the applications of the generalized Stirling number pairs. A unified approach to generalized Stirling number sequences and generalized Stirling functions can be found in He [34, 35] with a brief introduction shown in Exercise 6.2.

In [57], Roman presented an interesting approach to the inverse relations based on the theory of Sheffer sequences and the related umbral calculus (cf. [57, Theorem 5.5.1]). We now extend Roman's result to the general case as follows.

Theorem 6.15 *Let $p_n(x)$ be associated with $f(t)$ and let $q_n(x)$ be associated with $l(t)$. Then for any invertible series $h(t)$ we have the inverse pair*

$$y_n = \sum_{k=0}^{n} \frac{\langle h(t)(l(t))^k | p_n(x)\rangle}{c_k} x_k \quad \text{and} \quad x_n = \sum_{k=0}^{n} \frac{\langle h(t)^{-1}(f(t))^k | q_n(x)\rangle}{c_k} y_k\,.$$

Proof Now, let $p_n(x) = \sum_{j=0}^{n} a_{n,j}x^j$, then we have

$$\frac{1}{c_k}\langle h(t)(l(t))^k | p_n(x)\rangle = \frac{1}{c_k}\left\langle h(t)(l(t))^k \middle| \sum_{j=0}^{n} a_{n,j}x^j \right\rangle$$

$$= \frac{1}{c_k}\sum_{j=0}^{n} a_{n,j}\langle h(t)(l(t))^k | x^j\rangle = \frac{1}{c_k}\sum_{j=0}^{n} a_{n,j}\left[\frac{t^j}{c_j}\right](h(t)(l(t))^k)\,. \tag{6.3.4}$$

Since $(p_n(x))_{n\in\mathbb{N}}$ is the Sheffer sequence for the pair $(1, f(t))$, then according to Theorem 6.12, $a_{n,j}$ is the (n,k) entry of the Riordan array $(1, \bar{f}(t))$. By using the summation rule (6.2.2), (6.3.4) equals

$$\frac{1}{c_k}\left[\frac{t^n}{c_n}\right] h(\bar{f}(t))(l(\bar{f}(t)))^k = \left[\frac{t^n}{c_n}\right] h(\bar{f}(t))\frac{(l(\bar{f}(t)))^k}{c_k}\,,$$

which indicates that the quantity $\frac{1}{c_k}\langle h(t)(l(t))^k | p_n(x)\rangle$ is in fact the (n,k) entry of the Riordan array $(h(\bar{f}(t)), l(\bar{f}(t)))$. Similarly, we can prove that $\frac{1}{c_k}\langle h(t)^{-1}(f(t))^k|$ $q_n(x)\rangle$ is the (n,k) entry of the inverse array $\left(\frac{1}{h(\bar{l}(t))}, f(\bar{l}(t))\right)$. Therefore, the desired result follows. □

Thus, it can be found that Roman's result is equivalent to giving a Riordan array $R = (h(\bar{f}(t)), l(\bar{f}(t)))$ and its inverse $R^{-1} = \left(\frac{1}{h(\bar{l}(t))}, f(\bar{l}(t))\right)$, and is trivial in the view of the theory of Riordan arrays. In other words, by using the Riordan approach, Theorem 6.15 can be reformulated as follows.

Theorem 6.16 *For any invertible series $h(t)$ and any delta series $f(t)$ and $l(t)$, we can construct two Riordan arrays which are inverse to each other:*

$$R = (h(\bar{f}(t)), l(\bar{f}(t))) \quad \text{and} \quad R^{-1} = \left(\frac{1}{h(\bar{l}(t))}, f(\bar{l}(t))\right).$$

Suppose the (n, k)-th entries of R and R^{-1} are $a_{n,k}$ and $b_{n,k}$, respectively, then we have the following inverse relation:

$$y_n = \sum_{k=0}^{n} a_{n,k} x_k \quad \text{and} \quad x_n = \sum_{k=0}^{n} b_{n,k} y_k .$$

Because the examples presented in [57, Sect. 5.5] are all special cases of Theorem 6.15, they can also be deduced from Theorem 6.16. Another two examples will be represented in formulas (6.4.20) and (6.4.21).

Connection constants problem

The connection constants problem is to determine the connection constants $a_{n,k}$ in the expression

$$r_n(x) = \sum_{k=0}^{n} a_{n,k} s_k(x),$$

where $(r_n(x))_{n\in\mathbb{N}}$ and $(s_n(x))_{n\in\mathbb{N}}$ are sequences of polynomials. Roman showed how to use the theory of umbral calculus to give an explicit solution of this problem when the sequences involve Sheffer sequences (cf. [56, Theorem 8.5] and [57, Sect. 5.1]). In this subsection, we solve this problem by using the Riordan array approach, which gives an equivalent result of [56, Theorem 8.5] and is easier to follow.

Theorem 6.17 *Let $(s_n(x))_{n\in\mathbb{N}}$ be the Sheffer for $(g(t), f(t))$ and let $(r_n(x))_{n\in\mathbb{N}}$ be the Sheffer for $(h(t), l(t))$. Suppose $r_n(x) = \sum_{k=0}^{n} a_{n,k} s_k(x)$, then $a_{n,k}$ is the (n, k) entry of the Riordan array*

$$\left(\frac{g(\bar{l}(t))}{h(\bar{l}(t))}, f(\bar{l}(t))\right).$$

Proof We prove the theorem by the technique of matrix representation. Let

$$S[x] = (s_0(x), s_1(x), \ldots)^T, \quad R[x] = (r_0(x), r_1(x), \ldots)^T, \quad X = (1, x, x^2, \ldots)^T .$$

According to Theorem 6.12, we have

$$S[x] = \left(\frac{1}{g(\bar{f}(t))}, \bar{f}(t)\right) X, \quad R[x] = \left(\frac{1}{h(\bar{l}(t))}, \bar{l}(t)\right) X .$$

Then $X = (g(t), f(t))S[x]$. By (6.2.3), we have

$$R[x] = \left(\frac{1}{h(\bar{l}(t))}, \bar{l}(t)\right)(g(t), f(t))S[x] = \left(\frac{g(\bar{l}(t))}{h(\bar{l}(t))}, f(\bar{l}(t))\right)S[x],$$

which completes the proof. □

Corollary 6.3 *Let $s_n(x)$ be the Sheffer sequence for the pair $(g(t), f(t))$, and suppose $x^n = \sum_{k=0}^{n} a_{n,k}s_k(x)$, then $a_{n,k}$ is the (n,k) entry of the Riordan array $(g(t), f(t))$.*

Proof It can be obtained from Theorem 6.17 directly by considering $(x^n)_{n\in\mathbb{N}}$ is the Sheffer for $(1, t)$. It can also be verified alternatively as follows. Let $s_n(x) = \sum_{k=0}^{n} d_{n,k}x^k$. Then according to Theorem 6.12, $d_{n,k}$ is the (n,k) entry of the Riordan array $(1/g(\bar{f}(t)), \bar{f}(t))$, which is the inverse of the Riordan array $(g(t), f(t))$. Thus, the result is obtained immediately. □

Determinantal approach to Sheffer sequences

In 1999, Costabile [21] proposed a determinantal definition for the classical Bernoulli polynomials, which was used by Costabile, Dell'Accio, and Gualtieri [22] to obtain some well-known properties of these polynomials. Yang [76] and Costabile and Longo [23] further generalized independently the results of [21, 22] to define the determinantal definition for Appell sequences.

Since the Bernoulli polynomial sequence is an Appell sequence, and all the Appell sequences are in fact Sheffer sequences, it is natural to extend the determinantal theory to Sheffer sequences. In this subsection, we will study this problem by using the relations between Sheffer sequences and Riordan arrays presented in previous subsections. It should be noted that Yang [75] established a different determinantal representation for Sheffer sequences by using Riordan approach.

We first show that Sheffer sequences can be expressed by determinants.

Theorem 6.18 *Let $(s_n(x))_{n\in\mathbb{N}}$ be the Sheffer for $(g(t), f(t))$, then we have*

$$s_0(x) = \frac{1}{a_{0,0}}, \tag{6.3.5}$$

$$s_n(x) = \frac{(-1)^n}{a_{0,0}a_{1,1}\cdots a_{n,n}} \begin{vmatrix} 1 & x & x^2 & \cdots & x^{n-1} & x^n \\ a_{0,0} & a_{1,0} & a_{2,0} & \cdots & a_{n-1,0} & a_{n,0} \\ 0 & a_{1,1} & a_{2,1} & \cdots & a_{n-1,1} & a_{n,1} \\ 0 & 0 & a_{2,2} & \cdots & a_{n-1,2} & a_{n,2} \\ \vdots & \vdots & \ddots & \ddots & \vdots & \vdots \\ 0 & 0 & 0 & \cdots & a_{n-1,n-1} & a_{n,n-1} \end{vmatrix}$$

$$= \frac{(-1)^n}{a_{0,0}a_{1,1}\cdots a_{n,n}} \det\begin{pmatrix} X_{n+1} \\ S_{n\times(n+1)} \end{pmatrix}, \tag{6.3.6}$$

where $X_{n+1} = (1, x, x^2, \ldots, x^n)$, $S_{n\times(n+1)} = (a_{j-1,i-1})_{1\le i\le n, 1\le j\le n+1}$, *and* $a_{n,k}$ *is the* (n, k) *entry of the Riordan array* $(g(t), f(t))$.

Proof Let $x^n = \sum_{k=0}^{n} a_{n,k} s_k(x)$, then by Corollary 6.3, $a_{n,k}$ is the (n, k) entry of the Riordan array $(g(t), f(t))$. The above identity leads to the following system of infinite equations in the unknown $s_n(x)$ for $n = 0, 1, 2, \ldots$:

$$\begin{cases} a_{0,0}s_0(x) = 1\,, \\ a_{1,0}s_0(x) + a_{1,1}s_1(x) = x\,, \\ a_{2,0}s_0(x) + a_{2,1}s_1(x) + a_{2,2}s_2(x) = x^2\,, \\ \cdots \\ a_{n,0}s_0(x) + a_{n,1}s_1(x) + a_{n,2}s_2(x) + \cdots + a_{n,n}s_n(x) = x^n\,, \\ \cdots \end{cases}$$

Applying Cramer's rule to the first $n + 1$ equations gives

$$s_n(x) = \frac{1}{a_{0,0}a_{1,1}\cdots a_{n,n}} \begin{vmatrix} a_{0,0} & 0 & 0 & \cdots & 0 & 1 \\ a_{1,0} & a_{1,1} & 0 & \cdots & 0 & x \\ a_{2,0} & a_{2,1} & a_{2,2} & \ddots & 0 & x^2 \\ \vdots & \vdots & \vdots & \ddots & \vdots & \vdots \\ a_{n-1,0} & a_{n-1,1} & a_{n-1,2} & \cdots & a_{n-1,n-1} & x^{n-1} \\ a_{n,0} & a_{n,1} & a_{n,2} & \cdots & a_{n,n-1} & x^n \end{vmatrix}.$$

Then, bringing the $(n + 1)$-th column to the first place by n transpositions of adjacent columns, and noting that the determinant of a square matrix is the same as that of its transpose, we obtain the result. □

For the Appell sequences, Theorem 6.18 reduces to the following corollary.

Corollary 6.4 *Let* $c_n = n!$ *and* $(s_n(x))_{n\in\mathbb{N}}$ *be the Appell sequence for* $g(t)$*, where*

$$g(t) = \sum_{k=0}^{\infty} g_k \frac{t^k}{k!}\,,$$

then we have $s_0(x) = 1/g_0$ *and*

$$s_n(x) = \frac{(-1)^n}{g_0^{n+1}} \begin{vmatrix} 1 & x & x^2 & \cdots & x^{n-1} & x^n \\ g_0 & \binom{1}{0}g_1 & \binom{2}{0}g_2 & \cdots & \binom{n-1}{0}g_{n-1} & \binom{n}{0}g_n \\ 0 & g_0 & \binom{2}{1}g_1 & \cdots & \binom{n-1}{1}g_{n-2} & \binom{n}{1}g_{n-1} \\ 0 & 0 & g_0 & \cdots & \binom{n-1}{2}g_{n-3} & \binom{n}{2}g_{n-2} \\ \vdots & \vdots & \ddots & \ddots & \vdots & \vdots \\ 0 & 0 & 0 & \cdots & g_0 & \binom{n}{n-1}g_1 \end{vmatrix} = \frac{(-1)^n}{g_0^{n+1}} \det\begin{pmatrix} X_{n+1} \\ S_{n\times(n+1)} \end{pmatrix},$$

where

$$X_{n+1} = (1, x, x^2, \dots, x^n) \quad \textit{and} \quad S_{n\times(n+1)} = \left(\binom{j-1}{i-1} g_{j-i}\right)_{1\le i\le n, 1\le j\le n+1} . \tag{6.3.7}$$

Proof The element $a_{n,k}$ in (6.3.6) is now the (n,k) entry of the exponential Riordan array $[g(t), t]$. Thus, we have

$$a_{n,k} = \left[\frac{t^n}{n!}\right] g(t)\frac{t^k}{k!} = \frac{n!}{k!}[t^{n-k}]g(t) = \binom{n}{k} g_{n-k} ,$$

from which the desired result can be obtained. Note that this is the main result of Costabile and Longo [23] and Yang [76]. □

Let us consider the associated sequences. If $c_n = n!$ and $(s_n(x))_{n\in\mathbb{N}}$ is associated with $f(t)$, where $f(t) = \sum_{j=1}^{\infty} f_j t^j/j!$, then in the determinantal representation of $s_n(x)$, $a_{n,k}$ is the (n,k) entry of the Riordan array $[1, f(t)]$. By definition, we have

$$a_{n,k} = \left[\frac{t^n}{n!}\right] \frac{1}{k!}(f(t))^k = \left[\frac{t^n}{n!}\right] \frac{1}{k!}\left(\sum_{j=1}^{\infty} f_j \frac{t^j}{j!}\right)^k = B_{n,k}(f_1, f_2, \dots) , \tag{6.3.8}$$

where $B_{n,k}(f_1, f_2, \dots)$ are the *exponential partial Bell polynomials* (cf. [19, Sect. 3.3]).

Moreover, for any fixed sequence $(c_n)_{n\in\mathbb{N}}$, the next corollary holds.

Corollary 6.5 *Let $(s_n(x))_{n\in\mathbb{N}}$ be associated with $f(t)$, then we have*

$$s_0(x) = 1 ,$$

$$s_n(x) = \frac{(-1)^{n+1}}{a_{0,0}a_{1,1}\cdots a_{n,n}} \begin{vmatrix} x & x^2 & \cdots & x^{n-1} & x^n \\ a_{1,1} & a_{2,1} & \cdots & a_{n-1,1} & a_{n,1} \\ 0 & a_{2,2} & \cdots & a_{n-1,2} & a_{n,2} \\ \vdots & \ddots & \ddots & \vdots & \vdots \\ 0 & 0 & \cdots & a_{n-1,n-1} & a_{n,n-1} \end{vmatrix}$$

$$= \frac{(-1)^{n+1}}{a_{0,0}a_{1,1}\cdots a_{n,n}} \det\begin{pmatrix} \widetilde{X}_n \\ \widetilde{S}_{(n-1)\times n} \end{pmatrix} ,$$

where $\widetilde{X}_n = (x, x^2, \dots, x^n)$, $\widetilde{S}_{(n-1)\times n} = (a_{j,i})_{1\le i\le n-1, 1\le j\le n}$, *and* $a_{n,k}$ *is the* (n,k) *entry of the Riordan array* $(1, f(t))$.

Proof For the determinant on the right-hand side of (6.3.6), the entries of the second row can be determined:

$$a_{n,0} = \left[\frac{t^n}{c_n}\right] \frac{(f(t))^0}{c_0} = \frac{c_n}{c_0}[t^n]1 = \delta_{n,0} .$$

Thus, (6.3.6) turns into

$$s_n(x) = \frac{(-1)^n}{a_{0,0}a_{1,1}\cdots a_{n,n}} \begin{vmatrix} 1 & x & x^2 & \cdots & x^{n-1} & x^n \\ 1 & 0 & 0 & \cdots & 0 & 0 \\ 0 & a_{1,1} & a_{2,1} & \cdots & a_{n-1,1} & a_{n,1} \\ 0 & 0 & a_{2,2} & \cdots & a_{n-1,2} & a_{n,2} \\ \vdots & \vdots & \ddots & \ddots & \vdots & \vdots \\ 0 & 0 & 0 & \cdots & a_{n-1,n-1} & a_{n,n-1} \end{vmatrix}.$$

Expanding the above determinant with respect to the second row, we obtain the result. □

Now, let us show that if a polynomial sequence is defined by (6.3.5) and (6.3.6), then it is a Sheffer sequence. To do this, the following lemma presented in [22, Lemma 3] and [69, Theorem 4.20] is required.

Lemma 6.1 *Let H_n be the determinant of an upper Hessenberg matrix of order n denoted by*

$$H_n = \begin{vmatrix} h_{1,1} & h_{1,2} & h_{1,3} & \cdots & h_{1,n-1} & h_{1,n} \\ h_{2,1} & h_{2,2} & h_{2,3} & \cdots & h_{2,n-1} & h_{2,n} \\ 0 & h_{3,2} & h_{3,3} & \cdots & h_{3,n-1} & h_{3,n} \\ 0 & 0 & h_{4,3} & \cdots & h_{4,n-1} & h_{4,n} \\ \vdots & \vdots & \ddots & \ddots & \vdots & \vdots \\ 0 & 0 & 0 & \cdots & h_{n,n-1} & h_{n,n} \end{vmatrix},$$

and define $H_0 = 1$, then the following recurrence holds:

$$H_n = \sum_{k=0}^{n-1} (-1)^{n-k-1} q_k(n) h_{k+1,n} H_k\,,$$

where $q_{n-1}(n) = 1$ and $q_k(n) = \prod_{j=k+2}^{n} h_{j,j-1}$ for $k = 0, 1, 2, \ldots, n-2$.

Based on Theorem 6.12 and Lemma 6.1, we establish the next theorem.

Theorem 6.19 *Let $(s_n(x))_{n\in\mathbb{N}}$ be the sequence of polynomials defined by (6.3.5) and (6.3.6), where $a_{n,k}$ is the (n, k) entry of the Riordan array $(g(t), f(t))$, then*

$$s_n(x) = \sum_{k=0}^{n} b_{n,k} x^k\,, \tag{6.3.9}$$

where $b_{n,k}$ is the (n, k) entry of the Riordan array $(1/g(\bar{f}(t)), \bar{f}(t))$, and the sequence $(s_n(x))_{n\in\mathbb{N}}$ is the Sheffer for $(g(t), f(t))$.

Proof Denote the determinant in (6.3.6) by H_{n+1}. Then according to Lemma 6.1, we have

$$\begin{aligned}
H_{n+1} &= \sum_{k=0}^{n}(-1)^{n-k}q_k(n+1)h_{k+1,n+1}H_k \\
&= (-1)^n q_0(n+1)h_{1,n+1}H_0 + \sum_{k=1}^{n-1}(-1)^{n-k}q_k(n+1)h_{k+1,n+1}H_k \\
&\quad + q_n(n+1)h_{n+1,n+1}H_n ,
\end{aligned}$$

where $h_{1,n+1} = x^n$, $h_{k+1,n+1} = a_{n,k-1}$ for $k = 1, 2, \ldots, n$, $q_n(n+1) = 1$, and $q_k(n+1) = a_{k,k} \cdots a_{n-1,n-1}$ for $k = 0, 1, 2, \ldots, n-1$. Multiplying both sides by

$$\frac{(-1)^n}{a_{0,0}a_{1,1}\cdots a_{n,n}}$$

gives us the following recurrence

$$\begin{aligned}
s_n(x) &= \frac{x^n}{a_{n,n}} - \sum_{k=1}^{n-1}\frac{a_{n,k-1}}{a_{n,n}}s_{k-1}(x) - \frac{a_{n,n-1}}{a_{n,n}}s_{n-1}(x) \\
&= \frac{1}{a_{n,n}}\left(x^n - \sum_{k=0}^{n-1}a_{n,k}s_k(x)\right),
\end{aligned}$$

which is equivalent to

$$x^n = \sum_{k=0}^{n}a_{n,k}s_k(x).$$

Let $b_{n,k}$ be the (n,k) entry of the Riordan array $(1/g(\bar{f}(t)), \bar{f}(t))$, then Eq. (6.3.9) holds, and by Theorem 6.12, $(s_n(x))_{n\in\mathbb{N}}$ is the Sheffer for $(g(t), f(t))$. □

Combining Theorems 6.18 and 6.19, the following determinantal definition of Sheffer sequences can be established.

Definition 6.4 The sequence $(s_n(x))_{n\in\mathbb{N}}$ is the Sheffer for the pair $(g(t), f(t))$ if and only if the polynomials $s_n(x)$ are defined by (6.3.5) and (6.3.6), where $a_{n,k}$ is the (n,k) entry of the Riordan array $(g(t), f(t))$.

General properties of Sheffer sequences

From the determinantal definition of Sheffer sequences and the theory of Riordan arrays, some basic properties of Sheffer sequences can be derived once again. We first rederive the conjugate representation and the generating function for Sheffer sequences (cf. [56, Theorems 5.3 and 5.4]).

Theorem 6.20 *Let $(s_n(x))_{n\in\mathbb{N}}$ be the Sheffer for $(g(t), f(t))$, then we have the conjugate representation*

$$s_n(x) = \sum_{k=0}^{n} \frac{\langle g(\bar{f}(t))^{-1}(\bar{f}(t))^k | x^n \rangle}{c_k} x^k \tag{6.3.10}$$

and its generating function

$$\sum_{n=0}^{\infty} s_n(x) \frac{t^n}{c_n} = \frac{1}{g(\bar{f}(t))} \varepsilon_x(\bar{f}(t)), \tag{6.3.11}$$

where $\varepsilon_x(t) = \sum_{k=0}^{\infty} x^k t^k / c_k$ *is the generalized exponential series.*

Proof By Definition 6.4, the polynomials $s_n(x)$ can be expressed by (6.3.5) and (6.3.6), where $a_{n,k}$ is the (n,k) entry of the Riordan array $(g(t), f(t))$. Thus, by Eq. (6.3.9), $s_n(x) = \sum_{k=0}^{n} b_{n,k} x^k$, where $b_{n,k}$ is the (n,k) entry of the Riordan array $(1/g(\bar{f}(t)), \bar{f}(t))$. According to the definition of Riordan arrays, we have

$$\begin{aligned} b_{n,k} &= \left[\frac{t^n}{c_n}\right] \frac{1}{g(\bar{f}(t))} \frac{(\bar{f}(t))^k}{c_k} = \frac{1}{c_k} \left[\frac{t^n}{c_n}\right] g(\bar{f}(t))^{-1} (\bar{f}(t))^k \\ &= \frac{1}{c_k} \langle g(\bar{f}(t))^{-1} (\bar{f}(t))^k | x^n \rangle, \end{aligned}$$

which leads to the conjugate representation (6.3.10). Next, by (6.2.1), we have

$$\begin{aligned} \sum_{n=0}^{\infty} s_n(x) \frac{t^n}{c_n} &= \sum_{n=0}^{\infty} \left(\sum_{k=0}^{n} b_{n,k} x^k \right) \frac{t^n}{c_n} \\ &= \sum_{k=0}^{\infty} x^k \sum_{n=k}^{\infty} b_{n,k} \frac{t^n}{c_n} = \frac{1}{g(\bar{f}(t))} \sum_{k=0}^{\infty} \frac{(x\bar{f}(t))^k}{c_k} \\ &= \frac{1}{g(\bar{f}(t))} \varepsilon_x(\bar{f}(t)), \end{aligned}$$

which gives us the generating function (6.3.11). □

We next establish the operator characterization of Sheffer sequences (cf. [56, Theorem 5.7]).

Theorem 6.21 *Let* $(s_n(x))_{n\in\mathbb{N}}$ *be the Sheffer for* $(g(t), f(t))$*. Then we have*

$$f(t) s_n(x) = \frac{c_n}{c_{n-1}} s_{n-1}(x). \tag{6.3.12}$$

Proof Similar to the proof of Theorem 6.20, since $(s_n(x))_{n\in\mathbb{N}}$ is the Sheffer for $(g(t), f(t))$, we have $s_n(x) = \sum_{k=0}^{n} b_{n,k} x^k$, where $b_{n,k}$ is the (n,k) entry of the Riordan array $(1/g(\bar{f}(t)), \bar{f}(t))$. Set $f(t) = \sum_{j=1}^{\infty} f_j t^j / c_j$. By Eqs. (6.2.11) and (6.3.2), we have

$$
\begin{aligned}
f(t)s_n(x) &= \sum_{k=1}^{n} b_{n,k} \sum_{j=1}^{k} \frac{f_j}{c_j} \frac{c_k}{c_{k-j}} x^{k-j} = \sum_{l=0}^{n-1} \left(\sum_{k=l+1}^{n} b_{n,k} \frac{f_{k-l}}{c_{k-l}} \frac{c_k}{c_l} \right) x^l \\
&= \sum_{l=0}^{n-1} \frac{c_n}{c_{n-1}} a_{n-1,l} x^l = \frac{c_n}{c_{n-1}} s_{n-1}(x),
\end{aligned}
$$

which completes the proof. □

Finally, by the determinantal definition, we reprove the umbral composition of two Sheffer sequences [56, Theorem 8.4].

Theorem 6.22 *Let $(s_n(x))_{n\in\mathbb{N}}$ be the Sheffer for $(g(t), f(t))$ and $(r_n(x))_{n\in\mathbb{N}}$ be the Sheffer for $(h(t), l(t))$. Then $(r_n(s(x)))_{n\in\mathbb{N}}$ is the Sheffer for the pair $(g(t)h(f(t)), l(f(t)))$.*

Proof Based on the definition, we have

$$
s_n(x) = \sum_{k=0}^{n} b_{n,k} x^k \quad \text{and} \quad r_n(x) = \sum_{k=0}^{n} d_{n,k} x^k,
$$

where $b_{n,k}$ is the (n, k) entry of the Riordan array $(1/g(\bar{f}(t)), \bar{f}(t))$, and $d_{n,k}$ is the (n, k) entry of the Riordan array $(1/h(\bar{l}(t)), \bar{l}(t))$. Then

$$
r_n(s(x)) = \sum_{k=0}^{n} d_{n,k} s_k(x) = \sum_{k=0}^{n} d_{n,k} \sum_{j=0}^{k} b_{k,j} x^j = \sum_{j=0}^{n} \left(\sum_{k=j}^{n} d_{n,k} b_{k,j} \right) x^j.
$$

Denote the coefficient of x^j by $p_{n,j}$. According to the product rule of two Riordan arrays (6.2.3), $p_{n,j}$ is the (n, j) entry of the Riordan array

$$
\left(\frac{1}{h(\bar{l}(t))g(\bar{f}(\bar{l}(t)))}, \bar{f}(\bar{l}(t)) \right).
$$

Therefore, by Theorem 6.12, $(r_n(s(x)))_{n\in\mathbb{N}}$ is the Sheffer sequence for the pair $(g(t)h(f(t)), l(f(t)))$. □

Determinant consisting of entries of Riordan arrays

By the relations between the cofactor of x^k in (6.3.6) and the coefficient $b_{n,k}$ in (6.3.9), we obtain the next theorem, which shows that the entries of a Riordan array can be expressed by a *determinant* consisting of the entries of its inverse.

Theorem 6.23 *Let $a_{n,k}$ be the (n, k) entry of the Riordan array $(g(t), f(t))$ and $b_{n,k}$ be the (n, k) entry of the Riordan array $(1/g(\bar{f}(t)), \bar{f}(t))$. Then*

$$b_{n,k} = \frac{(-1)^{n-k}}{a_{k,k}\cdots a_{n,n}} \begin{vmatrix} a_{k+1,k} & a_{k+2,k} & \cdots & a_{n-1,k} & a_{n,k} \\ a_{k+1,k+1} & a_{k+2,k+1} & \cdots & a_{n-1,k+1} & a_{n,k+1} \\ 0 & a_{k+2,k+2} & \cdots & a_{n-1,k+2} & a_{n,k+2} \\ \vdots & \ddots & \ddots & \vdots & \vdots \\ 0 & 0 & \cdots & a_{n-1,n-1} & a_{n,n-1} \end{vmatrix}.$$

As an instance, let us consider a classical Riordan array $(f(t), t)$ and its inverse $(g(t), t)$, where $f(t) = \sum_{k=0}^{\infty} f_k t^k$ and $g(t) = 1/f(t) = \sum_{k=0}^{\infty} g_k t^k$. The (n, k) entries of these two arrays are

$$a_{n,k} = [t^n]f(t)t^k = [t^{n-k}]f(t) = f_{n-k} \quad \text{and} \quad b_{n,k} = g_{n-k}\,,$$

respectively. Thus, using Theorem 6.23 and then replacing n by $n+k$, we obtain finally the determinantal expression:

$$g_n = \frac{(-1)^n}{f_0^{n+1}} \begin{vmatrix} f_1 & f_2 & f_3 & \cdots & f_{n-1} & f_n \\ f_0 & f_1 & f_2 & \cdots & f_{n-2} & f_{n-1} \\ 0 & f_0 & f_1 & \cdots & f_{n-3} & f_{n-2} \\ 0 & 0 & f_0 & \cdots & f_{n-4} & f_{n-3} \\ \vdots & \vdots & \ddots & \ddots & \vdots & \vdots \\ 0 & 0 & 0 & \cdots & f_0 & f_1 \end{vmatrix} = (-1)^n f_0^{-n-1} \det(S_{n\times n})\,.$$

where $S_{n\times n} = (f_{j-i+1})_{1\le i\le n, 1\le j\le n}$, and $f_k := 0$ for $k < 0$. This is a result given in [19, p. 157, Exercise 5].

6.4 Special Riordan Arrays and Sheffer Sequences

In this section, we apply the results presented in Sects. 6.2 and 6.3 to some well-known Riordan arrays, such as the Pascal matrix, the Stirling matrices of the first kind and the second kind, and the Lah matrix, as well as some well-known Sheffer sequences including those of higher-order Bernoulli polynomials, higher-order Euler polynomials, Hermite polynomials, Laguerre polynomials, Gegenbauer polynomials, Chebyshev polynomials, Jacobi polynomials, and q-Bernoulli polynomials.

The Pascal matrix

Consider the classical Riordan array $(\frac{1}{1-t}, \frac{t}{1-t})$. The (n, k) entry is

$$[t^n]\frac{1}{1-t}\left(\frac{t}{1-t}\right)^k = \binom{n}{k},$$

then $(\frac{1}{1-t}, \frac{t}{1-t})$ is the well-known *Pascal matrix*. The corresponding row generating functions are $(x+1)^n$, which form the Sheffer sequence for $(\frac{1}{1+t}, \frac{t}{1+t})$.

Since $\bar{f}(t) = t/(1+t)$, then $A(t) = t/\bar{f}(t) = 1+t$, and (6.2.12) gives the relation

$$\binom{n+1}{k+1} = \binom{n}{k} + \binom{n}{k+1}.$$

Next, because $g'(t) = 1/(1-t)^2$, we have

$$\tilde{d}_{n-1,k} = [t^{n-1}]g'(t)(f(t))^k = [t^{n-1-k}](1-t)^{-k-2} = \binom{n}{k+1}.$$

Thus, in view of $f_k = 1$, we deduce from (6.2.13) and (6.2.14) that

$$\binom{n+1}{k+1} = \sum_{l=k}^{n}(n-l+1)\binom{l-1}{k-1},$$
$$\binom{n}{k} = \sum_{l=k}^{n}\binom{l-1}{k-1}.$$

Finally, since the coefficient of t^j in $\bar{f}(t)$ is $\bar{f}_j = (-1)^{j-1}$, then by Eq. (6.2.15), the following recurrence holds:

$$\binom{n}{k} = \sum_{l=k}^{n}(-1)^{l-k}\binom{n+1}{l+1}. \tag{6.4.1}$$

The inverse of the array $(\frac{1}{1-t}, \frac{t}{1-t})$ is $(\frac{1}{1+t}, \frac{t}{1+t})$, whose (n,k) entry is $(-1)^{n-k}\binom{n}{k}$. Now, the row generating functions are $(x-1)^n$, which form the Sheffer sequence for $(\frac{1}{1-t}, \frac{t}{1-t})$. Additionally, making use of Theorem 6.23 and doing some computation, we obtain the determinantal representation of the *binomial coefficients*:

$$\binom{n}{k} = \begin{vmatrix} \binom{k+1}{k} & \binom{k+2}{k} & \cdots & \binom{n-1}{k} & \binom{n}{k} \\ \binom{k+1}{k+1} & \binom{k+2}{k+1} & \cdots & \binom{n-1}{k+1} & \binom{n}{k+1} \\ 0 & \binom{k+2}{k+2} & \cdots & \binom{n-1}{k+2} & \binom{n}{k+2} \\ \vdots & \ddots & \ddots & \vdots & \vdots \\ 0 & 0 & \cdots & \binom{n-1}{n-1} & \binom{n}{n-1} \end{vmatrix}.$$

The Stirling matrices of both kinds

Let us consider the exponential Riordan array $[1, \ln(1+t)]$, which is the Stirling matrix of the first kind. Then, its (n,k) entry is

$$\left[\frac{t^n}{n!}\right]\frac{1}{k!}(\ln(1+t))^k = s(n,k)\,,$$

i.e., the *signed Stirling number of the first kind*. $s(n,k) = (-1)^{n-k}\left[{n \atop k}\right]$. The row generating functions are $\sum_{k=0}^{n} s(n,k)x^k = (x)_n$, which are the falling factorials defined by $(x)_0 = 1$ and $(x)_n = x(x-1)\cdots(x-n+1)$ for $n = 1, 2, \ldots$, and form the sequence associated with $\mathrm{e}^t - 1$ [57, Sect. 4.1.2].

By (6.2.6), the generating function of the A-sequence of $[1, \ln(1+t)]$ is

$$A(t) = \frac{t}{\bar{f}(t)} = \frac{t}{\mathrm{e}^t - 1} = \sum_{j=0}^{\infty} B_j \frac{t^j}{j!}\,,$$

where B_j are the *Bernoulli numbers*. Then Eq. (6.2.16) reduces to

$$\left[{n+1 \atop k+1}\right] = \sum_{j=0}^{\infty}(-1)^j \frac{n+1}{k+1}\binom{k+j}{j}\left[{n \atop k+j}\right]B_j\,.$$

Additionally, using $f_k = (-1)^{k-1}(k-1)!$, we obtain from (6.2.17) and (6.2.18) that

$$\left[{n \atop k}\right] = \sum_{l=k}^{n}\binom{n-1}{l-1}\left[{l-1 \atop k-1}\right](n-l)!\,,$$

$$k\left[{n \atop k}\right] = \sum_{l=k}^{n}\binom{n}{l-1}\left[{l-1 \atop k-1}\right](n-l)!\,.$$

Finally, since $\bar{f}(t) = \mathrm{e}^t - 1$, then $\bar{f}_j = 1$, and by Eq. (6.2.19), we have

$$(n+1)\left[{n \atop k}\right] = \sum_{l=k}^{n}(-1)^{l-k}\binom{l+1}{k}\left[{n+1 \atop l+1}\right].$$

The inverse of the exponential array $[1, \ln(1+t)]$ is $[1, \mathrm{e}^t - 1]$, which is the Stirling matrix of the second kind, and its (n,k) entry is

$$\left[\frac{t^n}{n!}\right]\frac{1}{k!}(\mathrm{e}^t - 1)^k = \left\{{n \atop k}\right\},$$

that is, the *Stirling number of the second kind*. The row generating functions are $\sum_{k=0}^{n}\left\{{n \atop k}\right\}x^k$, which are called the *exponential polynomials* and denoted by $\phi_n(x)$. The sequence $(\phi_n(x))_{n\in\mathbb{N}}$ is associated with $\ln(1+t)$ [57, Sect. 4.1.3].

The generating function of the A-sequence of the array $[1, \mathrm{e}^t - 1]$ is

$$A(t)=\frac{t}{\bar{f}(t)}=\frac{t}{\ln(1+t)}=\sum_{j=0}^{\infty}b_j(0)\frac{t^j}{j!},$$

where $b_j(0)$ are the *Bernoulli numbers of the second kind* [57, p. 114], and they are also called the *Cauchy numbers of the first kind* (cf. Example 4.3, [19, p. 294], and [48]). Thus, we have

$$\left\{\begin{matrix} n+1 \\ k+1 \end{matrix}\right\}=\sum_{j=0}^{\infty}\frac{n+1}{k+1}\binom{k+j}{j}\left\{\begin{matrix} n \\ k+j \end{matrix}\right\}b_j(0).$$

Next, because $f_k=1$, Eqs. (6.2.17) and (6.2.18) lead us at once to

$$\left\{\begin{matrix} n \\ k \end{matrix}\right\}=\sum_{l=k}^{n}\binom{n-1}{l-1}\left\{\begin{matrix} l-1 \\ k-1 \end{matrix}\right\},$$

$$k\left\{\begin{matrix} n \\ k \end{matrix}\right\}=\sum_{l=k}^{n}\binom{n}{l-1}\left\{\begin{matrix} l-1 \\ k-1 \end{matrix}\right\}.$$

Now, $\bar{f}(t)=\ln(1+t)$, so $\bar{f}_j=(-1)^{j-1}(j-1)!$, and Eq. (6.2.19) gives the recurrence

$$(n+1)\left\{\begin{matrix} n \\ k \end{matrix}\right\}=\sum_{l=k}^{n}(-1)^{l-k}(l-k)!\binom{l+1}{k}\left\{\begin{matrix} n+1 \\ l+1 \end{matrix}\right\}.$$

Finally, let us consider the related determinants. Set $p_n(x)=\left(\frac{x}{a}\right)_n$, then $(p_n(x))_{n\in\mathbb{N}}$ is associated with $f(t)=\mathrm{e}^{at}-1$ (cf. [57, Sect. 1.2]). The (n,k) entry of the exponential Riordan array $[1,\mathrm{e}^{at}-1]$ is

$$a_{n,k}=\left[\frac{t^n}{n!}\right]\frac{1}{k!}(\mathrm{e}^{at}-1)^k=\left[\frac{t^n}{n!}\right]\sum_{n=k}^{\infty}\left\{\begin{matrix} n \\ k \end{matrix}\right\}\frac{a^nt^n}{n!}=a^n\left\{\begin{matrix} n \\ k \end{matrix}\right\},$$

which indicates that

$$\left(\frac{x}{a}\right)_n=\frac{(-1)^{n+1}}{a^{\binom{n+1}{2}}}\begin{vmatrix} x & x^2 & \cdots & x^{n-1} & x^n \\ a\left\{\begin{smallmatrix} 1 \\ 1 \end{smallmatrix}\right\} & a^2\left\{\begin{smallmatrix} 2 \\ 1 \end{smallmatrix}\right\} & \cdots & a^{n-1}\left\{\begin{smallmatrix} n-1 \\ 1 \end{smallmatrix}\right\} & a^n\left\{\begin{smallmatrix} n \\ 1 \end{smallmatrix}\right\} \\ & a^2\left\{\begin{smallmatrix} 2 \\ 2 \end{smallmatrix}\right\} & \cdots & a^{n-1}\left\{\begin{smallmatrix} n-1 \\ 2 \end{smallmatrix}\right\} & a^n\left\{\begin{smallmatrix} n \\ 2 \end{smallmatrix}\right\} \\ & & \ddots & \vdots & \vdots \\ & & & a^{n-1}\left\{\begin{smallmatrix} n-1 \\ n-1 \end{smallmatrix}\right\} & a^n\left\{\begin{smallmatrix} n \\ n-1 \end{smallmatrix}\right\} \end{vmatrix}.$$

Note that setting $a=1$ and $x=n$ further leads to the determinantal representation of $n!$. The determinantal representation of the exponential polynomials $\phi_n(x)$ is

$$\phi_n(x) = (-1)^{n+1} \begin{vmatrix} x & x^2 & \cdots & x^{n-1} & x^n \\ s(1,1) & s(2,1) & \cdots & s(n-1,1) & s(n,1) \\ & s(2,2) & \cdots & s(n-1,2) & s(n,2) \\ & & \ddots & \vdots & \vdots \\ & & & s(n-1,n-1) & s(n,n-1) \end{vmatrix}.$$

When $x = 1$, the above formula gives the determinantal representation of the *Bell numbers* $\mathcal{B}_n = \phi_n(1) = \sum_{k=0}^{n} \left\{ {n \atop k} \right\}$ (cf. [19, Sect. 5.4]). Moreover, by Theorem 6.23, we have

$$\begin{bmatrix} n \\ k \end{bmatrix} = \begin{vmatrix} \left\{ {k+1 \atop k} \right\} & \left\{ {k+2 \atop k} \right\} & \cdots & \left\{ {n-1 \atop k} \right\} & \left\{ {n \atop k} \right\} \\ \left\{ {k+1 \atop k+1} \right\} & \left\{ {k+2 \atop k+1} \right\} & \cdots & \left\{ {n-1 \atop k+1} \right\} & \left\{ {n \atop k+1} \right\} \\ 0 & \left\{ {k+2 \atop k+2} \right\} & \cdots & \left\{ {n-1 \atop k+2} \right\} & \left\{ {n \atop k+2} \right\} \\ \vdots & \ddots & \ddots & \vdots & \vdots \\ 0 & 0 & \cdots & \left\{ {n-1 \atop n-1} \right\} & \left\{ {n \atop n-1} \right\} \end{vmatrix}$$

and

$$\left\{ {n \atop k} \right\} = \begin{vmatrix} \left[{k+1 \atop k} \right] & \left[{k+2 \atop k} \right] & \cdots & \left[{n-1 \atop k} \right] & \left[{n \atop k} \right] \\ \left[{k+1 \atop k+1} \right] & \left[{k+2 \atop k+1} \right] & \cdots & \left[{n-1 \atop k+1} \right] & \left[{n \atop k+1} \right] \\ 0 & \left[{k+2 \atop k+2} \right] & \cdots & \left[{n-1 \atop k+2} \right] & \left[{n \atop k+2} \right] \\ \vdots & \ddots & \ddots & \vdots & \vdots \\ 0 & 0 & \cdots & \left[{n-1 \atop n-1} \right] & \left[{n \atop n-1} \right] \end{vmatrix}.$$

A generalized Pascal matrix

A generalized Pascal matrix $P[x]$ has been introduced by Call and Velleman in [13]:

Definition 6.5 Let x be a non-zero real number. The $n \times n$ *generalized Pascal matrix* is defined by

$$(P_n[x])_{i,j} := \begin{cases} \binom{i}{j} x^{i-j}, & 0 \le j \le i \le n-1, \\ 0, & \text{otherwise}, \end{cases} \tag{6.4.2}$$

where $n \in \mathbb{N}$, and $P_n[0] := I$. Here $P_n[x]$ is also called the *Pascal matrix function.*

$P_n \equiv P_n[1]$ is clearly the classical Pascal matrix. Many well-known and new properties of *Bernoulli numbers* and *Bernoulli polynomials* and *Euler numbers* and *Euler polynomials* were obtained through this matrix approach, for instance, see [13, 14, 17, 37, 41, 50]. An extension of the above definition to the shifted generalized Pascal matrix follows.

Definition 6.6 Let x be a non-zero real number. The $n \times n$ *shifted generalized Pascal matrix* $P_{n,r}[x]$ is defined by

$$(P_{n,r}[x])_{i,j} := \begin{cases} \binom{i+r}{j+r} x^{i-j}, & 0 \le j \le i \le n-r-1, \\ 0, & \text{otherwise}, \end{cases} \tag{6.4.3}$$

where $n \in \mathbb{N}$ and $n \geq r+1$. In particular, $P_{n,0}[x] = P_n[x]$.

We also define the following *shifted Stirling matrices of the first kind and the second kind.*

Definition 6.7 Denote the Stirling numbers of the first kind and the second kind by $(-1)^{n-k}\left[{n \atop k}\right]$ and $\left\{{n \atop k}\right\}$ $(n, k \geq 0)$, respectively, where $\left[{n \atop k}\right]$ are unsigned Stirling numbers of the first kind. Then the $n \times n$ shifted Stirling matrices of the first kind and the second kind are defined by

$$[s_{n,r}]_{i,j} := (-1)^{i-j}\left[{r+i+1 \atop r+j+1}\right] \quad \text{and} \quad [S_{n,r}]_{i,j} := \left\{{r+i+1 \atop r+j+1}\right\}, \tag{6.4.4}$$

respectively.

Note that if $r = 0$, then $s_{n,0} = s_n$, $S_{n,0} = S_n$, the Stirling matrices of the first kind and the second kind.

Definition 6.8 A special *Vandermonde matrix* is defined by

$$V_r(x) = \begin{bmatrix} 1 & 1 & \cdots & 1 \\ x & x+1 & \cdots & x+r-1 \\ x^2 & (x+1)^2 & \cdots & (x+r-1)^2 \\ \vdots & \vdots & \ddots & \vdots \\ x^{r-1} & (x+1)^{r-1} & \cdots & (x+r-1)^{r-1} \end{bmatrix}. \tag{6.4.5}$$

Let A and B be any $m \times m$ and $n \times n$ square matrices, respectively. We use the notation $\oplus$ for the direct sum of matrices A and B:

$$A \oplus B = \begin{bmatrix} A & 0 \\ 0 & B \end{bmatrix}. \tag{6.4.6}$$

The following two lemmas are cited from [17].

Lemma 6.2 *For $n \geq 1$, there hold*

$$(a)\ P_n([1] \oplus S_{n-1}) = S_n,$$
$$(b)\ ([1] \oplus s_{n-1})P_n^{-1} = s_n.$$

Lemma 6.3 *Let x be real numbers, and $r \geq 1$. There holds*

$$V_r(x) = (P_r[x-1]S_r)D_rP_r^T,$$

where $D_r := \operatorname{diag}(0!, 1!, \ldots, (r-1)!)$.

Theorem 6.24 *Let $P_{n,r}[x]$ be defined by* (6.4.3). *Then*

$$(P_{n,r}[x])^{-1} = P_{n,r}[-x]. \tag{6.4.7}$$

Proof It is known that $P_n[x]P_n[-x] = I_n$, the identity, and $P_n[x]$ can be written as

$$P_n[x] = \begin{bmatrix} P_r[x] & O \\ A & P_{n,r}[x] \end{bmatrix}$$

for some $(n-r)\times r$ matrix A, where O is a zero matrix. Hence, we have

$$P_n[x]P_n[-x] = \begin{bmatrix} P_r[x] & O \\ A & P_{n,r}[x] \end{bmatrix}\begin{bmatrix} P_r[-x] & O \\ B & P_{n,r}[-x] \end{bmatrix} = \begin{bmatrix} I_r & O \\ O & I_{n-r} \end{bmatrix},$$

which implies

$$P_{n,r}[x]P_{n,r}[-x] = AO + P_{n,r}[x]P_{n,r}[-x] = I_{n-r}.$$

Therefore, $(P_{n,r}[x])^{-1} = P_{n,r}[-x]$. □

Similarly, we have

Theorem 6.25 *Let $S_{n,r}$ be defined by* (6.4.4). *Then there holds*

$$S_{n,r}^{-1} = s_{n,r} \tag{6.4.8}$$

for all positive integers n and r.

Theorem 6.26 *Suppose $0 \le r \le n-1$. There hold*

$$\begin{aligned} &(a)\ \ P_{n,r}S_{n-1,r-1} = S_{n,r}\,, \\ &(b)\ \ s_{n-1,r-1}(P_{n,r})^{-1} = s_{n-1,r-1}P_{n,r}[-1] = s_{n,r}\,, \end{aligned}$$

where $P_{n,r} = P_{n,r}[1]$.

Proof From Lemma 6.2, we have

$$S_n = P_n([1]\oplus S_{n-1}) = \begin{bmatrix} P_r & O \\ A & P_{n,r} \end{bmatrix}\begin{bmatrix} [1]\oplus S_{r-1} & O \\ B & S_{n-1,r-1} \end{bmatrix}.$$

Noting

$$S_n = \begin{bmatrix} S_r & O \\ C & S_{n,r} \end{bmatrix}$$

and comparing with the previous equation yields

$$S_{n,r} = AO + P_{n,r}S_{n-1,r-1} = P_{n,r}S_{n-1,r-1},$$

i.e., (a). It is clear that (b) can be proved from (a) by Theorems 6.24 and 6.25. □

From (a) of Theorem 6.26, we also have

$$(I_r \oplus P_{n-r})(I_r \oplus ([1] \oplus S_{n-r-1})) = (I_r \oplus S_{n-r})$$

for the Stirling numbers of the second kind, and

$$(I_r \oplus ([1] \oplus s_{n-r-1}))(I_r \oplus P_{n-r}[-1]) = (I_r \oplus s_{n-r})$$

for the Stirling numbers of the first kind.

From the definition of the forward difference, we immediately have the following inverse relation:

$$\Delta^r f(x) = \sum_{k=0}^{r} (-1)^{r-k} \binom{r}{k} f(x+k) \Leftrightarrow f(x+r) = \sum_{k=0}^{r} \binom{r}{k} \Delta^k f(x). \quad (6.4.9)$$

By using $P_r[x]$ defined by (6.4.2), we present the above equations in matrix form:

$$[f(x),\ \Delta f(x),\ \ldots,\ \Delta^{r-1} f(x)]^T = P_r[-1][f(x),\ f(x+1),\ \ldots,\ f(x+r-1)]^T.$$

Hence we may generalize (6.4.8) to the following result by using $P_r[n]P_r[-n] = I$.

Theorem 6.27 *Let $P_r[x]$ be defined by* (6.4.2). *Then we have the relationship*

$$[f(x), \Delta f(x), \ldots, \Delta^{r-1} f(x)]^T = P_r[-n][f(x), f(x+1), \ldots, f(x+r-1)]^T$$
$$\Leftrightarrow [f(x), f(x+1), \ldots, f(x+r-1)]^T = P_r[n][f(x), \Delta f(x), \ldots, \Delta^{r-1} f(x)]^T,$$

or equivalently,

$$\Delta^r f(x) = \sum_{k=0}^{r} (-1)^{r-k} \binom{r}{k} n^{r-k} f(x+k) \Leftrightarrow f(x+r) = \sum_{k=0}^{r} \binom{r}{k} n^{r-k} \Delta^k f(x).$$

If $f(n) = n^m$, $0 \le m \le r-1$, then

$$\begin{bmatrix} n^0 & n & \ldots & n^{r-1} \\ \Delta n^0 & \Delta n & \ldots & \Delta n^{r-1} \\ \Delta^2 n^0 & \Delta^2 n & \ldots & \Delta^2 n^{r-1} \\ \vdots & \vdots & \ddots & \vdots \\ \Delta^{r-1} n^0 & \Delta^{r-1} n & \ldots & \Delta^{r-1} n^{r-1} \end{bmatrix} = P_r[-1] V_r[n]^T, \quad (6.4.10)$$

where $V_r[n]$ is defined by (6.4.5).

Since $x^m = \sum_{r=0}^{m} \left\{ {m \atop r} \right\} (x)_r$ and noting that Newton's formula (cf. [44]) implies that

$$x^m = \sum_{r=0}^{m} \left. \frac{\Delta^r x^m}{r!} \right|_{x=0} (x)_r,$$

we have

$$\left.\frac{\Delta^r x^m}{r!}\right|_{x=0} = \left\{ \begin{matrix} m \\ r \end{matrix} \right\}. \tag{6.4.11}$$

Hence, from

$$\frac{\Delta^r n^m}{r!} = \left.\frac{\Delta^r (x+n)^m}{r!}\right|_{x=0} = \sum_{k=0}^{m} \binom{m}{k} n^{m-k} \left.\frac{\Delta^r x^k}{r!}\right|_{x=0},$$

we obtain

$$\frac{\Delta^r n^m}{r!} = \sum_{k=0}^{m} \binom{m}{k} n^{m-k} \left\{ \begin{matrix} k \\ r \end{matrix} \right\}. \tag{6.4.12}$$

The matrix form of (6.4.12) can be presented as below.

Theorem 6.28 *Let $P_r[x]$ and S_r be the Pascal matrix function and the Stirling matrix of the second kind, respectively. Then there exist the following matrix identities:*

$$D_r^{-1} \begin{bmatrix} n^0 & n & \dots & n^{r-1} \\ \Delta n^0 & \Delta n & \dots & \Delta n^{r-1} \\ \Delta^2 n^0 & \Delta^2 n & \dots & \Delta^2 n^{r-1} \\ \vdots & \vdots & \ddots & \vdots \\ \Delta^{r-1} n^0 & \Delta^{r-1} n & \dots & \Delta^{r-1} n^{r-1} \end{bmatrix} = (P_r[n-1]S_r)^T = (P_r[n]([1] \oplus S_{r-1}))^T ,$$

where $D_r = \mathrm{diag}(0!, 1!, \dots, (r-1)!)$.

Proof Noting $P_r[n] = P_r[n-1]P_r[1] = P_r[n-1]P_r$ as well as Lemmas 6.2 and 6.3, from (6.4.10), we have

$$\begin{aligned} &\begin{bmatrix} n^0 & n & \dots & n^{r-1} \\ \Delta n^0 & \Delta n & \dots & \Delta n^{r-1} \\ \Delta^2 n^0 & \Delta^2 n & \dots & \Delta^2 n^{r-1} \\ \vdots & \vdots & \ddots & \vdots \\ \Delta^{r-1} n^0 & \Delta^{r-1} n & \dots & \Delta^{r-1} n^{r-1} \end{bmatrix} = P_r[-1]V_r[n]^T \\ &= P_r[-1]\left(P_r[n-1]S_r)D_r P_r^T\right)^T = P_r[-1]P_r D_r \left(P_r[n]([1] \oplus S_{r-1})\right)^T \\ &= D_r \left(P_r[n]([1] \oplus S_{r-1})\right)^T = D_r \left(P_r[n-1]S_r\right)^T . \end{aligned}$$

Hence, the desired result follows. □

Let a, b, and x be real numbers and n a non-negative integer. Then, by using Newton's formula (cf. [44])

$$(x-a)^m = \sum_{r=0}^{m} \frac{1}{r!} \left[\Delta^r (x-a)^m\right]_{x=b} (x-b)_r .$$

The numbers, denoted by $S(m, r, b-a)$, in the sum before $(x-b)_r$ are referred to as the *Riordan-Stirling numbers* of the second kind with parameter a and b, and are defined in [53] and also in [46] for the case of $b=0$ where the numbers $S(m, r, -a)$ are called *non-central Stirling numbers*. Hence, we have

$$(x-a)^m = \sum_{r=0}^{m} S(m, r, -a)(x)_r, \quad S(m, r, b-a) = \left.\frac{\Delta^r (x-a)^m}{r!}\right|_{x=b}. \quad (6.4.13)$$

In particular,

$$S(m, r, -a) = \left.\frac{\Delta^r (x-a)^m}{r!}\right|_{x=0}.$$

From (6.4.13), there hold $S(0, 0, -a) = 1$, $S(m, 0, -a) = (-a)^m$, $S(0, r, -a) = 0$ for all $m, r \neq 0$, and

$$S(m+1, r, -a) = S(m, r-1, -a) + (r-a)S(m, r, -a).$$

Since $[\Delta^r (x-a)^m]_{x=0} = [\Delta^r x^m]_{x=-a}$, there holds

$$S(m, r, a) = \left.\frac{1}{r!}\Delta^r x^m\right|_{x=a},$$

which shows the numbers $S(m, r, -a)$ coincide with the numbers $A_{m,r}(a)$ presented in [54]. In the above sense, $S(m, r, b-a)$ is an extension of (6.4.11), the case of $n=0$ of (6.4.12).

The higher-order Bernoulli polynomials

The *higher-order Bernoulli polynomials* $B_n^{(\alpha)}(x)$ form the Appell sequence for $g(t) = \left(\frac{e^t-1}{t}\right)^\alpha$ (cf. [57, Sect. 2.2]), and the related Riordan array is $\left[\left(\frac{t}{e^t-1}\right)^\alpha, t\right]$. By computation, the (n, k) entry of this Riordan array is

$$[x^k]B_n^{(\alpha)}(x) = \left[\frac{t^n}{n!}\right]\left(\frac{t}{e^t-1}\right)^\alpha \frac{t^k}{k!} = \frac{n!}{k!}[t^{n-k}]\left(\frac{t}{e^t-1}\right)^\alpha = \binom{n}{k}B_{n-k}^{(\alpha)},$$

where $B_k^{(\alpha)} := B_k^{(\alpha)}(0)$ are the *higher-order Bernoulli numbers*. The corresponding inverse array is $\left[\left(\frac{e^t-1}{t}\right)^\alpha, t\right]$, and the (n, k) entry is $\binom{n}{k}B_{n-k}^{(-\alpha)}$.

The *potential polynomials* $P_n^{(z)} = P_n^{(z)}(f_1, \ldots, f_n)$ are defined by

$$1 + \sum_{n=1}^{\infty} P_n^{(z)} \frac{t^n}{n!} = \left(1 + \sum_{i=1}^{\infty} f_i \frac{t^i}{i!}\right)^z,$$

where z is a complex number (cf. [19, Sect. 3.5]). Denote $P_0^{(z)} := 1$, then according to [19, p. 141, Theorem B], we have

$$P_n^{(z)} = P_n^{(z)}(f_1, f_2, \ldots, f_n) = \sum_{k=0}^{n} (z)_k B_{n,k}(f_1, f_2, \ldots), \tag{6.4.14}$$

which gives the relation between the potential polynomials and the *exponential Bell polynomials*. Based on the potential polynomials and the identity [19, p. 136, Eq. (31')]

$$B_{n,k}\left(\frac{x_2}{2}, \frac{x_3}{3}, \ldots\right) = \frac{n!}{(n+k)!} B_{n+k,k}(0, x_2, x_3, \ldots),$$

we have

$$\begin{aligned}
\binom{n}{k} B_{n-k}^{(-\alpha)} &= \left[\frac{t^n}{n!}\right] \left(\frac{\mathrm{e}^t - 1}{t}\right)^{\alpha} \frac{t^k}{k!} \\
&= \frac{n!}{k!}[t^{n-k}] \left(\frac{\mathrm{e}^t - 1}{t}\right)^{\alpha} = \frac{n!}{k!}[t^{n-k}] \left(1 + \sum_{i=1}^{\infty} \frac{1}{i+1} \frac{t^i}{i!}\right)^{\alpha} \\
&= \binom{n}{k} P_{n-k}^{(\alpha)}\left(\frac{1}{2}, \frac{1}{3}, \frac{1}{4}, \cdots\right) = \binom{n}{k} \sum_{i=0}^{n-k} (\alpha)_i B_{n-k,i}\left(\frac{1}{2}, \frac{1}{3}, \frac{1}{4}, \cdots\right) \\
&= \binom{n}{k} \sum_{i=0}^{n-k} (\alpha)_i \frac{(n-k)!}{(n-k+i)!} B_{n-k+i,i}(0, 1, 1, \ldots) \\
&= \binom{n}{k} \sum_{i=0}^{n-k} \binom{\alpha}{i} \binom{n-k+i}{i}^{-1} S_2(n-k+i, i),
\end{aligned}$$

where $S_2(k+i, i)$ are the 2-associated Stirling numbers of the second kind, defined by

$$\sum_{n=k}^{\infty} S_2(n, k) \frac{t^n}{n!} = \frac{1}{k!} \left(\sum_{i=2}^{\infty} \frac{t^i}{i!}\right)^k$$

(cf. [19, p. 221, Exercise 7]). Thus, we establish the following expressions for the *higher-order Bernoulli polynomials*:

$$B_n^{(\alpha)}(x) = \sum_{k=0}^{n} \binom{n}{k} B_{n-k}^{(\alpha)} x^k = \sum_{k=0}^{n} \left\{ \binom{n}{k} \sum_{i=0}^{n-k} \binom{-\alpha}{i} \binom{n-k+i}{i}^{-1} S_2(n-k+i, i) \right\} x^k.$$

To give the determinantal representation of $B_n^{(\alpha)}(x)$, we should determine the coefficient of $t^k/k!$ in the invertible series $g(t)$:

$$g_k = \left[\frac{t^k}{k!}\right] g(t) = \left[\frac{t^k}{k!}\right] \left(\frac{e^t-1}{t}\right)^{\alpha} = \sum_{i=0}^{k} \binom{\alpha}{i} \binom{k+i}{i}^{-1} S_2(k+i, i)\,.$$

Thus, we have

$$g_0 = 1\,,\ g_1 = \frac{\alpha}{2}\,,\ g_2 = \frac{\alpha^2}{4} + \frac{\alpha}{12}\,,\ g_3 = \frac{\alpha^3}{8} + \frac{\alpha^2}{8}\,,$$
$$g_4 = \frac{\alpha^4}{16} + \frac{\alpha^3}{8} + \frac{\alpha^2}{48} - \frac{\alpha}{120}\,,\ \ldots,$$

and by Corollary 6.4, the determinantal representation of $B_n^{(\alpha)}(x)$ can be established. For example, when $n = 4$, it is

$$B_4^{(\alpha)}(x) = \begin{vmatrix} 1 & x & x^2 & x^3 & x^4 \\ 1 & \frac{1}{2}\alpha & \frac{1}{4}\alpha^2 + \frac{1}{12}\alpha & \frac{1}{8}\alpha^3 + \frac{1}{8}\alpha^2 & \frac{1}{16}\alpha^4 + \frac{1}{8}\alpha^3 + \frac{1}{48}\alpha^2 - \frac{1}{120}\alpha \\ 0 & 1 & \alpha & \frac{3}{4}\alpha^2 + \frac{1}{4}\alpha & \frac{1}{2}\alpha^3 + \frac{1}{2}\alpha^2 \\ 0 & 0 & 1 & \frac{3}{2}\alpha & \frac{3}{2}\alpha^2 + \frac{1}{2}\alpha \\ 0 & 0 & 0 & 1 & 2\alpha \end{vmatrix}$$
$$= x^4 - 2\alpha x^3 + \left(\frac{3\alpha^2}{2} - \frac{\alpha}{2}\right)x^2 + \left(-\frac{\alpha^3}{2} + \frac{\alpha^2}{2}\right)x + \frac{\alpha^4}{16} - \frac{\alpha^3}{8} + \frac{\alpha^2}{48} + \frac{\alpha}{120}\,.$$

Setting $\alpha = 1$ yields the *classical Bernoulli polynomials*. Now, the coefficient of $t^k/k!$ in $g(t)$ is

$$g_k = \left[\frac{t^k}{k!}\right] \frac{e^t-1}{t} = \left[\frac{t^k}{k!}\right] \sum_{i=0}^{\infty} \frac{t^i}{(i+1)i!} = \frac{1}{k+1}\,.$$

Applying Corollary 6.4 and multiplying the ith row by $i-2$ for $i = 3, 4, \ldots, n+1$, we obtain

$$B_n(x) = \frac{(-1)^n}{(n-1)!} \begin{vmatrix} 1 & x & x^2 & \cdots & x^{n-1} & x^n \\ 1 & \frac{1}{2} & \frac{1}{3} & \cdots & \frac{1}{n} & \frac{1}{n+1} \\ 0 & \binom{1}{0} & \binom{2}{0} & \cdots & \binom{n-1}{0} & \binom{n}{0} \\ 0 & 0 & \binom{2}{1} & \cdots & \binom{n-1}{1} & \binom{n}{1} \\ \vdots & \vdots & \ddots & \ddots & \vdots & \vdots \\ 0 & 0 & 0 & \cdots & \binom{n-1}{n-2} & \binom{n}{n-2} \end{vmatrix},$$

which was proposed by Costabile [21]. See also the paper due to Costabile, Dell'Accio, and Gualtieri [22].

The Riordan array $\left[\left(\frac{t}{e^t-1}\right)^{\alpha}, t\right] = \left[\binom{n}{k} B_{n-k}^{(\alpha)}\right]_{n,k\in\mathbb{N}}$ has the inverse $\left[\left(\frac{e^t-1}{t}\right)^{\alpha}, t\right] = \left[\binom{n}{k} B_{n-k}^{(-\alpha)}\right]_{n,k\in\mathbb{N}}$, so the following relation holds:

$$y_n = \sum_{k=0}^{n} \binom{n}{k} B_{n-k}^{(\alpha)} x_k \quad \text{and} \quad x_n = \sum_{k=0}^{n} \binom{n}{k} B_{n-k}^{(-\alpha)} y_k .$$

Particularly, when $\alpha = 1$, we obtain the Riordan array $\left[\frac{e^t-1}{t}, t\right]$, which has the (n,k) entry $\binom{n}{k}\frac{1}{n-k+1}$. Then the above inverse relation reduces to

$$y_n = \sum_{k=0}^{n} \binom{n}{k} B_{n-k} x_k \quad \text{and} \quad x_n = \sum_{k=0}^{n} \binom{n}{k} \frac{1}{n-k+1} y_k .$$

As an example, from

$$\sum_{k=0}^{n} \binom{n}{k} B_{n-k} x^k = B_n(x) ,$$

we have

$$\sum_{k=0}^{n} \binom{n}{k} \frac{1}{n-k+1} B_k(x) = x^n .$$

Similarly, from

$$\sum_{k=m}^{n} \binom{n}{k} B_{n-k} \left\{ {k \atop m} \right\} = \left[\frac{t^n}{n!}\right] \frac{t}{e^t-1} \frac{1}{m!} (e^t-1)^m$$
$$= \frac{n!}{m!}[t^{n-1}](e^t-1)^{m-1} = \frac{n}{m} \left\{ {n-1 \atop m-1} \right\},$$

we have

$$\sum_{k=m}^{n} \binom{n}{k} \frac{1}{n-k+1} \frac{k}{m} \left\{ {k-1 \atop m-1} \right\} = \frac{1}{m} \sum_{k=m}^{n} \binom{n}{k-1} \left\{ {k-1 \atop m-1} \right\} = \left\{ {n \atop m} \right\}.$$

The higher-order Euler polynomials

The higher-order Euler polynomials $E_n^{(\alpha)}(x)$ form the Appell sequence for $g(t) = \left(\frac{e^t+1}{2}\right)^\alpha$ (cf. [57, Sect. 2.3]). The corresponding Riordan array is $\left[\left(\frac{2}{e^t+1}\right)^\alpha, t\right]$, which has the (n,k) entry

$$[x^k]E_n^{(\alpha)}(x) = \left[\frac{t^n}{n!}\right] \left(\frac{2}{e^t+1}\right)^\alpha \frac{t^k}{k!} = \frac{n!}{k!}[t^{n-k}] \left(\frac{2}{e^t+1}\right)^\alpha = \binom{n}{k} E_{n-k}^{(\alpha)}(0) .$$

The inverse array is $\left[\left(\frac{e^t+1}{2}\right)^\alpha, t\right]$, and the (n,k) entry is $\binom{n}{k}E_{n-k}^{(-\alpha)}(0)$.

According to [19, Sect. 3.3, Theorems A and B], there holds

$$B_{k,i}\left(\frac{1}{2},\frac{1}{2},\ldots\right)=\frac{1}{2^i}\left\{{k \atop i}\right\}.$$

Then we obtain another expression of the (n,k) entry of the inverse array:

$$\begin{aligned}\binom{n}{k}E_{n-k}^{(-\alpha)}(0)&=\left[\frac{t^n}{n!}\right]\left(\frac{\mathrm{e}^t+1}{2}\right)^{\alpha}\frac{t^k}{k!}=\frac{n!}{k!}[t^{n-k}]\left(\frac{\mathrm{e}^t+1}{2}\right)^{\alpha}\\&=\frac{n!}{k!}[t^{n-k}]\left(1+\sum_{i=1}^{\infty}\frac{1}{2}\frac{t^i}{i!}\right)^{\alpha}=\binom{n}{k}P_{n-k}^{(\alpha)}\left(\frac{1}{2},\frac{1}{2},\ldots\right)\\&=\binom{n}{k}\sum_{i=0}^{n-k}(\alpha)_i B_{n-k,i}\left(\frac{1}{2},\frac{1}{2},\ldots\right)=\binom{n}{k}\sum_{i=0}^{n-k}(\alpha)_i 2^{-i}\left\{{n-k \atop i}\right\}.\end{aligned}$$

Therefore, the explicit expression of the *higher-order Euler polynomials* is

$$E_n^{(\alpha)}(x)=\sum_{k=0}^{n}\binom{n}{k}E_{n-k}^{(\alpha)}(0)x^k=\sum_{k=0}^{n}\binom{n}{k}\sum_{i=0}^{n-k}(-\alpha)_i 2^{-i}\left\{{n-k \atop i}\right\}x^k.$$

The coefficient of $t^k/k!$ in $g(t)$ is

$$g_k=\left[\frac{t^k}{k!}\right]g(t)=\left[\frac{t^k}{k!}\right]\left(\frac{\mathrm{e}^t+1}{2}\right)^{\alpha}=\sum_{i=0}^{k}(\alpha)_i 2^{-i}\left\{{k \atop i}\right\}.$$

By computation, the first few coefficients are

$$g_0=1\,,\ g_1=\frac{\alpha}{2}\,,\ g_2=\frac{\alpha^2}{4}+\frac{\alpha}{4}\,,\ g_3=\frac{\alpha^3}{8}+\frac{3\alpha^2}{8}\,,\ g_4=\frac{\alpha^4}{16}+\frac{3\alpha^3}{8}+\frac{3\alpha^2}{16}-\frac{\alpha}{8}\,,\ \ldots\,,$$

and the determinantal representation of $E_n^{(\alpha)}(x)$ can be established. For example, we have

$$\begin{aligned}E_4^{(\alpha)}(x)&=\begin{vmatrix}1 & x & x^2 & x^3 & x^4\\ 1 & \frac{1}{2}\alpha & \frac{1}{4}\alpha^2+\frac{1}{4}\alpha & \frac{1}{8}\alpha^3+\frac{3}{8}\alpha^2 & \frac{1}{16}\alpha^4+\frac{3}{8}\alpha^3+\frac{3}{16}\alpha^2-\frac{1}{8}\alpha\\ 0 & 1 & \alpha & \frac{3}{4}\alpha^2+\frac{3}{4}\alpha & \frac{1}{2}\alpha^3+\frac{3}{2}\alpha^2\\ 0 & 0 & 1 & \frac{3}{2}\alpha & \frac{3}{2}\alpha^2+\frac{3}{2}\alpha\\ 0 & 0 & 0 & 1 & 2\alpha\end{vmatrix}\\&=x^4-2\alpha x^3+\left(\frac{3\alpha^2}{2}-\frac{3\alpha}{2}\right)x^2+\left(-\frac{\alpha^3}{2}+\frac{3\alpha^2}{2}\right)x+\frac{\alpha^4}{16}-\frac{3\alpha^3}{8}+\frac{3\alpha^2}{16}+\frac{\alpha}{8}.\end{aligned}$$

When $\alpha=1$, we obtain the *classical Euler polynomials* $E_n(x)$. Since

$$g_k = \left[\frac{t^k}{k!}\right] g(t) = \left[\frac{t^k}{k!}\right]\left(1+\sum_{i=1}^{\infty}\frac{1}{2}\frac{t^i}{i!}\right),$$

then $g_0 = 1$ and $g_k = 1/2$ for $k \geq 1$, and we have

$$E_n(x) = \left(-\frac{1}{2}\right)^n \begin{vmatrix} 1 & x & x^2 & \cdots & x^{n-1} & x^n \\ 2 & 1 & 1 & \cdots & 1 & 1 \\ 0 & 2 & \binom{2}{1} & \cdots & \binom{n-1}{1} & \binom{n}{1} \\ 0 & 0 & 2 & \cdots & \binom{n-1}{2} & \binom{n}{2} \\ \vdots & \vdots & \ddots & \ddots & \vdots & \vdots \\ 0 & 0 & 0 & \cdots & 2 & \binom{n}{n-1} \end{vmatrix},$$

which can also be found in [23, Sect. 4.3] and [76, Corollary 2].

The Riordan array $\left[\frac{2}{e^t+1}, t\right] = \left[\binom{n}{k} E_{n-k}(0)\right]_{n,k\in\mathbb{N}}$ has the inverse $\left[\frac{e^t+1}{2}, t\right]$. It can be deduced directly that the (n, k) entry of the inverse is $\frac{1}{2}\binom{n}{k}$ for $n \neq k$ and 1 for $n = k$. Thus, the following inverse pair can be established:

$$y_n = \sum_{k=0}^{n}\binom{n}{k} E_{n-k}(0) x_k \quad \text{and} \quad x_n = \frac{1}{2}\sum_{k=0}^{n-1}\binom{n}{k} y_k + y_n .$$

For example, from the fact

$$\sum_{k=0}^{n}\binom{n}{k} E_{n-k}(0) x^k = E_n(x),$$

we have

$$\sum_{k=0}^{n-1}\frac{1}{2}\binom{n}{k} E_k(x) + E_n(x) = x^n .$$

Let us study the connection constants between $(B_n^{(\alpha)}(x))_{n\in\mathbb{N}}$ and $(E_n^{(\alpha)}(x))_{n\in\mathbb{N}}$. Suppose

$$B_n^{(\alpha)}(x) = \sum_{k=0}^{n} a_{n,k} E_k^{(\beta)}(x), \quad \beta \in \mathbb{N},$$

then $a_{n,k}$ is the (n, k) entry of the Riordan array $\left[\left(\frac{2}{e^t+1}\right)^{-\beta}\left(\frac{t}{e^t-1}\right)^{\alpha}, t\right]$. By computation, we have

$$a_{n,k} = \left[\frac{t^n}{n!}\right]\left(\frac{2}{e^t+1}\right)^{-\beta}\left(\frac{t}{e^t-1}\right)^{\alpha}\frac{t^k}{k!} = \frac{n!}{k!}[t^{n-k}]\sum_{i=0}^{\infty}E_i^{(-\beta)}(0)\frac{t^i}{i!}\sum_{j=0}^{\infty}B_j^{(\alpha)}(0)\frac{t^j}{j!}$$

$$= \binom{n}{k}\sum_{i=0}^{n-k}\binom{n-k}{i}E_i^{(-\beta)}(0)B_{n-k-i}^{(\alpha)}(0)\,.$$

Since

$$\sum_{i=0}^{\infty}E_i^{(-\beta)}(0)\frac{t^i}{i!} = \left(\frac{2}{e^t+1}\right)^{-\beta} = \frac{1}{2^\beta}(e^t+1)^\beta = \frac{1}{2^\beta}\sum_{m=0}^{\beta}\binom{\beta}{m}e^{mt}$$

$$= \frac{1}{2^\beta}\sum_{m=0}^{\beta}\binom{\beta}{m}\sum_{i=0}^{\infty}\frac{m^i t^i}{i!} = \sum_{i=0}^{\infty}\left(\frac{1}{2^\beta}\sum_{m=0}^{\beta}\binom{\beta}{m}m^i\right)\frac{t^i}{i!}\,,$$

we have

$$E_i^{(-\beta)}(0) = \frac{1}{2^\beta}\sum_{m=0}^{\beta}\binom{\beta}{m}m^i$$

and

$$a_{n,k} = \binom{n}{k}\sum_{i=0}^{n-k}\binom{n-k}{i}\frac{1}{2^\beta}\sum_{m=0}^{\beta}\binom{\beta}{m}m^i B_{n-k-i}^{(\alpha)}(0) = \frac{1}{2^\beta}\binom{n}{k}\sum_{m=0}^{\beta}\binom{\beta}{m}B_{n-k}^{(\alpha)}(m)\,,$$

which gives

$$B_n^{(\alpha)}(x) = \frac{1}{2^\beta}\sum_{k=0}^{n}\left\{\binom{n}{k}\sum_{m=0}^{\beta}\binom{\beta}{m}B_{n-k}^{(\alpha)}(m)\right\}E_k^{(\beta)}(x)\,. \tag{6.4.15}$$

Next, suppose

$$E_n^{(\alpha)}(x) = \sum_{k=0}^{n}b_{n,k}B_k^{(\beta)}(x)\,, \quad \beta\in\mathbb{N}\,,$$

then $b_{n,k}$ is the (n,k) entry of the Riordan array $\left[\left(\frac{t}{e^t-1}\right)^{-\beta}\left(\frac{2}{e^t+1}\right)^{\alpha}, t\right]$. By computation,

$$b_{n,k} = \left[\frac{t^n}{n!}\right]\left(\frac{t}{e^t-1}\right)^{-\beta}\left(\frac{2}{e^t+1}\right)^{\alpha}\frac{t^k}{k!} = \binom{n}{k}\sum_{i=0}^{n-k}\binom{n-k}{i}B_i^{(-\beta)}(0)E_{n-k-i}^{(\alpha)}(0)\,.$$

According to [57, p. 99], $\left\{ {n \atop k} \right\} = \binom{n}{k}B_{n-k}^{(-k)}(0)$, where $\left\{ {n \atop k} \right\}$ are the Stirling numbers of the second kind, then $\left\{ {i+\beta \atop \beta} \right\} = \binom{i+\beta}{\beta}B_i^{(-\beta)}(0)$, and we have

$$b_{n,k} = \binom{n}{k} \sum_{i=0}^{n-k} \binom{n-k}{i} \binom{i+\beta}{\beta}^{-1} \left\{ \begin{matrix} i+\beta \\ \beta \end{matrix} \right\} E_{n-k-i}^{(\alpha)}(0) ,$$

which indicates

$$E_n^{(\alpha)}(x) = \sum_{k=0}^{n} \left\{ \binom{n}{k} \sum_{i=0}^{n-k} \binom{n-k}{i} \binom{i+\beta}{\beta}^{-1} \left\{ \begin{matrix} i+\beta \\ \beta \end{matrix} \right\} E_{n-k-i}^{(\alpha)}(0) \right\} B_k^{(\beta)}(x) . \tag{6.4.16}$$

Note that relations (6.4.15) and (6.4.16) generalize the results of Cheon [14] and Srivastava and Pintér [67]. The readers are referred to [72] for another proof of these two relations. By specifying the parameters α and β, various elegant identities can be obtained, which can also be found in [14, 67, 72].

The Hermite polynomials

From [57, Sect. 4.2.1], the Hermite polynomials $H_n^{(v)}(x)$ of variance v form the Appell sequence for $g(t) = \mathrm{e}^{vt^2/2}$. By using Theorem 6.12, one may see that $H_{n,k}^{(v)} := [x^k] H_n^{(v)}(x)$ is the (n, k) entry of the Riordan array $[\exp(-vt^2/2), t]$:

$$H_{n,k}^{(v)} = \left[\frac{t^n}{n!} \right] \mathrm{e}^{-\frac{vt^2}{2}} \frac{t^k}{k!} = \frac{n!}{k!} [t^{n-k}] \mathrm{e}^{-\frac{vt^2}{2}} = \begin{cases} 0 , & \text{if } n-k \text{ is odd} , \\ \frac{n!}{k!} \frac{(-\frac{v}{2})^{\frac{n-k}{2}}}{(\frac{n-k}{2})!} , & \text{if } n-k \text{ is even} . \end{cases} \tag{6.4.17}$$

Therefore, the explicit expression of the *Hermite polynomials* is

$$\begin{aligned} H_n^{(v)}(x) = & \sum_{\substack{k=0 \\ n-k \text{ even}}}^{n} \frac{n!}{k!} \frac{(-\frac{v}{2})^{\frac{n-k}{2}}}{(\frac{n-k}{2})!} x^k = \sum_{\substack{k=0 \\ k \text{ even}}}^{n} \frac{n!}{(n-k)!} \frac{(-\frac{v}{2})^{\frac{k}{2}}}{(\frac{k}{2})!} x^{n-k} \\ = & \sum_{k=0}^{[\frac{n}{2}]} \left(-\frac{v}{2} \right)^k \frac{(n)_{2k}}{k!} x^{n-2k} . \end{aligned}$$

The inverse of the Riordan array $[\exp(-vt^2/2), t]$ is $[\exp(vt^2/2), t]$, whose (n, k) entry can be obtained easily from (6.4.17) by replacing v with $-v$.

Moreover, for the invertible series $g(t)$, we have

$$g_k = \left[\frac{t^k}{k!} \right] g(t) = \left[\frac{t^k}{k!} \right] \mathrm{e}^{\frac{vt^2}{2}} = \begin{cases} 0 & \text{if } k \text{ is odd} , \\ \frac{k!(\frac{v}{2})^{\frac{k}{2}}}{(\frac{k}{2})!} & \text{if } k \text{ is even} , \end{cases}$$

and the first few coefficients are

$$g_0 = 1 , \; g_2 = v , \; g_4 = 3v^2 , \; g_6 = 15v^3 , \; g_8 = 105v^4 , \; g_{10} = 945v^5 , \; \ldots .$$

By Corollary 6.4, the determinantal representation of $H_n^{(v)}(x)$ can be obtained. For example, for $n = 4$ we have

$$H_4^{(v)}(x) = \begin{vmatrix} 1 & x & x^2 & x^3 & x^4 \\ 1 & 0 & v & 0 & 3v^2 \\ 0 & 1 & 0 & 3v & 0 \\ 0 & 0 & 1 & 0 & 6v \\ 0 & 0 & 0 & 1 & 0 \end{vmatrix} = x^4 - 6vx^2 + 3v^2 .$$

From [1, Eq. (22.5.19)] and [52, Sect. 103 and 104], it is known that the *probabilists' Hermite polynomials* are defined by $He_n(x) := H_n^{(1)}(x)$, and the physicists' Hermite polynomials are defined by $H_n(x) = 2^{\frac{n}{2}} H_n^{(1)}(\sqrt{2}x)$, so the corresponding explicit expressions and determinantal representations can be established directly.

The Laguerre polynomials

According to [57, Section 4.3.1], the *Laguerre polynomials* $L_n^{(\alpha)}(x)$ of order α form the Sheffer sequence for the pair $((1-t)^{-\alpha-1}, t/(t-1))$, and the corresponding Riordan array is $[(1-t)^{-\alpha-1}, t/(t-1)]$. It is interesting to notice that the inverse of the Riordan array $[(1-t)^{-\alpha-1}, t/(t-1)]$ is just itself (cf. [40]). The (n, k) entry of this Riordan array is

$$\begin{aligned}[x^k]L_n^{(\alpha)}(x) &= \left[\frac{t^n}{n!}\right] \frac{1}{k!}\left(\frac{-1}{t-1}\right)^{\alpha+1}\left(\frac{t}{t-1}\right)^k = (-1)^k \frac{n!}{k!}[t^{n-k}](1-t)^{-(\alpha+k+1)} \\ &= (-1)^n \frac{n!}{k!}\binom{-\alpha-1-k}{n-k} = (-1)^k \frac{n!}{k!}\binom{n+\alpha}{n-k},\end{aligned}$$

which implies that

$$L_n^{(\alpha)}(x) = \sum_{k=0}^{n} \frac{n!}{k!}\binom{n+\alpha}{n-k}(-x)^k .$$

Moreover, by Theorem 6.18, the determinant of $L_n^{(\alpha)}(x)$ can be established. For example, when $n = 4$, we have

$$\begin{aligned}L_4^{(\alpha)}(x) &= \begin{vmatrix} 1 & x & x^2 & x^3 & x^4 \\ 1 & \alpha+1 & (\alpha+2)_2 & (\alpha+3)_3 & (\alpha+4)_4 \\ 0 & -1 & -2(\alpha+2) & -3(\alpha+3)_2 & -4(\alpha+4)_3 \\ 0 & 0 & 1 & 3(\alpha+3) & 6(\alpha+4)_2 \\ 0 & 0 & 0 & -1 & -4(\alpha+4) \end{vmatrix} \\ &= x^4 - 4(\alpha+4)x^3 + 6(\alpha+4)_2 x^2 - 4(\alpha+4)_3 x + (\alpha+4)_4 .\end{aligned}$$

Note that many papers, including [1, Eq. (22.9.15)] and [52, Sect. 113], denote the Laguerre polynomials by $L_n^{(\alpha)}(x)/n!$.

From the exponential Riordan array $[(1-t)^{-\alpha-1}, t/(t-1)]$, it is known that

$$f(t) = \bar{f}(t) = \frac{t}{t-1} = -\sum_{k=1}^{\infty} t^k,$$

and the $(n-1,k)$ entry of the array $[(\alpha+1)(1-t)^{-\alpha-2}, t/(t-1)]$ is

$$\tilde{d}_{n-1,k} = (\alpha+1)(-1)^k \frac{(n-1)!}{k!}\binom{n+\alpha}{n-1-k}.$$

Then, by using (6.2.17), (6.2.18), and (6.2.19) with some computation, we obtain

$$\begin{aligned}\binom{n+\alpha}{n-k} &= \sum_{l=k}^{n}\binom{l-1+\alpha}{l-k} = \sum_{l=k}^{n}(-1)^{l-k}\binom{n+1+\alpha}{n-l}\\ &= \frac{\alpha+1}{n}\binom{n+\alpha}{n-1-k} + \frac{k}{n}\sum_{l=k}^{n}(n-l+1)\binom{l-1+\alpha}{l-k}.\end{aligned}$$

Moreover, when $\alpha = -1$, the Riordan array $[(1-t)^{-\alpha-1}, t/(t-1)]$ reduces to $[1, t/(t-1)]$, and its (n,k) entry becomes

$$a_{n,k} = (-1)^k \frac{n!}{k!}\binom{n-1}{k-1} = (-1)^k L(n,k),$$

where $L(n,k)$ are the *Lah numbers*. The generating function of the A-sequence of this array is $A(t) = t-1$, then (6.2.16) gives

$$L(n+1,k+1) = \frac{n+1}{k+1}L(n,k) + (n+1)L(n,k+1).$$

Additionally, by Theorem 6.23, the following determinantal representation for the Lah numbers holds:

$$L(n,k) = \begin{vmatrix} L(k+1,k) & L(k+2,k) & \cdots & L(n-1,k) & L(n,k) \\ L(k+1,k+1) & L(k+2,k+1) & \cdots & L(n-1,k+1) & L(n,k+1) \\ 0 & L(k+2,k+2) & \cdots & L(n-1,k+2) & L(n,k+2) \\ \vdots & \ddots & \ddots & \vdots & \vdots \\ 0 & 0 & \cdots & L(n-1,n-1) & L(n,n-1) \end{vmatrix}.$$

The polynomials of the Gegenbauer case

In this case, $c_n = 1/\binom{-\lambda}{n}$. Let $(s_n(x))_{n\in\mathbb{N}}$ be the Sheffer sequence for the pair $(g(t), f(t))$ where

$$g(t)=\left(\frac{2}{1+\sqrt{1-t^2}}\right)^{\lambda_0} \quad \text{and} \quad f(t)=\frac{-t}{1+\sqrt{1-t^2}}.$$

Then $\bar{f}(t)=-2t/(1+t^2)$, $g(\bar{f}(t))=(1+t^2)^{\lambda_0}$, and the generating function is

$$\sum_{n=0}^{\infty}\binom{-\lambda}{n}s_n(x)t^n=(1+t^2)^{\lambda-\lambda_0}(1-2xt+t^2)^{-\lambda},$$

from which, it can be seen that $s_n(x)$ is related to the *Gegenbauer polynomials*, and in fact, when $\lambda_0=\lambda$, the polynomials $\binom{-\lambda}{n}s_n(x)$ are the Gegenbauer polynomials $C_n^{(\lambda)}(x)$ (cf., for example, [1, Eq. (22.9.3)], [52, Sect. 143], and [56, pp. 97–99]).

The coefficient $[x^k]s_n(x)$ is the (n,k) entry of the Riordan array $R=((1+t^2)^{-\lambda_0}, -2t/(1+t^2))$. Thus, we have

$$[x^k]s_n(x)=\left[\frac{t^n}{c_n}\right]\frac{(1+t^2)^{-\lambda_0}}{c_k}\left(\frac{-2t}{1+t^2}\right)^k=\frac{c_n}{c_k}(-2)^k[t^{n-k}]\sum_{j=0}^{\infty}\binom{-\lambda_0-k}{j}t^{2j}$$

$$=\begin{cases}0, & \text{if } n-k \text{ is odd},\\ \frac{c_n}{c_k}(-2)^k\binom{-\lambda_0-k}{\frac{n-k}{2}}, & \text{if } n-k \text{ is even},\end{cases}$$

which further gives the explicit expression of $s_n(x)$:

$$s_n(x)=\sum_{\substack{k=0\\ n-k \text{ even}}}^{n}\frac{\binom{-\lambda}{k}}{\binom{-\lambda}{n}}\binom{-\lambda_0-k}{\frac{n-k}{2}}(-2)^kx^k$$

$$=\sum_{k=0}^{[\frac{n}{2}]}\frac{\binom{-\lambda}{n-2k}\binom{-\lambda_0-n+2k}{k}}{\binom{-\lambda}{n}}(-2x)^{n-2k}. \tag{6.4.18}$$

The inverse of R is $\left(\left(\frac{2}{1+\sqrt{1-t^2}}\right)^{\lambda_0}, \frac{-t}{1+\sqrt{1-t^2}}\right)$. Using the equation

$$(1+\sqrt{1-z})^{-\alpha}=\sum_{i=0}^{\infty}\frac{\alpha(\alpha+2i-1)_{i-1}}{2^{\alpha+2i}i!}z^i \tag{6.4.19}$$

(cf. [56, p. 94, Eq. (9.1)]), we determine the (n,k) entry of this array:

$$\begin{aligned} a_{n,k} &= \left[\frac{t^n}{c_n}\right] \frac{1}{c_k} \left(\frac{2}{1+\sqrt{1-t^2}}\right)^{\lambda_0} \left(\frac{-t}{1+\sqrt{1-t^2}}\right)^k \\ &= \frac{c_n}{c_k} 2^{\lambda_0} (-1)^k [t^{n-k}] \left(1+\sqrt{1-t^2}\right)^{-\lambda_0-k} \\ &= \begin{cases} 0 & \text{if } n-k \text{ is odd}\,, \\ \frac{c_n}{c_k} \frac{(-1)^k}{2^n} \frac{\lambda_0+k}{\lambda_0+n} \binom{\lambda_0+n}{\frac{n-k}{2}} & \text{if } n-k \text{ is even}\,. \end{cases} \end{aligned}$$

In particular, $a_{n,n} = (-\frac{1}{2})^n$. Thus, the determinant of $s_n(x)$ can be obtained. For example, when $n = 4$, it is

$$\begin{aligned} s_4(x) &= 1024 \begin{vmatrix} 1 & x & x^2 & x^3 & x^4 \\ 1 & 0 & \frac{\lambda_0}{2\lambda(\lambda+1)} & 0 & \frac{3\lambda_0(\lambda_0+3)}{4\lambda(\lambda+1)(\lambda+2)(\lambda+3)} \\ 0 & -\frac{1}{2} & 0 & \frac{-3(\lambda_0+1)}{4(\lambda+1)(\lambda+2)} & 0 \\ 0 & 0 & \frac{1}{4} & 0 & \frac{3(\lambda_0+2)}{4(\lambda+2)(\lambda+3)} \\ 0 & 0 & 0 & -\frac{1}{8} & 0 \end{vmatrix} \\ &= 16x^4 - \frac{48(\lambda_0+2)}{(\lambda+2)(\lambda+3)} x^2 + \frac{12\lambda_0(\lambda_0+1)}{\lambda(\lambda+1)(\lambda+2)(\lambda+3)}\,. \end{aligned}$$

As mentioned above, when $\lambda_0 = \lambda$, $\binom{-\lambda}{n} s_n(x) = C_n^{(\lambda)}(x)$. Therefore, we have the explicit expression and determinantal representation of $C_n^{(\lambda)}(x)$. For example, there holds

$$\begin{aligned} C_4^{(\lambda)}(x) &= \frac{128\lambda(\lambda+1)(\lambda+2)(\lambda+3)}{3} \begin{vmatrix} 1 & x & x^2 & x^3 & x^4 \\ 1 & 0 & \frac{1}{2(\lambda+1)} & 0 & \frac{3}{4(\lambda+1)(\lambda+2)} \\ 0 & -\frac{1}{2} & 0 & \frac{-3}{4(\lambda+2)} & 0 \\ 0 & 0 & \frac{1}{4} & 0 & \frac{3}{4(\lambda+3)} \\ 0 & 0 & 0 & -\frac{1}{8} & 0 \end{vmatrix} \\ &= \frac{2\lambda(\lambda+1)(\lambda+2)(\lambda+3)}{3} x^4 - 2\lambda(\lambda+1)(\lambda+2)x^2 + \frac{\lambda(\lambda+1)}{2}\,. \end{aligned}$$

Moreover, according to [1, Eqs. (22.5.34) and (22.5.36)], it is known that the *Chebyshev polynomials of the second kind* $U_n(x) = C_n^{(1)}(x)$ and the *Legendre polynomials* $P_n(x) = C_n^{(1/2)}(x)$. Thus, the expressions and determinants of $U_n(x)$ and $P_n(x)$ can be established immediately.

Using the (n, k) entries of the Riordan array $((1+t^2)^{-\lambda_0}, -2t/(1+t^2))$ and its inverse $\left(\left(\frac{2}{1+\sqrt{1-t^2}}\right)^{\lambda_0}, \frac{-t}{1+\sqrt{1-t^2}}\right)$, we establish the following inverse pair:

$$y_n = \sum_{k=0}^{[\frac{n}{2}]} \frac{\binom{-\lambda}{n-2k}\binom{-\lambda_0-n+2k}{k}}{\binom{-\lambda}{n}}(-2)^{n-2k}x_{n-2k}\,,$$

$$x_n = \sum_{k=0}^{[\frac{n}{2}]} \left(-\frac{1}{2}\right)^n \frac{\binom{-\lambda}{n-2k}\binom{\lambda_0+n}{k}}{\binom{-\lambda}{n}} \frac{\lambda_0+n-2k}{\lambda_0+n} y_{n-2k}\,.$$

For example, from the expression (6.4.18) of the polynomials $s_n(x)$, we have

$$x^n = \sum_{k=0}^{[\frac{n}{2}]} \left(-\frac{1}{2}\right)^n \frac{\binom{-\lambda}{n-2k}\binom{\lambda_0+n}{k}}{\binom{-\lambda}{n}} \frac{\lambda_0+n-2k}{\lambda_0+n} s_{n-2k}\,.$$

The polynomials of the Chebyshev case

Let us consider the polynomials of the Chebyshev case specifically. Now, $c_n = (-1)^n$, which can be obtained from $1/\binom{-\lambda}{n}$ by setting $\lambda = 1$. Let $(t_n(x))_{n\in\mathbb{N}}$ be the Sheffer sequence for the pair $(g(t), f(t))$, where

$$g(t) = \frac{1}{\sqrt{1-t^2}} \quad \text{and} \quad f(t) = \frac{-t}{1+\sqrt{1-t^2}}\,.$$

Since $g(\bar{f}(t)) = (1+t^2)/(1-t^2)$, the generating function of $(t_n(x))_{n\in\mathbb{N}}$ is

$$\sum_{n=0}^{\infty}(-1)^n t_n(x)t^n = \frac{1-t^2}{1-2xt+t^2}$$

(cf. [56, pp. 99–100]).

The corresponding Riordan array is $\left(\frac{1-t^2}{1+t^2}, \frac{-2t}{1+t^2}\right)$, which has the (n,k) entry

$$\begin{aligned}[x^k]t_n(x) &= \left[\frac{t^n}{c_n}\right]\frac{1}{c_k}\frac{1-t^2}{1+t^2}\left(\frac{-2t}{1+t^2}\right)^k \\ &= \frac{c_n}{c_k}(-2)^k[t^{n-k}]\left(\frac{1}{(1+t^2)^{k+1}} - \frac{t^2}{(1+t^2)^{k+1}}\right).\end{aligned}$$

Because

$$\begin{aligned}\frac{1}{(1+t^2)^{k+1}} - \frac{t^2}{(1+t^2)^{k+1}} &= \sum_{i=0}^{\infty}\binom{-k-1}{i}t^{2i} - \sum_{i=0}^{\infty}\binom{-k-1}{i}t^{2i+2} \\ &= 1+\sum_{i=1}^{\infty}\left(\binom{-k-1}{i}-\binom{-k-1}{i-1}\right)t^{2i} = 1+\sum_{i=1}^{\infty}(-1)^i\frac{k+2i}{i}\binom{k+i-1}{i-1}t^{2i}\,,\end{aligned}$$

we have

$$[x^k]t_n(x) = \begin{cases} 0\,, & \text{if } n-k \text{ is odd}\,,\\ (-2)^n\,, & \text{if } n-k=0\,,\\ \frac{c_n}{c_k}(-2)^k\frac{(-1)^{\frac{n-k}{2}}n}{\frac{n-k}{2}}\binom{k+\frac{n-k}{2}-1}{\frac{n-k}{2}-1}\,, & \text{if } n-k \text{ is even and } n\neq k\,,\end{cases}$$

from which the explicit expression of $t_n(x)$ can be established:

$$\begin{aligned} t_n(x) &= \sum_{\substack{k=0\\ n-k \text{ even}}}^{n-2}(-1)^{n-k}(-2)^k\frac{(-1)^{\frac{n-k}{2}}n}{\frac{n-k}{2}}\binom{k+\frac{n-k}{2}-1}{\frac{n-k}{2}-1}x^k+(-2)^n x^n\\ &= \sum_{k=1}^{[\frac{n}{2}]}(-1)^k\frac{n}{k}\binom{n-k-1}{k-1}(-2x)^{n-2k}+(-2)^n x^n\\ &= \sum_{k=0}^{[\frac{n}{2}]}(-1)^k\frac{n}{n-k}\binom{n-k}{k}(-2x)^{n-2k}\,. \end{aligned}$$

The inverse Riordan array is $\left(\frac{1}{\sqrt{1-t^2}}, \frac{-t}{1+\sqrt{1-t^2}}\right)$. To determine the (n,k) entry of the inverse, we first compute

$$\frac{1}{\sqrt{1-t^2}} = \sum_{i=0}^{\infty}\binom{-\frac{1}{2}}{i}(-1)^i t^{2i} = \sum_{i=0}^{\infty}\binom{2i}{i}\frac{1}{2^{2i}}t^{2i}\,,$$

then by Eq. (6.4.19), we have

$$\begin{aligned} a_{n,k} &= \left[\frac{t^n}{c_n}\right]\frac{1}{c_k}\frac{1}{\sqrt{1-t^2}}\left(\frac{-t}{1+\sqrt{1-t^2}}\right)^k = (-1)^n[t^{n-k}]\frac{1}{\sqrt{1-t^2}}\left(\frac{1}{1+\sqrt{1-t^2}}\right)^k\\ &= (-1)^n[t^{n-k}]\sum_{l=0}^{\infty}\sum_{j=0}^{l}\binom{2l-2j}{l-j}\frac{k(k+2j-1)_{j-1}}{2^{k+2l}j!}t^{2l}\\ &= \begin{cases} 0 & \text{if } n-k \text{ is odd}\,,\\ \left(-\frac{1}{2}\right)^n\sum_{j=0}^{\frac{n-k}{2}}\binom{n-k-2j}{\frac{n-k}{2}-j}\binom{k+2j}{j}\frac{k}{k+2j} & \text{if } n-k \text{ is even}\,.\end{cases} \end{aligned}$$

In particular, $a_{n,n} = (-\frac{1}{2})^n$. By Theorem 6.18, the determinantal representation of $t_n(x)$ can be obtained. For example, when $n = 4$, it is

$$t_4(x) = 1024\begin{vmatrix} 1 & x & x^2 & x^3 & x^4\\ 1 & 0 & \frac{1}{2} & 0 & \frac{3}{8}\\ 0 & -\frac{1}{2} & 0 & \frac{-3}{8} & 0\\ 0 & 0 & \frac{1}{4} & 0 & \frac{1}{4}\\ 0 & 0 & 0 & -\frac{1}{8} & 0 \end{vmatrix} = 16x^4 - 16x^2 + 2\,.$$

According to [52, Sect. 155] and [1, Eq. (22.9.9)], the polynomials $t_n(x)$ are related to $T_n(x)$, the *Chebyshev polynomials of the first kind*. In fact, $T_0(x) = t_0(x) = 1$ and $T_n(x) = (-1)^n t_n(x)/2$ for $n \geq 1$. Thus, the expression and determinantal representation of $T_n(x)$ can be established immediately.

If $c_n = (-1)^n$, the Sheffer sequence $(u_n(x))_{n\in\mathbb{N}}$ for the pair $\left(\frac{2-2\sqrt{1-t^2}}{t^2}, \frac{-t}{1+\sqrt{1-t^2}}\right)$ has the generating function

$$\sum_{n=0}^{\infty}(-1)^n u_n(x)t^n = \frac{1}{1-2xt+t^2},$$

and satisfies $(-1)^n u_n(x) = U_n(x)$, where $U_n(x)$ are the *Chebyshev polynomials of the second kind*. Note that $(u_n(x))_{n\in\mathbb{N}}$ also coincides with the $\lambda = \lambda_0 = 1$ case of the sequence $(s_n(x))_{n\in\mathbb{N}}$ proposed in the previous subsection, so we obtain the following expression directly:

$$u_n(x) = \sum_{k=0}^{[\frac{n}{2}]}\binom{-n+2k-1}{k}(-2x)^{n-2k}.$$

It is interesting to determine the connection constants between $(t_n(x))_{n\in\mathbb{N}}$ and $(u_n(x))_{n\in\mathbb{N}}$. Suppose $t_n(x) = \sum_{k=0}^{n} b_{n,k}u_k(x)$, then by Theorem 6.17, $b_{n,k}$ is the (n,k) entry of the Riordan array $(1-t^2, t)$ with respect to $c_n = (-1)^n$. By computation, we have $b_{n,k} = \delta_{n,k} - \delta_{n,k+2}$, then

$$t_n(x) = u_n(x) - u_{n-2}(x).$$

Next, suppose $u_n(x) = \sum_{k=0}^{n} c_{n,k}t_k(x)$; then $c_{n,k}$ is the (n,k) entry of the array $(1/(1-t^2), t)$. Thus, $c_{n,k} = 1$ if $n-k$ is even and 0 if $n-k$ is odd, which implies

$$u_n(x) = \sum_{\substack{k=0\\ n-k \text{ even}}}^{n} t_k(x) = \sum_{k=0}^{[\frac{n}{2}]} t_{n-2k}(x).$$

Now, we give an inverse pair by Theorem 6.16. Let

$$f(t) = \frac{-t}{1+\sqrt{1-t^2}}, \quad l(t) = \frac{2-2\sqrt{1-t^2}}{t}, \quad h(t) = \frac{2-2\sqrt{1-t^2}}{t^2}$$

From these series, we have $\bar{f}(t) = \frac{-2t}{1+t^2}$ and $\bar{l}(t) = \frac{4t}{4+t^2}$. Then

$$h(\bar{f}(t)) = 1+t^2, \quad l(\bar{f}(t)) = -2t; \quad \frac{1}{h(\bar{l}(t))} = \frac{4}{4+t^2}, \quad f(\bar{l}(t)) = -\frac{t}{2}.$$

The (n,k) entry of $(h(\bar{f}(t)), l(\bar{f}(t)))$ is

$$\begin{aligned} a_{n,k} &= \left[\frac{t^n}{c_n}\right](1+t^2)\frac{(-2)^k t^k}{c_k} = \frac{c_n}{c_k}(-2)^k[t^{n-k}](1+t^2) \\ &= \frac{c_n}{c_k}(-2)^k(\delta_{n,k}+\delta_{n,k+2}), \end{aligned}$$

that is,

$$a_{n,k} = \begin{cases} (-2)^n, & \text{if } n=k, \\ \frac{c_n}{c_k}(-2)^k, & \text{if } n=k+2, \\ 0, & \text{else}. \end{cases}$$

The (n,k) entry of $\left(\frac{1}{h(\bar{l}(t))}, f(\bar{l}(t))\right)$ is

$$\begin{aligned} b_{n,k} &= \left[\frac{t^n}{c_n}\right]\frac{4}{4+t^2}\left(-\frac{1}{2}\right)^k\frac{t^k}{c_k} = \frac{c_n}{c_k}\left(-\frac{1}{2}\right)^k[t^{n-k}]\sum_{i=0}^{\infty}\frac{(-1)^i}{2^{2i}}t^{2i} \\ &= \begin{cases} 0, & \text{if } n-k \text{ is odd}; \\ \frac{c_n}{c_k}(-1)^k(-1)^{\frac{n-k}{2}}\frac{1}{2^n}, & \text{if } n-k \text{ is even}. \end{cases} \end{aligned}$$

By Theorem 6.16, the following inverse relation holds:

$$y_n = (-2)^n x_n + \frac{c_n}{c_{n-2}}(-2)^{n-2}x_{n-2},$$

$$x_n = \sum_{\substack{k=0 \\ n-k \text{ even}}}^{n} \frac{c_n}{c_k}(-1)^k(-1)^{\frac{n-k}{2}}\frac{1}{2^n}y_k = \sum_{k=0}^{[\frac{n}{2}]}\frac{c_n}{c_{n-2k}}(-1)^{n-k}\frac{1}{2^n}y_{n-2k},$$

which, by setting $y_n/c_n \to y_n$ and $(-2)^n x_n/c_n \to x_n$, reduces to

$$y_n = x_n + x_{n-2} \quad \text{and} \quad x_n = \sum_{k=0}^{[\frac{n}{2}]}(-1)^k y_{n-2k}. \tag{6.4.20}$$

The polynomials of the Jacobi case

Define the rising factorials $\langle x\rangle_n$ by $\langle x\rangle_0 = 1$ and $\langle x\rangle_n = x(x+1)\cdots(x+n-1)$ for $n = 1, 2, \ldots$, and set

$$c_n = \frac{\langle 1+\alpha\rangle_n n!}{\left\langle\frac{1+\alpha+\beta}{2}\right\rangle_n\left\langle\frac{2+\alpha+\beta}{2}\right\rangle_n} = \frac{4^n(\alpha+n)_n n!}{(\alpha+\beta+2n)_{2n}}.$$

Let $(J_n(x))_{n\in\mathbb{N}}$ be the Sheffer sequence for the pair $(g(t), f(t))$, where

$$g(t)=\left(\frac{2}{1+\sqrt{1+2t}}\right)^{1+\alpha+\beta} \quad \text{and} \quad f(t)=\frac{t}{1+t+\sqrt{1+2t}};$$

then $\bar{f}(t)=2t/(1-t)^2$, $g(\bar{f}(t))=(1-t)^{1+\alpha+\beta}$, and the generating function of $J_n(x)$ is

$$\sum_{n=0}^{\infty} J_n(x)\frac{t^n}{c_n}=(1-t)^{-1-\alpha-\beta}{}_2F_1\left[\begin{matrix}\frac{1+\alpha+\beta}{2},\ \frac{2+\alpha+\beta}{2}\\ 1+\alpha\end{matrix}\middle|\ \frac{2xt}{(1-t)^2}\right],$$

where the hypergeometric series is defined by

$${}_pF_q\left[\begin{matrix}a_1,a_2,\cdots,a_p\\ b_1,b_2,\cdots,b_q\end{matrix}\middle|\ z\right]=\sum_{k=0}^{\infty}\frac{\langle a_1\rangle_k\langle a_2\rangle_k\cdots\langle a_p\rangle_k}{\langle b_1\rangle_k\langle b_2\rangle_k\cdots\langle b_q\rangle_k}\frac{z^k}{k!}.$$

Thus, it can be seen that the polynomials $J_n(x)$ are related to the *classical Jacobi polynomials* $P_n^{(\alpha,\beta)}(x)$ by the formula

$$J_n(x)=\frac{\langle 1+\alpha+\beta\rangle_n}{\langle 1+\alpha\rangle_n}c_nP_n^{(\alpha,\beta)}(x+1)=4^nn!\frac{(\alpha+\beta+n)_n}{(\alpha+\beta+2n)_{2n}}P_n^{(\alpha,\beta)}(x+1)$$

(cf. [52, Sect. 132] and [56, Sect. 10]).

Now, the related Riordan array is $((1-t)^{-(1+\alpha+\beta)}, 2t/(1-t)^2)$, and its (n,k) entry is

$$[x^k]J_n(x)=\left[\frac{t^n}{c_n}\right]\frac{1}{c_k}\frac{1}{(1-t)^{1+\alpha+\beta}}\left(\frac{2t}{(1-t)^2}\right)^k=\frac{c_n}{c_k}2^k\binom{\alpha+\beta+n+k}{n-k}.$$

Thus, the following expression holds:

$$J_n(x)=\sum_{k=0}^{n}\binom{n}{k}\frac{(\alpha+\beta+n+k)_{n-k}(\alpha+n)_{n-k}}{(\alpha+\beta+2n)_{2n-2k}}2^{2n-k}x^k.$$

The inverse Riordan array is

$$\left(\left(\frac{2}{1+\sqrt{1+2t}}\right)^{1+\alpha+\beta},\frac{t}{1+t+\sqrt{1+2t}}\right),$$

which has the (n,k) entry

$$\begin{aligned} a_{n,k} &= \left[\frac{t^n}{c_n}\right]\frac{1}{c_k}\left(\frac{2}{1+\sqrt{1+2t}}\right)^{1+\alpha+\beta}\left(\frac{2t}{(1+\sqrt{1+2t})^2}\right)^k \\ &= \frac{c_n}{c_k}2^{1+\alpha+\beta+k}[t^{n-k}](1+\sqrt{1+2t})^{-(1+\alpha+\beta+2k)} \\ &= \frac{c_n}{c_k}(-1)^{n-k}\frac{(1+\alpha+\beta+2k)(\alpha+\beta+2n)_{n-k-1}}{2^n(n-k)!} \\ &= \frac{(-1)^{n-k}}{2^n}\frac{c_n}{c_k}\frac{1+\alpha+\beta+2k}{1+\alpha+\beta+2n}\binom{1+\alpha+\beta+2n}{n-k}. \end{aligned}$$

In particular, $a_{n,n} = 1/2^n$. Then the determinantal representation of $J_n(x)$ can be established. For example, we have

$$\begin{aligned} J_4(x) &= 1024\begin{vmatrix} 1 & x & x^2 & x^3 & x^4 \\ 1 & -\frac{2(\alpha+1)}{\alpha+\beta+2} & \frac{4(\alpha+2)_2}{(\alpha+\beta+3)_2} & -\frac{8(\alpha+3)_3}{(\alpha+\beta+4)_3} & \frac{16(\alpha+4)_4}{(\alpha+\beta+5)_4} \\ 0 & \frac{1}{2} & -\frac{2(\alpha+2)}{\alpha+\beta+4} & \frac{6(\alpha+3)_2}{(\alpha+\beta+5)_2} & -\frac{16(\alpha+4)_3}{(\alpha+\beta+6)_3} \\ 0 & 0 & \frac{1}{4} & -\frac{3(\alpha+3)}{2(\alpha+\beta+6)} & \frac{6(\alpha+4)_2}{(\alpha+\beta+7)_2} \\ 0 & 0 & 0 & \frac{1}{8} & -\frac{\alpha+4}{\alpha+\beta+8} \end{vmatrix} \\ &= 16x^4 + \frac{128(\alpha+4)}{\alpha+\beta+8}x^3 + \frac{384(\alpha+4)_2}{(\alpha+\beta+8)_2}x^2 + \frac{512(\alpha+4)_3}{(\alpha+\beta+8)_3}x + \frac{256(\alpha+4)_4}{(\alpha+\beta+8)_4}. \end{aligned}$$

Moreover, by the relations between $J_n(x)$ and $P_n^{(\alpha,\beta)}(x)$, the expression and determinantal representation of the Jacobi polynomials $P_n^{(\alpha,\beta)}(x)$ can also be obtained.

Using Theorem 6.16, we give here an inverse relation. Let

$$f(t) = \frac{1+t-\sqrt{1+2t}}{t}, \quad l(t) = \frac{2t}{1+\sqrt{1+2t}}, \quad h(t) = \left(\frac{2}{1+\sqrt{1+2t}}\right)^{1+\alpha+\beta}.$$

By computation, we have $\bar{f}(t) = \frac{2t}{(1-t)^2}$ and $\bar{l}(t) = \frac{t^2+2t}{2}$. Then

$$\begin{aligned} h(\bar{f}(t)) &= (1-t)^{1+\alpha+\beta}, \quad l(\bar{f}(t)) = \frac{2t}{1-t}, \\ \frac{1}{h(\bar{l}(t))} &= \left(\frac{2+t}{2}\right)^{1+\alpha+\beta}, \quad f(\bar{l}(t)) = \frac{t}{2+t}. \end{aligned}$$

The (n,k) entry of $(h(\bar{f}(t)), l(\bar{f}(t)))$ is

$$a_{n,k} = \left[\frac{t^n}{c_n}\right]\frac{1}{c_k}(1-t)^{1+\alpha+\beta}\left(\frac{2t}{1-t}\right)^k = 2^k\frac{c_n}{c_k}\binom{1+\alpha+\beta-k}{n-k}(-1)^{n-k},$$

and the (n,k) entry of $\left(\frac{1}{h(\bar{l}(t))}, f(\bar{l}(t))\right)$ is

$$b_{n,k} = \left[\frac{t^n}{c_n}\right]\frac{1}{c_k}\left(\frac{2+t}{2}\right)^{1+\alpha+\beta}\left(\frac{t}{2+t}\right)^k = \frac{c_n}{c_k}\binom{1+\alpha+\beta-k}{n-k}\frac{1}{2^n}.$$

Therefore, we have the inverse pair

$$y_n = \sum_{k=0}^{n} \frac{c_n}{c_k}\binom{1+\alpha+\beta-k}{n-k}(-1)^{n-k}2^k x_k,$$

$$x_n = \sum_{k=0}^{n} \frac{c_n}{c_k}\binom{1+\alpha+\beta-k}{n-k}\frac{1}{2^n}y_k.$$

Upon setting $y_n/c_n \to y_n$, $2^n x_n/c_n \to x_n$ and $-2-\alpha-\beta \to p$, this inverse pair reduces to

$$y_n = \sum_{k=0}^{n}\binom{n+p}{n-k}x_k \quad \text{and} \quad x_n = \sum_{k=0}^{n}\binom{n+p}{n-k}(-1)^{n-k}y_k, \tag{6.4.21}$$

which is the $c = 1$ case of [54, p. 69, Table 2.6, Entry 5].

The polynomials of the q-case

Let

$$c_n = \frac{(1-q)(1-q^2)\cdots(1-q^n)}{(1-q)^n} = \frac{(q;q)_n}{(1-q)^n} = n!_q,$$

which is the q-analog of $n!$, and let

$$\begin{bmatrix} n \\ k \end{bmatrix}_q = \frac{c_n}{c_k c_{n-k}} = \frac{(1-q)\cdots(1-q^n)}{(1-q)\cdots(1-q^k)(1-q)\cdots(1-q^{n-k})} = \frac{(q;q)_n}{(q;q)_k(q;q)_{n-k}},$$

which are called the *q-binomial coefficients* (cf. [2, Chap. 10]).

Define the sequence $([x]_{a,n})_{n\in\mathbb{N}}$ by

$$[x]_{a,0} = 1 \quad \text{and} \quad [x]_{a,n} = (x-a)(x-qa)\cdots(x-q^{n-1}a);$$

then according to [56, p. 108], $([x]_{a,n})_{n\in\mathbb{N}}$ is the Appell sequence for $\varepsilon_a(t)$, and the related Riordan array is $(1/\varepsilon_a(t), t)$. Using

$$\frac{1}{\varepsilon_a(t)} = \sum_{k=0}^{\infty}\frac{(1-q)^k}{(1-q)(1-q^2)\cdots(1-q^k)}q^{\binom{k}{2}}(-at)^k = \sum_{k=0}^{\infty}q^{\binom{k}{2}}(-at)^k\frac{1}{c_k},$$

we can determine the (n,k) entry of this array:

$$[x^k][x]_{a,n} = \left[\frac{t^n}{c_n}\right] \frac{1}{\varepsilon_a(t)} \frac{t^k}{c_k} = \frac{c_n}{c_k}[t^{n-k}] \sum_{i=0}^{\infty} q^{\binom{i}{2}} (-at)^i \frac{1}{c_i} = \begin{bmatrix} n \\ k \end{bmatrix}_q q^{\binom{n-k}{2}} (-a)^{n-k},$$

which indicates that

$$[x]_{a,n} = \sum_{k=0}^{n} \begin{bmatrix} n \\ k \end{bmatrix}_q (-a)^{n-k} q^{\binom{n-k}{2}} x^k . \tag{6.4.22}$$

By computation, the (n, k) entry of the inverse array $(\varepsilon_a(t), t)$ is

$$a_{n,k} = \left[\frac{t^n}{c_n}\right] g(t)\frac{t^k}{c_k} = \frac{c_n}{c_k}[t^{n-k}]\varepsilon_a(t) = \frac{c_n}{c_k}\frac{a^{n-k}}{c_{n-k}} = \begin{bmatrix} n \\ k \end{bmatrix}_q a^{n-k} .$$

Thus, we obtain from Theorem 6.18 the determinantal representation. For example, there holds

$$\begin{aligned}[x]_{a,4} &= \begin{vmatrix} 1 & x & x^2 & x^3 & x^4 \\ 1 & a & a^2 & a^3 & a^4 \\ 0 & 1 & (q+1)a & (q^2+q+1)a^2 & (q^3+q^2+q+1)a^3 \\ 0 & 0 & 1 & (q^2+q+1)a & (q^2+q+1)(q^2+1)a^2 \\ 0 & 0 & 0 & 1 & (q^3+q^2+q+1)a \end{vmatrix} \\ &= (x-a)(x-qa)(x-q^2a)(x-q^3a) .\end{aligned}$$

The inverse of the Riordan array $(1/\varepsilon_a(t), t)$ is $(\varepsilon_a(t), t)$. Thus, based on their (n, k) entries, the following inverse relation holds:

$$y_n = \sum_{k=0}^{n} \begin{bmatrix} n \\ k \end{bmatrix}_q (-a)^{n-k} q^{\binom{n-k}{2}} x_k \quad \text{and} \quad x_n = \sum_{k=0}^{n} \begin{bmatrix} n \\ k \end{bmatrix}_q a^{n-k} y_k .$$

As an example, from (6.4.22), we have

$$x^n = \sum_{k=0}^{n} \begin{bmatrix} n \\ k \end{bmatrix}_q a^{n-k} [x]_{a,k} .$$

Define the *q-Bernoulli polynomials* $\mathfrak{B}_n(x)$ by the generating function

$$\sum_{k=0}^{\infty} \mathfrak{B}_k(x) \frac{t^k}{c_k} = \frac{t}{\varepsilon_1(t) - 1} \varepsilon_x(t) .$$

Then $\mathfrak{B}_n(x)$ form the Appell sequence for $(\varepsilon_1(t) - 1)/t$, and $[x^k]\mathfrak{B}_n(x)$ is the (n, k) entry of the Riordan array $(t/(\varepsilon_1(t) - 1), t)$. Since

$$\frac{t}{\varepsilon_1(t)-1}=\frac{1}{\varepsilon_1(t)}\sum_{k=0}^{\infty}\frac{\mathfrak{B}_k(1)t^k}{c_k}=\sum_{i=0}^{\infty}\frac{q^{\binom{i}{2}}(-t)^i}{c_i}\sum_{j=0}^{\infty}\frac{\mathfrak{B}_j(1)t^j}{c_j}$$
$$=\sum_{n=0}^{\infty}\sum_{i=0}^{n}\begin{bmatrix}n\\ i\end{bmatrix}_q q^{\binom{i}{2}}(-1)^i\mathfrak{B}_{n-i}(1)\frac{t^n}{c_n},$$

we have

$$[x^k]\mathfrak{B}_n(x)=\left[\frac{t^n}{c_n}\right]\frac{t}{\varepsilon_1(t)-1}\frac{t^k}{c_k}=\begin{bmatrix}n\\ k\end{bmatrix}_q\sum_{i=0}^{n-k}\begin{bmatrix}n-k\\ i\end{bmatrix}_q q^{\binom{i}{2}}(-1)^i\mathfrak{B}_{n-k-i}(1),$$

which implies that

$$\mathfrak{B}_n(x)=\sum_{k=0}^{n}\begin{bmatrix}n\\ k\end{bmatrix}_q\sum_{i=0}^{n-k}\begin{bmatrix}n-k\\ i\end{bmatrix}_q q^{\binom{i}{2}}(-1)^i\mathfrak{B}_{n-k-i}(1)x^k$$
$$=\sum_{k=0}^{n}\begin{bmatrix}n\\ k\end{bmatrix}_q\sum_{i=k}^{n}\begin{bmatrix}n-k\\ i-k\end{bmatrix}_q q^{\binom{i-k}{2}}(-1)^{i-k}\mathfrak{B}_{n-i}(1)x^k$$
$$=\sum_{i=0}^{n}\begin{bmatrix}n\\ i\end{bmatrix}_q\mathfrak{B}_{n-i}(1)\sum_{k=0}^{i}\begin{bmatrix}i\\ k\end{bmatrix}_q(-1)^{i-k}q^{\binom{i-k}{2}}x^k=\sum_{i=0}^{n}\begin{bmatrix}n\\ i\end{bmatrix}_q\mathfrak{B}_{n-i}(1)[x]_i,$$

where the last step comes from (6.4.22). By computation, the (n,k) entry of the inverse array $((\varepsilon_1(t)-1)/t,t)$ is

$$a_{n,k}=\left[\frac{t^n}{c_n}\right]\frac{\varepsilon_1(t)-1}{t}\frac{t^k}{c_k}=\frac{c_n}{c_k}[t^{n-k}]\frac{\varepsilon_1(t)-1}{t}$$
$$=\frac{c_n}{c_kc_{n-k+1}}=\frac{n!_q}{k!_q(n-k+1)!_q},$$

which, together with Theorem 6.18, gives the determinantal representation of $\mathfrak{B}_n(x)$. For example, we have

$$\mathfrak{B}_4(x)=\begin{vmatrix}1 & x & x^2 & x^3 & x^4\\ 1 & \frac{1}{q+1} & \frac{1}{q^2+q+1} & \frac{1}{q^3+q^2+q+1} & \frac{1}{q^4+q^3+q^2+q+1}\\ 0 & 1 & 1 & 1 & 1\\ 0 & 0 & 1 & \frac{q^2+q+1}{q+1} & q^2+1\\ 0 & 0 & 0 & 1 & q^2+1\end{vmatrix}$$
$$=x^4-(q^2+1)x^3+\frac{(q^2+1)q^2}{q+1}x^2-\frac{(q-1)q^3}{q+1}x$$
$$+\frac{(q^6-q^4-2q^3-q^2+1)q^4}{(q+1)^2(q^2+q+1)(q^4+q^3+q^2+q+1)}.$$

Note that when $q \to 1$, we obtain the corresponding result for the classical Bernoulli polynomials.

6.5 Double Riordan Arrays and Sheffer Polynomial Pairs

A Riordan array is an infinite lower triangular matrix defined by two generating functions g and f, where $g(t)$ is an invertible series and $f(t)$ is a delta series. The kth column of the matrix has the generating function gf^k. In a double Riordan array, there are two generating functions h_1 and h_2 instead of the one function f so that the columns starting from the left of the array have the generating functions alternately using h_1 and h_2. When g is an even function and h_1 and h_2 are odd functions this class of Riordan arrays form a group, called the double Riordan group, where $g(t)$ is an invertible series and $h_1(t)$ and $h_2(t)$ are delta series.

The *ECO* (an abbreviation of enumeration of combinatorial objects) is a method for enumerating some classes of combinatorial objects. The basic idea of this method is to use an operator that performs a "local expansion" on the objects (cf. [3]). Inspired by the Fibonacci tree shown in the recent book, *Catalan Numbers*, by Richard Stanley, we present a combinatorial way to construct double Riordan arrays by using the ECO technique. We will give some recursive constructions for the double Riordan group. We also develop the sequence characterizations of double Riordan arrays and examine the subgroups. As an extension of the Pascal-Fibonacci triangle, the compression forms of double Riordan arrays are defined. Those results are also extended to the case of higher-order Riordan arrays. The *pairs of Sheffer polynomials* and *pairs of summation formulas* associated with double Riordan arrays are defined and discussed. The major results of this section are from the reference [38].

ECO method and Riordan arrays

Let R be a Riordan array $(d(t), h(t)) = (d_{n,k})_{n,k\geq 0}$ with the generating functions $A(t)$ and $Z(t)$ of its A-sequence and Z-sequence, respectively. Then,

$$d(t) = \frac{d_{0,0}}{1 - tZ(h(t))} \quad \text{and} \quad h(t) = tA(h(t)). \tag{6.5.1}$$

We may write (6.5.1) as

$$\frac{d(t) - d_{0,0}}{t} = d(t)Z(h(t)) \quad \text{and} \quad \frac{d(t)h^n(t)}{t} = d(t)h^{n-1}(t)A(h(t)). \tag{6.5.2}$$

Denote the *upper shift matrix* by U, i.e.,

$$U = (\delta_{i+1,j})_{i,j\geq 0} = \begin{bmatrix} 0 & 1 & 0 & 0 & 0 & \cdots \\ 0 & 0 & 1 & 0 & 0 & \cdots \\ 0 & 0 & 0 & 1 & 0 & \cdots \\ 0 & 0 & 0 & 0 & 1 & \cdots \\ \vdots & \vdots & \vdots & \vdots & \vdots & \ddots \end{bmatrix}$$

and

$$P = \begin{bmatrix} z_0 & a_0 & 0 & 0 & 0 & \cdots \\ z_1 & a_1 & a_0 & 0 & 0 & \cdots \\ z_2 & a_2 & a_1 & a_0 & 0 & \cdots \\ z_3 & a_3 & a_2 & a_1 & a_0 & \cdots \\ \vdots & \vdots & \vdots & \vdots & \vdots & \ddots \end{bmatrix} = \left(Z(t), A(t), tA(t), t^2A(t), \ldots\right), \tag{6.5.3}$$

where the rightmost expression is the representation of P by using its column generating functions. Let M be a matrix. Then UM is the matrix obtained from M by removing the first row of M and elevating the remaining rows of M up one row. Thus, (6.5.2) can be written in a matrix form by using upper shift matrix U:

$$U(d(t), h(t)) = (d(t), h(t))P \tag{6.5.4}$$

because its left-hand side and right-hand side are the same. In fact, due to (6.5.2) and the fundamental theorem of Riordan arrays (cf. [60]) we can write

$$\text{LHS} = U(d(t), h(t)) = \left(\frac{d(t) - d_{0,0}}{t}, \frac{d(t)h(t)}{t}, \frac{d(t)h^2(t)}{t}, \frac{d(t)h^3(t)}{t}, \cdots\right)$$

and

$$\begin{aligned} \text{RHS} &= (d(t), h(t))P \\ &= \left(d(t)Z(h(t)), d(t)A(h(t)), d(t)h(t)A(h(t)), d(t)h^2(t)A(h(t)), \cdots\right), \end{aligned}$$

where the Riordan array $(d(t), h(t))$ can be written as

$$(d(t), d(t)h(t), d(t)h^2(t), \ldots)$$

in terms of the generating functions of the columns of $(d(t), h(t))$. Thus, two sides of (6.5.4) are equal. Conversely, if $R = (d_{n,k})_{n,k\geq 0}$ is a lower triangle matrix satisfying $UR = RP$, where P is defined by (6.5.3), then we have

$$\begin{aligned} d_{n+1,0} &= z_0 d_{n,0} + z_1 d_{n,1} + \cdots + z_n d_{n,n} \quad for\ n \geq 0, \\ d_{n+1,k+1} &= a_0 d_{n,k} + a_1 d_{n,k+1} + \cdots + a_{n-k} d_{n,n}, \quad for\ n \geq k \geq 0, \end{aligned}$$

which implies that R is a Riordan array because of the sequence characterization of Riordan arrays shown in (6.5.1). Hence, we call $P = (p_{n,k})_{n,k\geq 0}$ the *characterization matrix* and have the following matrix characterization of Riordan arrays. If $p_{n,k} \in \mathbb{N}_0$, then P is also named production matrix in [25]. The above presentation can be surveyed in the following theorem.

Theorem 6.29 *Let $d(t) \in \mathcal{F}_0$, $h(t) \in \mathcal{F}_1$, and $U = (\delta_{i+1,j})_{i,j\geq 0}$. Then a lower triangle matrix $R = (d_{n,k})_{n,k\geq 0} = (d(t), h(t))$ is a Riordan array if and only if*

$$UR = RP, \tag{6.5.5}$$

or equivalently,

$$P = R^{-1}UR, \tag{6.5.6}$$

where P is defined by (6.5.3).

Another proof of (6.5.6) by using a different approach can be found on Page 148–149 of Barry [8].

If there exists a solution of $MS_M = UM$, then its solution S_M is called a Stieltjes matrix of M (cf. for example [61]). If $M = R$, a proper Riordan array, then it is a lower triangular matrix with non-zero entries on the main diagonal. Thus, its Stieltjes matrix S_R always exists and it is unique. In other words, Equation (6.5.5) has a unique solution $P = R^{-1}UR$ when R is a proper Riordan array. Therefore, the characterization matrix P defined by (6.5.5) is a Stieltjes matrix. An application of the Stieltjes matrix in the LDL^T decomposition of Hankel matrix is presented in [51].

If the Stieltjes matrix $P = (p_{n,k})_{n,k\geq 0}$ has non-negative integer entries, i.e., $p_{n,k} \in \mathbb{N}_0$, then P is also called a production matrix in [25]. Hence, Theorem 6.29 is reduced to Proposition 3.2 of [25]. More precisely, denoting the so-called *ECO matrix* induced by P by $A_P := (u^T, u^T P, u^T P^2, \ldots)^T \equiv (d_{n,k})_{n,k\geq 0}$, [25] shows

$$d_{n+1,k} = \sum_{j\geq 0} d_{n,j} p_{j,k} = d_{n,0} p_{0,k} + d_{n,1} p_{1,k} + d_{n,2} p_{2,k} + \cdots \tag{6.5.7}$$

for $n \geq 0$, and $d_{0,k} = \delta_{0,k}$, the Kronecker delta. Thus, Proposition 3.2 of [25] presents that A_P is a Riordan array if and only if P can be written as in (6.5.3). One may see that Theorem 6.29 is an extension of Proposition 3.2 of [25] for an arbitrary Stieltjes matrix P with the form (6.5.3). In other words, in Theorem 6.29, we do not need the entries of P to be non-negative.

The P-matrix of a Riordan array R and R itself can be interpreted by a methodology of the enumeration of combinatorial objects abbreviated as *ECO*. The ECO method is a constructive method to produce all the objects of a given class, according to the growth of a certain parameter (in terms of the size) of the objects. A complete description of the ECO method and its applications for the enumeration of several classes of combinatorial objects is given in [3]. The roots of the ECO method can be traced back to the paper [18], where the authors study Baxter permutations: for the first time, a combinatorial construction is presented and described by means of a generating tree, as it usually happens for every ECO construction. If an ECO construction is sufficiently regular, then it is often possible to be described by using a succession rule, whose definition was first introduced by Julian West in [74]. Some algebraic properties of succession rules have been determined in [28].

A *succession rule* is a formal system consisting of an axiom (a), $a \in \mathbb{N}$, and a set of productions

$$\left\{(k_j) \to (e_1(k_j))(e_2(k_j)) \cdots (e_j(k_j)) : j \in \mathbb{N}\right\},$$

where $e_i : \mathbb{N} \to \mathbb{N}$ for deriving the successors $(e_1(k))(e_2(k)) \cdots (e_k(k))$ of any given label (k), $k \in \mathbb{N}$. Thus, we may write a succession rule, denoted by Ω, as

$$\Omega : \begin{cases} (a) \\ (k) \to (e_1(k))(e_2(k)) \cdots (e_k(k)). \end{cases} \tag{6.5.8}$$

(a), (k), and $(e_i(k))$ $(a, k, e_i(k) \in \mathbb{N})$ are called the labels of Ω. The succession rule can be presented as a *generating rooted tree* called a *generating tree*, whose vertices are the labels of Ω, where (a) is the label of the root, and each node labeled (k) has k sons labeled by $(e_1(k)), \ldots, (e_k(k))$, respectively. A study concerning the connection between generating trees and Riordan arrays can be found in [49]. Denote by f_n the number of nodes at level n in the generating tree determined by Ω. We call $\{f_n\}_{n\geq 0}$ the sequence $\{f_n\}_{n\geq 0}$ associated with Ω. In [24] the authors represent the succession rule Ω by an infinite matrix $P = (p_{i,j})_{i,j\geq 0}$, called the *production matrix* of the succession rule Ω, where $p_{i,j}$ is the number of labels ℓ_j produced by label ℓ_i, and $\{(\ell_k)\}_{k\geq 0}$ is the label set of Ω with the axiom label ℓ_0. Simply speaking, if Ω containing root (a) satisfies succession rule $(a) \to (a) \ldots (a)(a+1) \ldots (a+1) \ldots (a+m) \ldots (a+m)$, which we shorten it to $(a)^{n_1}(a+1)^{n_2} \ldots (a+m)^{n_m}$, then the first row of P is $(n_1, n_2, \ldots, n_m, 0, \ldots)$, where n_j $(j = 1, 2, \ldots, m)$ may be zero, say $n_k = 0$ if $(a+k)$ is not a label. Similarly, we may write other rows of P row by row.

In [25] the following result is proved, which can be viewed as a particular case of (6.5.6) when the A- and Z- sequences are non-negative sequences. The case of not non-negative A- and/or Z-sequence is discussed in [36].

Theorem 6.30 *[25] Let P be an infinite production matrix and let A_P be the ECO matrix induced by P satisfying* (6.5.6), *i.e., $A_P P = U A_P$. Then A_P is a Riordan matrix if and only if P is of the form*

$$P = \begin{bmatrix} z_0 & a_0 & 0 & 0 & 0 & \cdots \\ z_1 & a_1 & a_0 & 0 & 0 & \cdots \\ z_2 & a_2 & a_1 & a_0 & 0 & \cdots \\ z_3 & a_3 & a_2 & a_1 & a_0 & \cdots \\ \vdots & \vdots & \vdots & \vdots & \vdots & \ddots \end{bmatrix}, \tag{6.5.9}$$

where $z_j, a_j \geq 0$ *for* $j = 0, 1, \ldots$.

As an example, let us consider the Fibonacci tree F_1 shown in Horibe [42], which satisfies the following succession rule:

$$\Omega : \begin{cases} (a) \\ a(k) \to a(k)a(k+1). \end{cases} \tag{6.5.10}$$

The P matrix obtained from the succession rule is

$$P = \begin{bmatrix} 1 & 1 & 0 & 0 & 0 & \cdots \\ 0 & 1 & 1 & 0 & 0 & \cdots \\ 0 & 0 & 1 & 1 & 0 & \cdots \\ 0 & 0 & 0 & 1 & 1 & \cdots \\ \vdots & \vdots & \vdots & \vdots & \vdots & \ddots \end{bmatrix}. \tag{6.5.11}$$

From (6.5.6), one may find that the corresponding Riordan array R is the Pascal matrix $(1/(1-t), t/(1-t))$:

$$R = \begin{bmatrix} 1 & 0 & 0 & 0 & 0 & \cdots \\ 1 & 1 & 0 & 0 & 0 & \cdots \\ 1 & 2 & 1 & 0 & 0 & \cdots \\ 1 & 3 & 3 & 1 & 0 & \cdots \\ \vdots & \vdots & \vdots & \vdots & \vdots & \ddots \end{bmatrix}. \tag{6.5.12}$$

The sum of the numbers along the parallel lines in the direction of $(1, 1)$ are the *Fibonacci numbers*, $\{1, 1, 2, 3, 5, 8, \ldots\}$.

Succession rules can generate a wider class of lower triangle matrices that do not belong to Riordan array class. For instance, in Stanley [68], the Fibonacci tree F_2 is defined to be the rooted tree with root v such that the root has degree one, the child of every vertex of degree one has degree two, and the two children of every vertex of degree two have degrees one and two.

The succession rule Ω for the Fibonacci tree is

$$\Omega : \begin{cases} (a_{2l}) \to (a_{2l})(a_{2l+1}) \\ (a_{2l+1}) \to (a_{2l+2}), \end{cases} \tag{6.5.13}$$

for $\ell = 0, 1, 2, \ldots$. If we count the numbers of (a_k) on the nth level, $n \geq 0$, generated from the above succession rule and present them as the entry $d_{n,k}$ in a matrix, then this succession rule generates the following lower triangle matrix $\hat{R}$:

$$\hat{R} = \begin{bmatrix} 1\,0\,0\,0\,0\,0 & \dots \\ 1\,1\,0\,0\,0\,0 & \dots \\ 1\,1\,1\,0\,0\,0 & \dots \\ 1\,1\,2\,1\,0\,0 & \dots \\ 1\,1\,3\,2\,1\,0 & \dots \\ 1\,1\,4\,3\,3\,1 & \dots \\ \vdots\,\vdots\,\vdots\,\vdots\,\vdots\,\vdots & \ddots \end{bmatrix}. \tag{6.5.14}$$

The matrix $\hat{R}$ is called the *Pascal-Fibonacci array*. The row sums of $\hat{R}$ are the Fibonacci numbers. Obviously, $\hat{R}$ is not a Riordan array. We will see that it is the compression of a double Riordan array, which is a type of quasi-double Riordan array. We now explain what is a double Riordan array and what is its compression.

In the construction of a Riordan array, one multiplier function h is used to multiply a column to obtain the next column. Suppose alternating rules are applied to generate an infinite matrix similar to a Riordan array. To consider this case, one may use two multiplier functions, denoted by h_1 and h_2, respectively. Davenport, Shapiro, and Woodson [27] require that d be an even function and that h_1 and h_2 be odd functions so that they can develop an analog of the fundamental theorem and thus obtain a group structure. Let $d(t) = \sum_{k=0}^{\infty} d_{2k}t^{2k}$, $h_1(t) = \sum_{k=0}^{\infty} h_{1,2k+1}t^{2k+1}$, and $h_2(t) = \sum_{k=0}^{\infty} h_{2,2k+1}t^{2k+1}$, where $h_{1,1}, h_{2,1} \neq 0$. Then the double Riordan matrix in terms of $d(t)$, $h_1(t)$, and $h_2(t)$, denoted by $(d; h_1, h_2)$, is defined by the generating function of its columns as

$$(d, dh_1, dh_1h_2, dh_1^2h_2, dh_1^2h_2^2, \dots).$$

We now define a formal power series $\sqrt{h_1h_2}$ from h_1 and h_2. Denote $g = \sqrt{h_1h_2}$, $g(t) = \sum_{k=0}^{\infty} g_k t^{2k+1}$. Then

$$g_0 = \sqrt{h_{1,1}h_{2,1}}, \quad g_1 = \frac{1}{g_0}(h_{1,1}h_{2,3} + h_{1,3}h_{2,1}),$$
$$g_n = \frac{1}{g_0}\left(\sum_{j=0}^{n} h_{1,2(n-j)+1}h_{2,2j+1} - \sum_{j=1}^{n-1} g_j g_{n-j}\right), \tag{6.5.15}$$

where g_n can be found recursively for all $n \geq 2$. It can be seen that $g^2 = h_1h_2$ and $g = \sqrt{h_1h_2} \in \mathcal{F}_1$. There are two cases of the *first fundamental theorem of double Riordan arrays (FTDRA)* (cf. [27]):

$$(d; h_1, h_2)A(t) = B(t), \tag{6.5.16}$$

where for $A(t) = \sum_{k\geq 0} a_{2k}t^{2k}$ and $A(t) = \sum_{k\geq 0} a_{2k+1}t^{2k+1}$, we have

$$B(t) = dA(\sqrt{h_1h_2}) \quad \text{and} \quad B(t) = d\sqrt{h_1/h_2}A(\sqrt{h_1h_2}),$$

respectively, where $\sqrt{h_1/h_2} := h_2\sqrt{h_1h_2}$. Based on the fundamental theorem of double Riordan arrays, we may define a multiplication of two double Riordan arrays as

$$(d; h_1, h_2)(g; f_1, f_2) = (dg(\sqrt{h_1h_2}); \sqrt{h_1/h_2}f_1(\sqrt{h_1h_2}), \sqrt{h_2/h_1}f_2(\sqrt{h_1h_2})), \tag{6.5.17}$$

where $g(t) = \sum_{k=0}^{\infty} g_{2k}t^{2k}$, $f_1(t) = \sum_{k=0}^{\infty} f_{1,2k+1}t^{2k+1}$, and $f_2(t) = \sum_{k=0}^{\infty} f_{2,2k+1}\, t^{2k+1}$. The collection of all double Riordan arrays associated with the multiplication defined above form a group called the *double Riordan group*, which is denoted by $\mathcal{DR}$, where the identity is $(1; t, t)$. From [27], $\mathcal{S} := \{(d; h_1, h_2) \in \mathcal{DR} : d = 1\}$, $\mathcal{B}_1 := \{(d; h_1, h_2) \in \mathcal{DR} : h_1 = td\}$, $\mathcal{B}_2 := \{(d; h_1, h_2) \in \mathcal{DR} : h_2 = td\}$, and $\mathcal{A} := \{(d; t, t) \in \mathcal{DR}\}$ are *associated, type-1 Bell, type-2 Bell, and Appell subgroups* of $\mathcal{DR}$, respectively. The row sum generating function of a double Riordan array $(d; h_1, h_2)$ is

$$\begin{aligned}
&(d; h_1, h_2)\frac{1}{1-t}\\
&= d + dh_1 + dh_1h_2 + dh_1^2h_2 + dh_1^2h_2^2 + \cdots\\
&= d(1 + (h_1h_2) + (h_1h_2)^2 + \cdots) + dh_1(1 + (h_1h_2) + (h_1h_2)^2 + \cdots)\\
&= \frac{d}{1-h_1h_2} + \frac{dh_1}{1-h_1h_2} = \frac{d(1+h_1)}{1-h_1h_2}.
\end{aligned} \tag{6.5.18}$$

In [27], a normal subgroup of $\mathcal{DR}$ defined by $\mathcal{N} := \{(d; t, t) \in \mathcal{DR}\}$ is given. It can be seen that $\mathcal{DR}$ is the semidirect product of $\mathcal{N}$ and $\mathcal{A}$.

If $(d; h_1, h_2)$ is a double Riordan array, where $d(t) = \sum_{k=0}^{\infty} d_{2k}t^{2k}$, $h_1(t) = \sum_{k=0}^{\infty} h_{1,2k+1}t^{2k+1}$, and $h_2(t) = \sum_{k=0}^{\infty} h_{2,2k+1}t^{2k+1}$, then $(\hat{d}; \hat{h}_1, \hat{h}_2)$, where $\hat{d}(t) = \sum_{k=0}^{\infty} d_{2k}t^k$, $\hat{h}_1(t) = \sum_{k=0}^{\infty} h_{1,2k+1}t^{k+1}$, and $\hat{h}_2(t) = \sum_{k=0}^{\infty} h_{2,2k+1}t^{k+1}$, is the *compression* of the double Riordan array $(d; h_1, h_2)$. It can be seen that $\hat{R}$ shown above is the compression of the double Riordan array $(1/(1-t^2), t, t/(1-t^2))$.

In the next subsection, we will give the sequence characterization of double Riordan arrays, i.e., we will introduce A- and Z- sequences of double Riordan arrays in $\mathcal{DR}$ and its subgroups. After that, the relationship between a double Riordan array and its compression including their A- and Z- sequences will be discussed. Then, we define quasi Riordan arrays and discuss their A- and Z-sequence characterizations. Finally, we define Sheffer pairs related to double Riordan arrays.

Sequence characterization of double Riordan arrays

We now give a sequence characterization of double Riordan arrays.

Theorem 6.31 $(d; h_1, h_2) = (d_{n,k})_{n,k\geq 0}$ *is in* $\mathcal{DR}$ *with* $d_{n,n} \neq 0$ *if and only if there exist an* A_1*-sequence,* $\{a_{1,0}, a_{1,1}, a_{1,2}, \ldots\}$*, and* A_2*- sequence,* $\{a_{2,0} \neq 0, a_{2,1},$

$a_{2,2},\ldots\}$, *and a Z-sequence,* $\{z_0, z_1, z_2, \ldots\}$ *such that*

$$\begin{aligned} d_{n,2k-1} &= a_{1,0}d_{n-1,2k-2} + a_{1,1}d_{n-1,2k} + a_{1,2}d_{n-1,2k+2} + \cdots, \quad k \geq 1 \\ d_{n,2k} &= a_{2,0}d_{n-2,2k-2} + a_{2,1}d_{n-2,2k} + a_{2,2}d_{n-2,2k+2} + \cdots, \quad k \geq 1 \\ d_{n,0} &= z_0 d_{n-2,0} + z_1 d_{n-2,2} + z_2 d_{n-2,4} + \cdots \end{aligned} \tag{6.5.19}$$

for $n \geq 0$*. Furthermore, denoting the generating functions of the* A_1*-sequence, the* A_2*-sequence, and the Z-sequence by* $A_1(t) = \sum_{k\geq 0} a_{1,k}t^{2k}$, $A_2(t) = \sum_{k\geq 0} a_{2,k}t^{2k}$, *and* $Z(t) = \sum_{k\geq 0} z_k t^{2k}$*, respectively, we have equivalent forms of the three expressions in* (6.5.19) *as follows:*

$$h_1 = tA_1(\sqrt{h_1h_2}), \tag{6.5.20}$$

$$h_1h_2 = t^2A_2(\sqrt{h_1h_2}), \tag{6.5.21}$$

$$d(t) = \frac{d(0)}{1 - t^2Z(\sqrt{h_1h_2})}. \tag{6.5.22}$$

Proof Let $(d; h_1, h_2)$ be a double Riordan array. Then there exists another double Riordan array $(A_1; F_1, F_2)$ such that

$$(A_1; F_1, F_2) = (d; h_1, h_2)^{-1}\left(\frac{dh_1}{t}; h_1, h_2\right),$$

or equivalently,

$$(d; h_1, h_2)(A_1; F_1, F_2) = \left(\frac{dh_1}{t}; h_1, h_2\right).$$

Since the left-hand side of the above equation is

$$\left(dA_1(\sqrt{h_1h_2}); \sqrt{\frac{h_1}{h_2}}F_1(\sqrt{h_1h_2}), \sqrt{\frac{h_2}{h_1}}F_2(\sqrt{h_1h_2})\right),$$

we have

$$dA_1(\sqrt{h_1h_2}) = \frac{dh_1}{t}, \quad \sqrt{\frac{h_1}{h_2}}F_1(\sqrt{h_1h_2}) = h_1, \quad \sqrt{\frac{h_2}{h_1}}F_2(\sqrt{h_1h_2}) = h_2,$$

which implies

$$h_1 = tA_1(\sqrt{h_1h_2}), \quad F_1 = F_2 = t.$$

Denote $A_1 = \sum_{k\geq 0} a_{1,k}t^{2k}$. From formula (6.5.20), we have

$$[t^n]dh_1(h_1h_2)^{k-1} = [t^n]td\left(a_{1,0}(h_1h_2)^{k-1} + a_{1,1}(h_1h_2)^k + a_{1,2}(h_1h_2)^{k+1} + \cdots\right),$$

which implies

$$d_{n,2k-1} = a_{1,0}d_{n-1,2k-2} + a_{1,1}d_{n-1,2k} + a_{1,2}d_{n-1,2k+2} + \cdots .$$

Similarly, if $(d; h_1, h_2)$ is a double Riordan array, then there exists another double Riordan array $(A_2; F_1, F_2)$ such that

$$(A_2; F_1, F_2) = (d; h_1, h_2)^{-1}\left(\frac{dh_1h_2}{t^2}; h_1, h_2\right),$$

or equivalently,

$$(d; h_1, h_2)(A_2; F_1, F_2) = \left(\frac{dh_1h_2}{t^2}; h_1, h_2\right).$$

Thus, noting

$$(d; h_1, h_2)(A_2; F_1, F_2) = \left(dA_1(\sqrt{h_1h_2}); \sqrt{\frac{h_1}{h_2}}F_1(\sqrt{h_1h_2}), \sqrt{\frac{h_2}{h_1}}F_2(\sqrt{h_1h_2})\right),$$

we obtain (6.5.21) $h_1h_2 = t^2A_2\left(\sqrt{h_1h_2}\right)$ and $F_1 = F_2 = t$. From (6.5.21) there exists

$$[t^n]d(h_1h_2)^k = [t^n]t^2\left(a_{2,0}(h_1h_2)^{k-1} + a_{2,1}(h_1h_2)^k + a_{2,2}(h_1h_2)^{k+1} + \cdots\right),$$

which yields

$$d_{n,2k} = a_{2,0}d_{n-2,2k-2} + a_{2,1}d_{n-2,2k} + a_{2,2}d_{n-2,2k+2} + \cdots .$$

Let $(d_{n,k})_{n,k\geq 0}$ be any infinite lower triangular array with $d_{n,n} \neq 0$, and let its first column have an even generating function $d(t) = \sum_{k\geq 0} t^{2k}$. Then we denote $z_0 = d_{2,0}/d_{0,0}$ and use

$$d_{n,0} = z_0d_{n-2,0} + z_1d_{n-2,2} + z_2d_{n-2,4} + \cdots \tag{6.5.23}$$

to determine $z_1, z_2, \ldots$, uniquely and recursively. For instance,

$$z_1 = \frac{d_{4,0} - z_0d_{2,0}}{d_{2,2}} = \frac{d_{4,0}d_{0,0} - d_{2,0}^2}{d_{0,0}d_{2,2}},$$

etc. Finally, the equivalence between (6.5.22) and (6.5.23) can be seen from the equivalence between (6.5.23) and

$$[t^n]d = [t^n]t^2d\left(z_0 + z_1(h_1h_2) + z_2(h_1h_2)^2 + \cdots\right), \quad n \geq 2.$$

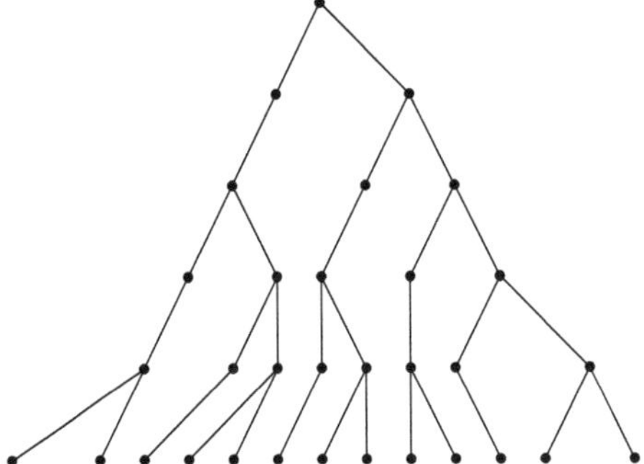

Fig. 6.1 The Fibonacci tree

The last expression implies

$$d(t) - d(0) = dt^2 Z(\sqrt{h_1 h_2}),$$

or equivalently, equation (6.5.22).

Conversely, we assume Z, A_1, and A_2 satisfying (6.5.19) or (6.5.20)–(6.5.22) are given. If the first column of a lower triangular matrix is determined by (6.5.23), then from conditions (6.5.21) $h_1 h_2$ can be determined uniquely. Substituting $h_1 h_2$ into $h_1 = tA_1(\sqrt{h_1 h_2})$ we obtain h_1, and h_2 can be obtained accordingly. Thus the given lower triangular matrix is the double Riordan array $(d; h_1, h_2) \in \mathcal{DR}$. This completes our proof. □

Equations (6.5.20) and (6.5.21) may be called *the second fundamental theorem of double Riordan arrays.*

Example 6.5 The Pascal-Fibonacci array $\hat{R}$ shown in (6.5.14) is the compression of the double Riordan array $(1/(1-t^2); t, t/(1-t^2))$, which represents the Fibonacci tree shown in Fig. 6.1. Thus $h_1 = t$, $h_2 = t/(1-t^2)$, and $d = 1/(1-t^2)$. and $\sqrt{h_1 h_2} = t/\sqrt{1-t^2}$. The compositional inverse of $\sqrt{h_1 h_2}$ is $t/\sqrt{1+t^2}$. Substituting $h_1 = t$ and $d = 1/(1-t^2)$ into Equations (6.5.20) and (6.5.22), we have $A_1 = 1$ and $Z = 1$. Equation (6.5.21) gives the generating functions of the A_2-sequence as

$$A_2(t) = \left.\frac{h_1 h_2}{t^2}\right|_{t=t/\sqrt{1+t^2}} = 1 + t^2.$$

Example 6.6 Kuznetkov, Pak, and Postnikov [47] consider ordered trees with no points at odd heights, which is converted to the *Dyck paths* with no valley at odd heights in Davenport, Shapiro, and Woodson [27]. Denote by $d_{n,k}$ the number of those paths with n edges that end at height k. Then the matrix $D = (d_{n,k})_{n,k\geq 0}$ [27] is

$$D = \begin{bmatrix} 1 & 0 & 0 & 0 & 0 & 0 & 0 & 0 & 0 & \dots \\ 0 & 1 & 0 & 0 & 0 & 0 & 0 & 0 & 0 & \dots \\ 1 & 0 & 1 & 0 & 0 & 0 & 0 & 0 & 0 & \dots \\ 0 & 2 & 0 & 1 & 0 & 0 & 0 & 0 & 0 & \dots \\ 2 & 0 & 2 & 0 & 1 & 0 & 0 & 0 & 0 & \dots \\ 0 & 4 & 0 & 3 & 0 & 1 & 0 & 0 & 0 & \dots \\ 4 & 0 & 5 & 0 & 3 & 0 & 1 & 0 & 0 & .. \\ 0 & 9 & 0 & 8 & 0 & 4 & 0 & 1 & 0 & \dots \\ 9 & 0 & 12 & 0 & 9 & 0 & 4 & 0 & 1 & \dots \\ \vdots & \vdots & \vdots & \vdots & \vdots & \vdots & \vdots & \vdots & \vdots & \ddots \end{bmatrix}. \tag{6.5.24}$$

It can be seen that

$$\begin{aligned} d_{n,2k} &= d_{n-2,2k-2} + d_{n-2,2k} + d_{n-2,2k+2}, \quad k \geq 1, \\ d_{n,2k-1} &= d_{n-1,2k-2} + d_{n-1,2k}, \quad k \geq 1, \\ d_{n,0} &= d_{n-2,0} + d_{n-2,2} \end{aligned}$$

for $n \geq 1$. Thus, $A_1 = 1 + t^2$, $A_2 = 1 + t^2 + t^4$, and $Z = 1 + t^2$.

Corollary 6.6 *Let $\mathcal{D} = (d; h_1, h_2) \in \mathcal{DR}$. Then $\mathcal{D}$ is in the Appell subgroup of $\mathcal{DR}$, i.e., $h_1 = h_2 = t$ if and only if the generating functions of its A_1- and A_2- sequences satisfy $A_1 = A_2 = 1$; $\mathcal{D}$ is in the associated subgroup of $\mathcal{DR}$, i.e., $d = 1$, if and only if the generating function of the Z-sequence satisfies $Z = 0$; $\mathcal{D}$ is in the type-1 Bell subgroup, i.e., $h_1 = td$, if and only if $A_1(\sqrt{h_1h_2}) = d(0) + th_1Z(\sqrt{h_1h_2})$; and $\mathcal{D}$ is in the type-2 Bell subgroup, i.e., $h_2 = td$, if and only if $A_2(\sqrt{h_1h_2}) = d(0)\frac{h_1}{t} + h_1h_2Z(\sqrt{h_1h_2})$.*

Proof If $\mathcal{D} = (d; t, t) \in \mathcal{DR}$, then from (6.5.20) and (6.5.21), we obtain $A_1 = 1$ and $A_2 = 1$, respectively. If $\mathcal{D} = (1, h_1, h_2) \in \mathcal{DR}$, then from (6.5.22), we have

$$1 = \frac{1}{1 - t^2Z(\sqrt{h_1h_2})},$$

which implies $Z = 0$. When $\mathcal{D} = (d; td, h_2) \in \mathcal{DR}$, we may use (6.5.22) to get

$$d(t) - d(0) = dt^2Z(\sqrt{h_1h_2}). \tag{6.5.25}$$

Noting $d = h_1/t = A_1(\sqrt{h_1h_2})$ from (6.5.20), we may immediately change (6.5.25) to

$$A_1(\sqrt{h_1h_2}) - d(0) = th_1Z(\sqrt{h_1h_2}),$$

i.e., $A_1(\sqrt{h_1h_2}) = d(0) + th_1Z(\sqrt{h_1h_2})$. Similarly, if $\mathcal{D} = (d; h_1, td) \in \mathcal{DR}$, then we use (6.5.25) again and change its left-hand side to

$$d(t) - d(0) = \frac{h_2}{t} - d(0) = \frac{t}{h_1}\frac{h_1 h_2}{t^2} - d(0) = \frac{t}{h_1} A_2(\sqrt{h_1 h_2}),$$

which yields $A_2(\sqrt{h_1 h_2}) = d(0)\frac{h_1}{t} + h_1 h_2 Z(\sqrt{h_1 h_2})$. □

Theorem 6.32 *Let $(d; h_1, h_2)$ be a double Riordan array, and let $\overline{h_{1,2}}$ be the compositional inverse of $\sqrt{h_1 h_2}$, i.e., $(\overline{h_{1,2}} \circ \sqrt{h_1 h_2})(t) = (\sqrt{h_1 h_2} \circ \overline{h_{1,2}})(t) = t$. Then it has a unique inverse*

$$(d; h_1, h_2)^{-1} = \left(\frac{1}{d(\overline{h_{1,2}})}; \overline{h_{1,2}} \frac{t}{h_1(\overline{h_{1,2}})}, \overline{h_{1,2}} \frac{t}{h_2(\overline{h_{1,2}})} \right). \tag{6.5.26}$$

Particularly,

$$(d; t, h_2)^{-1} = \left(\frac{1}{d(\overline{h_{1,2}})}; t, \overline{h_{1,2}} \frac{t}{h_2(\overline{h_{1,2}})} \right), \tag{6.5.27}$$

where $\overline{h_{1,2}} = \overline{th_2}$ is the compositional inverse of th_2.

Proof From (6.5.17),

$$\begin{aligned} &(d; h_1, h_2) \left(\frac{1}{d(\overline{h_{1,2}})}; \overline{h_{1,2}} \frac{t}{h_1(\overline{h_{1,2}})}, \overline{h_{1,2}} \frac{t}{h_2(\overline{h_{1,2}})} \right) \\ &= \left(d\frac{1}{d}; \sqrt{\frac{h_1}{h_2}} t \frac{\sqrt{h_1 h_2}}{h_1}, \sqrt{\frac{h_2}{h_1}} t \frac{\sqrt{h_1 h_2}}{h_2} \right) = (1; t, t), \end{aligned}$$

which implies (6.5.26). Equation (6.5.27) follows by substituting $h_1 = t$ into (6.5.26). □

It is well known that the A-sequence and Z-sequence of a Riordan array (d, h) can be easily represented by its inverse $(d, h)^{-1} = (1/d(\bar{h}), \bar{h}) =: (g, f)$, where $\bar{h}$ is the compositional inverse of h:

$$A = t/\bar{h} = t/f \quad \text{and} \quad Z(t) = (1 - (1/d(\bar{h}))/\bar{h} = (1 - g)/f.$$

We now present the generating functions of the A_1-, A_2-, and Z-sequences for double Riordan arrays in a similar way to the above expressions for Riordan arrays.

Corollary 6.7 *Let $(d; h_1, h_2)$ be a double Riordan array, and let $\overline{h_{1,2}}$ be the compositional inverse of $\sqrt{h_1 h_2}$, i.e., $(\overline{h_{1,2}} \circ \sqrt{h_1 h_2})(t) = (\sqrt{h_1 h_2} \circ \overline{h_{1,2}})(t) = t$. Denote the inverse of $(d; h_1, h_2)$ by $(g; f_1, f_2)$ presented as (6.5.26), i.e.,*

$$(g; f_1, f_2) = (d; h_1, h_2)^{-1} = \left(\frac{1}{d(\overline{h_{1,2}})}; \overline{h_{1,2}} \frac{t}{h_1(\overline{h_{1,2}})}, \overline{h_{1,2}} \frac{t}{h_2(\overline{h_{1,2}})} \right). \tag{6.5.28}$$

Then the generating functions of A_1-, A_2-, and Z- sequences of $(d; h_1, h_2)$ can be written as

$$A_1=\frac{t}{f_1},\quad A_2=\frac{t^2}{f_1f_2},\quad \textit{and}\quad Z=\frac{1-g/g(0)}{f_1f_2}. \tag{6.5.29}$$

Proof Substituting $t=\sqrt{h_1h_2}$ into (6.5.29) and noting (6.5.28), we have

$$\begin{aligned}
A_1(\sqrt{h_1h_2}) &= \frac{\sqrt{h_1h_2}}{t\sqrt{h_1h_2}/h_1},\\
A_2(\sqrt{h_1h_2}) &= \frac{h_1h_2}{t^2h_1h_2/(h_1h_2)},\\
Z(\sqrt{h_1h_2}) &= \frac{1-d(0)/d}{t^2(\sqrt{h_1h_2})^2/(h_1h_2)},
\end{aligned}$$

which implies (6.5.20), (6.5.21), and (6.5.22) with $d(0)=1$, respectively. □

Theorem 6.33 *Let A_1^*, A_2^*, and Z^* be the generating functions of the A_1-, A_2-, and Z- sequences of $(d;h_1,h_2)^{-1}$, where $(d;h_1,h_2)\in\mathcal{DS}$. Then we have*

$$A_1^*\left(\frac{t\overline{h_{1,2}}}{\sqrt{h_1\left(\overline{h_{1,2}}\right)h_2\left(\overline{h_{1,2}}\right)}}\right)=\frac{1}{A_1(t)}, \tag{6.5.30}$$

$$A_2^*\left(\frac{t\overline{h_{1,2}}}{\sqrt{h_1\left(\overline{h_{1,2}}\right)h_2\left(\overline{h_{1,2}}\right)}}\right)=\frac{1}{A_2(t)}, \tag{6.5.31}$$

$$\begin{aligned}
Z^*\left(\frac{t\overline{h_{1,2}}}{\sqrt{h_1\left(\overline{h_{1,2}}\right)h_2\left(\overline{h_{1,2}}\right)}}\right) &= \frac{1-d(\overline{h_{1,2}})}{t^2}\\
&= \frac{1-d\left(\frac{h_1\left(\overline{h_{1,2}}\right)}{A_1}\right)}{t^2}=-\frac{\overline{h_{1,2}}^2}{t^2}d(\overline{h_{1,2}})Z(t).
\end{aligned} \tag{6.5.32}$$

Proof Substituting $t=\overline{h_{1,2}}$ into (6.5.20)–(6.5.22) yields

$$\begin{aligned}
A_1(t) &= \frac{h_1(\overline{h_{1,2}})}{\overline{h_{1,2}}},\\
A_2(t) &= \frac{h_1(\overline{h_{1,2}})h_2(\overline{h_{1,2}})}{\overline{h_{1,2}}^2},\\
d(\overline{h_{1,2}})-1 &= \overline{h_{1,2}}^2 d(\overline{h_{1,2}})Z(t)\ (\textit{because}\ d(0)=1).
\end{aligned}$$

From (6.5.20)–(6.5.22), we have

$$\overline{h_{1,2}}\frac{t}{h_1(\overline{h_{1,2}})} = tA_1^*\left(\frac{\overline{h_{1,2}}^2 t^2}{h_1(\overline{h_{1,2}})h_2(\overline{h_{1,2}})}\right),$$

$$\frac{\overline{h_{1,2}}^2 t^2}{h_1(\overline{h_{1,2}})h_2(\overline{h_{1,2}})} = t^2 A_2^*\left(\frac{\overline{h_{1,2}}^2 t^2}{h_1(\overline{h_{1,2}})h_2(\overline{h_{1,2}})}\right),$$

$$\frac{1}{d(\overline{h_{1,2}})} = \frac{1}{1 - t^2 Z^*\left(\frac{\overline{h_{1,2}}^2 t^2}{h_1(\overline{h_{1,2}})h_2(\overline{h_{1,2}})}\right)},$$

or equivalently,

$$A_1^*\left(\sqrt{\frac{\overline{h_{1,2}}^2 t^2}{h_1(\overline{h_{1,2}})h_2(\overline{h_{1,2}})}}\right) = \frac{\overline{h_{1,2}}^2}{h_1(\overline{h_{1,2}})} = \frac{1}{A_1(t)},$$

$$A_2^*\left(\sqrt{\frac{\overline{h_{1,2}}^2 t^2}{h_1(\overline{h_{1,2}})h_2(\overline{h_{1,2}})}}\right) = \frac{\overline{h_{1,2}}}{h_1(\overline{h_{1,2}})h_2(\overline{h_{1,2}})} = \frac{1}{A_2(t)},$$

$$Z^*\left(\sqrt{\frac{\overline{h_{1,2}}^2 t^2}{h_1(\overline{h_{1,2}})h_2(\overline{h_{1,2}})}}\right) = \frac{1 - d(\overline{h_{1,2}})}{t^2} = \frac{1 - d\left(\frac{h_1(\overline{h_{1,2}})}{A_1}\right)}{t^2} = -\frac{\overline{h_{1,2}}^2}{t^2} d(\overline{h_{1,2}})Z(t).$$

This completes the proof. □

Example 6.7 $\hat{R}$ shown in (6.5.14) is the compression of the double Riordan array, the Pascal-Fibonacci array, $(1/(1-t^2), t, t/(1-t^2))$. The generating functions of its Z-, A_1-, and A_2- sequences are found in Example 6.1 as $A_1 = Z = 1$ and $A_2 = 1 + t^2$. Since $h_1 = t$ and $h_2 = t/(1-t^2)$, $\sqrt{h_1 h_2} = t/\sqrt{1-t^2}$. The compositional inverse of $\sqrt{h_1 h_2}$ is $\overline{h_{1,2}} = t/\sqrt{1+t^2}$. We have

$$h_1(\overline{h_{1,2}}) = \overline{h_{1,2}} = \frac{t}{\sqrt{1+t^2}},$$

$$h_2(\overline{h_{1,2}}) = \frac{\overline{h_{1,2}}}{1 - \overline{h_{1,2}}^2} = t\sqrt{1+t^2}.$$

Thus,

$$h_1(\overline{h_{1,2}})h_2(\overline{h_{1,2}}) = t^2.$$

From (6.5.26) or (6.5.27), the inverse of $(1/(1-t^2); t, t/(1-t^2))$ is $(1/(1+t^2); t, t/(1+t^2))$. By using (6.5.20)–(6.5.22), we may find the generating functions of Z-, A_1-, and A_2- sequences as follows:

$$Z^* = -1, \quad A_1^* = 1, \quad A_2^* = 1 - t^2.$$

Noting that

$$\frac{t\overline{h_{1,2}}}{h_1(\overline{h_{1,2}})h_2(\overline{h_{1,2}})} = \overline{h_{1,2}} = \frac{t}{\sqrt{1+t^2}},$$

we may check (6.5.30)–(6.5.32) hold because

$$A_1^*\left(\frac{t\overline{h_{1,2}}}{\sqrt{h_1\left(\overline{h_{1,2}}\right)h_2\left(\overline{h_{1,2}}\right)}}\right) = A_1^*(\overline{h_{1,2}}) = 1 = \frac{1}{A_1(t)},$$

$$A_2^*\left(\frac{t\overline{h_{1,2}}}{\sqrt{h_1\left(\overline{h_{1,2}}\right)h_2\left(\overline{h_{1,2}}\right)}}\right) = A_2^*(\overline{h_{1,2}}) = \frac{1}{1+t^2} = \frac{1}{A_2(t)},$$

and

$$Z^*\left(\frac{t\overline{h_{1,2}}}{\sqrt{h_1\left(\overline{h_{1,2}}\right)h_2\left(\overline{h_{1,2}}\right)}}\right) = Z^*(\overline{h_{1,2}}) = -1 = \frac{1-d(\overline{h_{1,2}})}{t^2}$$

$$= \frac{1-d\left(\frac{h_1\left(\overline{h_{1,2}}\right)}{A_1}\right)}{t^2} = -\frac{\overline{h_{1,2}}^2}{t^2}d(\overline{h_{1,2}})Z(t).$$

Double Riordan arrays and their compressions

In this subsection, we define an array called the compression array of a double Riordan array. The sequence characterization of the compression arrays is also given. We extend double Riordan arrays to high-order Riordan arrays and consider the second-order Riordan arrays as double Riordan arrays. Finally, the compressions of high-order Riordan arrays are defined accordingly.

Definition 6.9 Let $(d; h_1, h_2) = (d_{n,k})_{n,k\geq 0}$ be a double Riordan array. We define its compression array $(\hat{d}_{n,k})_{n,k\geq 0}$ as follows:

$$\hat{d}_{n,k} := d_{2n-k,k}, \quad n \geq k \geq 0. \tag{6.5.33}$$

We now study the structure of the compression of a double Riordan array starting from the following theorem.

Theorem 6.34 *Let $(d; h_1, h_2) = (d_{n,k})_{n,k\geq 0}$ be a double Riordan array, and let its compression array $(\hat{d}_{n,k})_{n,k\geq 0}$ be defined by* (6.5.33). *Then we have*

$$\hat{d}_{n,k} = \begin{cases} [t^n]\hat{d}(\hat{h}_1\hat{h}_2)^{k/2}, & \text{if } k \text{ is even} \\ [t^n]\hat{d}\hat{h}_1(\hat{h}_1\hat{h}_2)^{(k-1)/2}, & \text{if } k \text{ is odd,} \end{cases} \tag{6.5.34}$$

where

$$\hat{d}(t)=\sum_{k\geq 0} d_{2k}t^k,\quad \hat{h}_1(t)=\sum_{k\geq 0} h_{1,2k+1}t^{k+1},\quad \textit{and}\quad \hat{h}_2(t)=\sum_{k\geq 0} h_{2,2k+1}t^{k+1}.$$

Proof It is sufficient to prove the case of even k since the case of odd k can be proved similarly. If $k=2\ell$, Equation (6.5.33) shows

$$\hat{d}_{n,2\ell}=d_{2n-2\ell,2\ell}=[t^{2n-2\ell}]d(h_1h_2)^{\ell}=[t^{2n}]t^{2\ell}d(h_1h_2)\ell.$$

On the rightmost side, using the transformation $t^2\to t$ and noting $d(t^{1/2})=\hat{d}$ and $t^{1/2}h_i(t^{1/2})=\hat{h}_i$ $(i=1,2)$, we may write the above equation as

$$\hat{d}_{n,2\ell}=[t^n]\hat{d}(\hat{h}_1\hat{h}_2)\ell,$$

completing the proof. □

As an example, $\hat{R}=(1/(1-t);t,t/(1-t))$ shown in (6.5.14) is the compression of the double Riordan array $(1/(1-t^2);t,t/(1-t^2))$.

The sequence characterization of the compression of a double Riordan array is given in the following theorem.

Theorem 6.35 *Let $(d;h_1,h_2)=(d_{n,k})_{n,k\geq 0}$ be a double Riordan array, and let its compression be defined by $(\hat{d}_{n,k})_{n,k\geq 0}$, where $\hat{d}_{n,k}$ is shown in (6.5.33). Suppose the A_1-, A_2-, and Z sequences of $(d;h_1,h_2)$ are*

$$A_1=\{a_{1,0},a_{1,1},\ldots\},\quad A_2=\{a_{2,0},a_{2,1},\ldots\},\quad \textit{and}\quad Z=\{z_0,z_1,\ldots\}.$$

Then for $l=1,2,\ldots$ and $n=1,2,\ldots$,

$$\hat{d}_{n,2l-1}=a_{1,0}\hat{d}_{n-1,2l-2}+a_{1,1}\hat{d}_{n,2l}+a_{1,2}\hat{d}_{n+1,2l+2}+\cdots,\tag{6.5.35}$$
$$\hat{d}_{n,2l}=a_{1,0}\hat{d}_{n-2,2l-2}+a_{1,1}\hat{d}_{n-1,2l}+a_{1,2}\hat{d}_{n,2l+2}+\cdots,\tag{6.5.36}$$
$$\hat{d}_{n,0}=z_0\hat{d}_{n-1,0}+z_1\hat{d}_{n,2}+z_2\hat{d}_{n+1,4}+\cdots,\tag{6.5.37}$$

or equivalently,

$$\hat{h}_1=tA_1\left(\frac{\hat{h}_1\hat{h}_2}{t}\right),\tag{6.5.38}$$
$$\hat{h}_1\hat{h}_2=t^2A_2\left(\frac{\hat{h}_1\hat{h}_2}{t}\right),\tag{6.5.39}$$
$$\hat{d}=\frac{1}{1-tZ\left(\frac{\hat{h}_1\hat{h}_2}{t}\right)},\tag{6.5.40}$$

where $A_1(t)$, $A_2(t)$, and $Z(t)$ are the generating functions of the A_1-, A_2-, and Z-sequences of $(d;h_1,h_2)$.

Proof To find the sequence characterization of the compression of a double Riordan array, we consider the cases of $\hat{d}_{n,2l-1}$ and $\hat{d}_{n,2l}$ for $l = 1, 2, \ldots$ and $\hat{d}_{n,0}$, where $n \geq 0$. From (6.5.33) and (6.5.19), we have

$$\begin{aligned}
\hat{d}_{n,2l-1} &= d_{2(n-l)+1,2l-1} = a_{1,0}d_{2(n-l),2l-2} + a_{1,1}d_{2(n-l),2l} + a_{1,2}d_{2(n-l),2l+2} + \cdots \\
&= a_{1,0}\hat{d}_{n-1,2l-2} + a_{1,1}\hat{d}_{n,2l} + a_{1,2}\hat{d}_{n+1,2l+2} + \cdots, \\
\hat{d}_{n,2l} &= d_{2(n-l),2l} = a_{2,0}d_{2(n-l-1),2l-2} + a_{2,1}d_{2(n-l-1),2l} + a_{2,2}d_{2(n-l-1),2l+2} + \cdots \\
&= a_{2,0}\hat{d}_{n-2,2l-2} + a_{2,1}\hat{d}_{n-1,2l} + a_{2,2}\hat{d}_{n,2l+2} + \cdots, \\
\hat{d}_{n,0} &= d_{2n,0} = z_0 d_{2(n-1),0} + z_1 d_{2(n-1),2} + z_2 d_{2(n-1),4} + \cdots \\
&= z_0\hat{d}_{n-1,0} + z_1\hat{d}_{n,2} + z_2\hat{d}_{n+1,4} + \cdots.
\end{aligned}$$

Using the generating function representation, from Theorem 6.34 we may write (6.5.35) to be

$$\begin{aligned}
[t^n]\hat{d}\hat{h}_1(\hat{h}_1\hat{h}_2)^{\ell-1} &= [t^n]\hat{d}\left(ta_{1,0}(\hat{h}_1\hat{h}_2)^{\ell-1} + a_{1,1}(\hat{h}_1\hat{h}_2)^{\ell} + \frac{1}{t}a_{1,2}(\hat{h}_1\hat{h}_2)^{\ell+1} + \cdots\right) \\
&= [t^n]t\hat{d}A_1\left(\frac{\hat{h}_1\hat{h}_2}{t}\right)
\end{aligned}$$

for $n \geq 1$, which implies (6.5.38). Similarly, we may prove (6.5.39). To prove (6.5.40), we rewrite (6.5.37) for $n \geq 1$ as

$$[t^n]\hat{d} = [t^n]\hat{d}\left(tz_0 + z_1(\hat{h}_1\hat{h}_2) + \frac{1}{t}z_2(\hat{h}_1\hat{h}_2)^2 + \cdots\right).$$

Hence,

$$\hat{d} - 1 = t\hat{d}Z\left(\frac{\hat{h}_1\hat{h}_2}{t}\right),$$

which gives (6.5.40). □

Equation (6.5.35) can be represented as

$$\begin{aligned}
[t^n]\hat{d}\hat{h}_1(\hat{h}_1\hat{h}_2)^{\ell-1} &= \sum_{i\geq 0} a_{1,i}[t^{n-i-1}]\hat{d}(\hat{h}_1\hat{h}_2)^{\ell-1+i} \\
&= [t^n]\hat{d}(\hat{h}_1\hat{h}_2)^{\ell-1}\sum_{i\geq 0} a_{1,i}t^{i+1}(\hat{h}_1\hat{h}_2)^i.
\end{aligned}$$

The above expression has the following equivalent form by using the generating function of the columns of the *compression matrix* $(\hat{d}_{n,k})_{n,k\geq 0}$:

$$\hat{h}_1 = tA(t\hat{h}_1\hat{h}_2).$$

Example 6.8 Let $D = (d_{n,k})_{n,k\geq 0}$ be the double Riordan array of the Dyck path numbers studied in Example 6.6. Then its compression is

$$\hat{D} = (\hat{d}_{n,k})_{n,k\geq 0} = \begin{bmatrix} 1 & 0 & 0 & 0 & 0 & 0 & 0 & 0 & 0 & 0 & \dots \\ 1 & 1 & 0 & 0 & 0 & 0 & 0 & 0 & 0 & 0 & \dots \\ 2 & 2 & 1 & 0 & 0 & 0 & 0 & 0 & 0 & 0 & \dots \\ 4 & 4 & 2 & 1 & 0 & 0 & 0 & 0 & 0 & 0 & \dots \\ 9 & 9 & 5 & 3 & 1 & 0 & 0 & 0 & 0 & 0 & \dots \\ 21 & 21 & 12 & 8 & 3 & 1 & 0 & 0 & 0 & 0 & \dots \\ \vdots & \vdots & \vdots & \vdots & \vdots & \vdots & \vdots & \vdots & \vdots & \vdots & \ddots \end{bmatrix}.$$

It can be seen that

$$\begin{aligned} \hat{d}_{n,2l-1} &= \hat{d}_{n-1,2l-2} + \hat{d}_{n,2l}, \quad l \geq 1, \\ \hat{d}_{n,2l} &= d_{n-2,2l-2} + d_{n-1,2l} + \hat{d}_{n,2l+2}, \quad l \geq 1, \\ \hat{d}_{n,0} &= d_{n-1,0} + d_{n,2} \end{aligned}$$

for $n \geq 1$. Thus, $A_1 = 1 + t^2$, $A_2 = 1 + t^2 + t^4$, and $Z = 1 + t^2$.

The following theorem shows how to extend Riordan subgroups to double Riordan subgroups.

Theorem 6.36 *$\mathcal{L} := \{(d; t, h) \in \mathcal{DR}\}$ is a subgroup of $\mathcal{DR}$. Furthermore, if $\{(d, h) \in \mathcal{R}\}$ is in the Appell subgroup, associated subgroup, Bell subgroup of $\mathcal{R}$, then $\{(d; t, h) \in \mathcal{DR}\}$ is in the Appell subgroup, associated subgroup, and type-2 Bell subgroup of $\mathcal{DR}$, respectively.*

Proof Let $(d; t, h), (g; t, f) \in \mathcal{L}$. Then from (6.5.17)

$$\begin{aligned} (d; t, h)(g; t, f) &= (dg(\sqrt{th}); \sqrt{t/h}(\sqrt{th}), \sqrt{h/t}f(\sqrt{th})) \\ &= (dg(\sqrt{th}); t, \sqrt{h/t}f(\sqrt{th})), \end{aligned}$$

which is also in $\mathcal{L}$. $(1; t, t)$ is the identity element of $\mathcal{L}$ because $(d; t, h)(1; t, t) = (d; t, h)$. Hence, $\mathcal{L} \subset \mathcal{DR}$ is a subgroup of $\mathcal{DR}$.

If (d, h) is in the Appell subgroup, associated subgroup, or Bell subgroup of $\mathcal{R}$, i.e., $(d, h) = (d, t), (1, h)$, or (d, td), then $(d; t, h) = (d; t, t), (1; t, h)$, or $(d; t, td)$, respectively. Thus, the corresponding $(d; t, h)$ are in the Appell subgroup, associated subgroup, or type-2 Bell subgroup of $\mathcal{DR}$, respectively. □

Riordan arrays of kth order

The concept of double Riordan arrays can be further generalized.

Definition 6.10 Let $d(t) \in \mathcal{F}_0$ and $h_j(t) \in \mathcal{F}_1$, where $d(t) = \sum_{n\geq 0} d_{kn}t^{kn}$ and $h_j(t) = \sum_{n\geq 0} h_{j,kn+1}t^{kn+1}$, $1 \leq j \leq k$. Then the kth-order Riordan array with respect to d and h_j $(1 \leq j \leq k)$ is defined by

$$\begin{aligned}&(d; h_1, h_2, \ldots, h_k)\\ &\quad = (d; dh_1, dh_1h_2, \ldots, d\Pi_{j=1}^k h_j, dh_1\Pi_{j=1}^k h_j, dh_1h_2\Pi_{j=1}^k h_j, \ldots).\end{aligned}$$

We denote the entry of nth row and jth column as $d_{n,j}$ and write $(d; h_1, h_2, \ldots, h_k) = (d_{n,j})_{n\geq j\geq 0}$.

The fundamental theorem of kth-order Riordan arrays follows from their definition.

Theorem 6.37 *Let $(d; h_1, h_2, \ldots, h_k)$ be a kth-order Riordan array, and let $A_j(t) = t^j B_j(t)$, where $B_j(t) = \sum_{n\geq 0} b_{kn+j}t^{kn}$, $j = 0, 1, \ldots, k-1$. Then we have the following fundamental theorem of kth-order Riordan arrays (FTKRA):*

$$(d; h_1, h_2, \ldots, h_k)A_j(t) = \begin{cases} dB_0\left(\sqrt[k]{\Pi_{n=1}^k h_n(t)}\right), & \text{if } j = 0,\\ dh_1h_2\cdots h_jB_j\left(\sqrt[k]{\Pi_{n=1}^k h_n(t)}\right), & \text{if } 1\leq j\leq k-1.\end{cases}$$

Proof The proof is similar to that of the FTRA and is omitted. □

Similar to the double Riordan arrays, the kth-order Riordan arrays form a group, called the kth-order Riordan group, with respect to the following matrix multiplication: For kth-order Riordan arrays $(d; h_1, h_2, \ldots, h_k)$ and $(g; f_1, f_2, \ldots, f_k)$,

$$\begin{aligned}&(d; h_1, h_2, \ldots, h_k)(g; f_1, f_2, \ldots, f_k)\\ &= \left(dg\left(\sqrt[k]{\Pi_{n=1}^k h_n(t)}\right); \sqrt[k]{\frac{h_1^{k-1}}{h_2h_3\cdots h_k}}f_1\left(\sqrt[k]{\Pi_{n=1}^k h_n(t)}\right), \ldots,\right.\\ &\left.\sqrt[k]{\frac{h_k^{k-1}}{h_1h_2\cdots h_{k-1}}}f_k\left(\sqrt[k]{\Pi_{n=1}^k h_n(t)}\right)\right), \qquad (6.5.41)\end{aligned}$$

where $g(t) = \sum_{n\geq 0} g_{kn}t^{kn}$ and $f_j(t) = \sum_{n\geq 0} f_{j,kn+1}t^{kn+1}$, $1\leq j\leq k$. Formula (6.5.41) can be proved from the FTKRA. The proof is similar to the argument used for the cases of Riordan arrays and double Riordan arrays.

Theorem 6.38 *The kth-order Riordan arrays defined in Definition 6.10 associated with the multiplication defined by* (6.5.41) *form a group, called kth-order Riordan group. Particularly, for $k = 1$ and $k = 2$, the kth-order Riordan group reduces to the Riordan group and double Riordan group, respectively.*

Definition 6.11 The compression of a kth-order Riordan array is defined by $(\hat{d}; \hat{h}_1, \hat{h}_2, \ldots, \hat{h}_k) = (\hat{d}_{n,k})_{n,k\geq 0}$, where

$$\hat{d}_{n,j} = d_{kn-j,(k-1)j}, \quad n\geq j\geq 0, \qquad (6.5.42)$$

or equivalently,

$$\hat{d}(t) = \sum_{n\geq 0} d_{kn}t^n, \quad \hat{h}_j(t) = \sum_{n\geq 0} h_{j,kn+1}t^{n+1}.$$

Sheffer polynomial pairs associated with double Riordan arrays

Let $s_n(x) = \sum_{k=0}^{n} d_{n,k}x^k$. We have seen that if $d_{n,k}$ is the (n,k) entry of the Riordan array $[d(t), h(t)]$, then the polynomial sequence $(s_n(x))_{n\in\mathbb{N}}$ is the Sheffer for $(1/d(\bar{h}(t)), \bar{h}(t))$. Conversely, if the sequence $(s_n(x))_{n\in\mathbb{N}}$ is the Sheffer for $(d(t), h(t))$, then the coefficients $d_{n,k}$ are the entries of the Riordan array $[1/d(\bar{h}(t)), \bar{h}(t)]$.

Despite this apparently different definitions of Riordan arrays and Sheffer polynomial sequences, the two concepts are very similar and we have seen their intrinsic equivalence. It can be shown that the product of two Riordan arrays and the umbral composition of two Sheffer sequences formally follow the same rule. In fact, both the set of Sheffer sequences and the set of Riordan arrays form a group with an operation corresponding to the usual row-by-column product, and as groups they are isomorphic. The equivalence has also been known since the introduction of Riordan arrays: Sheffer-like polynomial sequences are a basic aspect of the $n!$- or classical umbral calculus, while classical Riordan arrays correspond to the same concept in the 1-umbral calculus. He et al. [40] present in an explicit way the stated equivalence and show some developments, which may be of interest to readers concerned with orthogonal polynomials, combinatorial sums, and combinatorial inversion.

In this subsection, we will see that some applications of Riordan arrays to Sheffer polynomials and summation formulas can be extended to the applications of double Riordan arrays to pairs of Sheffer polynomials and pairs of summation formulas.

Let $(g;\, f_1,\, f_2)$ be a double Riordan array, and let its inverse be

$$(d; h_1, h_2) = (g;\, f_1,\, f_2)^{-1} = \left(\frac{1}{g(\overline{f_{1,2}})};\, \overline{f_{1,2}}\frac{t}{f_1(\overline{f_{1,2}})},\, \overline{f_{1,2}}\frac{t}{f_2(\overline{f_{1,2}})}\right), \tag{6.5.43}$$

where $\overline{f_{1,2}}$ is the compositional inverse of $\sqrt{f_1 f_2}$. Inspired by Definition 6.3, we may give the definition of a pair of Sheffer sequences $(s_{1,n},\, s_{2,n})$.

Definition 6.12 Let $g(t)$ be a formal power series in $\mathcal{F}_0$ and let $f_1(t)$ and $f_2(t)$ be two formal power series in $\mathcal{F}_1$. We say that the sequence $(s_{1,n}(x), s_{2,n}(x))_{n\in\mathbb{N}}$ is a *Sheffer (polynomial) pair* associated with the triple $(g(t),\, f_1(t),\, f_2(t))$ if and only if

$$\sum_{k\geq 0} s_{1,k}(x)\frac{t^{2k}}{c_{2k}} + \sum_{k\geq 0} s_{2,k}(x)\frac{t^{2k+1}}{c_{2k+1}} = \frac{1}{d}\varepsilon_x(h_1(t)h_2(t)), \tag{6.5.44}$$

where $\varepsilon_x(uv) = \sum_{k\geq 0} x^{2k}(uv)^k/c_{2k} + u\sum_{k\geq 0} x^{2k+1}(uv)^k/c_{2k+1}$.

If the triple $(g,\, f_1,\, f_2)$ form a double Riordan array $(g;\, f_1,\, f_2)$ with the inverse $(d; h_1, h_2)$, then from Definition 6.12, we may show that

$$s_{1,n}(x) = \sum_{k=0}^{n} d_{2n,2k} \frac{x^{2k}}{c_{2k}} \quad \text{and} \quad s_{2,n}(x) = \sum_{k=0}^{n} d_{2n+1,2k+1} \frac{x^{2k+1}}{c_{2k+1}}, \tag{6.5.45}$$

where $d_{n,2k} = [t^n]d(h_1h_2)^k$ and $d_{n,2k+1} = [t^n]dh_1(h_1h_2)^k$.

As an important example, we will use the following result to construct a Sheffer pair.

Theorem 6.39 *Suppose $b > a$ and $b - a$ is an integer, then the double Riordan array*

$$(d_{n,k})_{n,k\geq 0} = \left(\frac{1}{1-t^2}; t, \frac{t^{2b-2a+1}}{(1-t^2)^b}\right)$$

possesses its entries as follows:

$$d_{2n,2k} = d_{2n+1,2k+1} = \binom{n+(a-1)k}{bk}$$

for $n, k = 0, 1, 2, \ldots$, and 0 otherwise.

Proof It is sufficient to exhibit $d_{2n,2k}$ since the derivation of $d_{2n+1,2k+1}$ is the same. From the definition of double Riordan array, we have

$$\begin{aligned} d_{2n,2k} &= [t^{2n}]\frac{1}{1-t^2}\left(\frac{t^{2b-2a+2}}{(1-t^2)^b}\right)^k \\ &= [t^{2n+2(a-b-1)k}]\frac{1}{(1-t^2)^{bk+1}} \\ &= [t^{2n+2(a-b-1)k}]\sum_{j\geq 0}\binom{-bk-1}{j}(-t^2)^j \\ &= \binom{-bk-1}{n+(a-b-1)k}(-1)^{n+(a-b-1)k} \\ &= \binom{n+(a-1)k}{n+(a-b-1)k} = \binom{n+(a-1)k}{bk}, \end{aligned}$$

which is the desired result. □

Example 6.9 From (6.5.45) and Theorem 6.39, the Sheffer pair associated with the triple $(1/(1-t^2), t, t^{b-a+1}/(1-t^2)^b)^{-1}$ is

$$s_{1,n}(x) = \sum_{k=0}^{n}\binom{n+(a-1)k}{bk}x^{2k} \quad \text{and} \quad s_{2,n}(x) = \sum_{k=0}^{n}\binom{n+(a-1)k}{bk}x^{2k+1}.$$

Particularly, if $a = b = 1$, then

$$s_{1,n}(x) = (1+x^2)^n \quad \text{and} \quad s_{2,n}(x) = x(1+x^2)^n.$$

Inspired by [39, 65], we now construct summation pairs. Let $(d_{n,k})_{n,k\geq 0} = (d; h_1, h_2)$ be a proper (i.e., $d(0) \neq 0$) or improper (i.e., $d(0) = 0$)) double Riordan array with $h_1, h_2 \in \mathcal{F}_1$, and let

$$f(t) = \sum_{k\geq 0} f_{2k}t^{2k} \quad \text{and} \quad g(t) = \sum_{k\geq 0} g_{2k+1}t^{2k+1}$$

be even and odd functions, respectively. The FTDRA shown in (6.5.16) provides a pair of summation formulas

$$\sum_{k\geq 0} d_{2n,2k} f_{2k} = [t^{2n}]df(\sqrt{h_1h_2}) \tag{6.5.46}$$

$$\sum_{k\geq 0} d_{2n+1,2k+1}g_{2k+1} = [t^{2n+1}]d\sqrt{\frac{h_1}{h_2}}g(\sqrt{h_1h_2}). \tag{6.5.47}$$

Let n (resp. m) be a variable, and let m (resp. n), a, and b be parameters. Then the following result provides many pairs of sums.

Theorem 6.40 *Suppose $b > a$ and $c > d$ and $b - a$ and $c - d$ are integers, then the double Riordan array (or improper double Riordan array when $m \neq 0$)*

$$(d_{n,k})_{n,k\geq 0} = \left(\frac{t^{2m}}{(1-t^2)^{m+1}}; \frac{t^{2b-2a+1}}{(1-t^2)^b}, \frac{t^{2c-2d+1}}{(1-t^2)^d}\right)$$

possesses its entries as follows:

$$d_{2n,2k} = \binom{n+(a+c-1)k}{m+(b+d)k} \quad \text{and} \quad d_{2n+1,2k+1} = \binom{n+a(k+1)+(c-1)k}{m+b(k+1)+dk} \tag{6.5.48}$$

for $n, k = 0, 1, 2, \ldots$, and 0 otherwise. Particularly, if $b = a = 0$, then

$$d_{2n,2k} = d_{2n+1,2k+1} = \binom{n+(c-1)k}{m+dk}$$

Proof We first show $d_{2n,2k}$ of (6.5.48). From the definition of double Riordan array, we have

$$
\begin{aligned}
d_{2n,2k} &= [t^{2n}]\frac{t^{2m}}{(1-t^2)^{m+1}}\left(\frac{t^{2b-2a+2d-2c+2}}{(1-t^2)^{b+d}}\right)^k \\
&= [t^{2n-2m+2(a-b+c-d-1)k}]\frac{1}{(1-t^2)^{m+1+(b+d)k}} \\
&= [t^{2n-2m+2(a-b+c-d-1)k}]\sum_{j\geq 0}\binom{-m-1-(b+d)k}{j}(-t^2)^j \\
&= \binom{-m-1-(b+d)k}{n-m+(a-b+c-d-1)k}(-1)^{n-m+(a-b+c-d-1)k} \\
&= \binom{n+(a+c-1)k}{n-m+(a-b+c-d-1)k} = \binom{n+(a+c-1)k}{m+(b+d)k}.
\end{aligned}
$$

We now derive $d_{2n+1,2k+1}$ of (6.5.47) using a similar argument. From the definition of double Riordan array,

$$
\begin{aligned}
d_{2n+1,2k+1} &= [t^{2n+1}]\frac{t^{2m}}{(1-t^2)^{m+1}}\frac{t^{2b-2a+1}}{(1-t^2)^b}\left(\frac{t^{2b-2a+2d-2c+2}}{(1-t^2)^{b+d}}\right)^k \\
&= [t^{2n-2m+2a-2b+2(a-b+c-d-1)k}]\frac{1}{(1-t^2)^{m+1-b+(b+d)k}} \\
&= [t^{2n-2m+2a-2b+2(a-b+c-d-1)k}]\sum_{j\geq 0}\binom{-m-1-b-(b+d)k}{j}(-t^2)^j \\
&= \binom{-m-1-b-(b+d)k}{n-m+a-b+(a-b+c-d-1)k}(-1)^{n-m+a-b+(a-b+c-d-1)k} \\
&= \binom{n+a+(a+c-1)k}{n-m+a-b+(a-b+c-d-1)k} \\
&= \binom{n+a(k+1)+(c-1)k}{m+b(k+1)+dk},
\end{aligned}
$$

completing the proof. □

Example 6.10 Suppose $a=b=c=d=1$, Theorem 6.40 gives the following entries of $(t^{2m}/(1-t^2)^{m+1}, t/(1-t^2), t/(1-t^2))$:

$$
d_{2n,2k} = \binom{n+k}{m+2k} \quad \text{and} \quad d_{2n+1,2k+1} = \binom{n+k+1}{m+2k+1}.
$$

From the generating function of *Catalan numbers* (cf., for example, [19, 33, 68])

$$
\sum_{k\geq 0}\frac{1}{k+1}\binom{2k}{k}t^k = \frac{1-\sqrt{1-4t}}{2t}
$$

we have

$$\sum_{k\geq 0}\frac{(-1)^k}{k+1}\binom{2k}{k}t^{2k}=\frac{\sqrt{1+4t^2}-1}{2t^2}\quad\text{and}\quad\sum_{k\geq 0}\frac{(-1)^k}{k+1}\binom{2k}{k}t^{2k+1}=\frac{\sqrt{1+4t^2}-1}{2t}.$$

Hence,

$$\left(\frac{t^{2m}}{(1-t^2)^{m+1}};\frac{t}{1-t^2},\frac{t}{1-t^2}\right)\frac{\sqrt{1+4t^2}-1}{2t^2}$$

$$=\frac{t^{2m}}{(1-t^2)^{m+1}}\left.\frac{\sqrt{1+4y^2}-1}{2y^2}\right|_{y=t/(1-t^2)}=\frac{t^{2m}}{(1-t^2)^{m+1}}\frac{\frac{1+t^2}{1-t^2}-1}{\frac{2t^2}{(1-t^2)^2}}=\frac{t^{2m}}{(1-t^2)^m}.$$

The above equations imply

$$\sum_{k\geq 0}d_{2n,2k}\frac{(-1)^k}{k+1}\binom{2k}{k}=[t^{2n}]\frac{t^{2m}}{(1-t^2)^m},$$

i.e.,

$$\sum_{k\geq 0}\frac{(-1)^k}{k+1}\binom{2k}{k}\binom{n+k}{m+2k}=\binom{n-1}{m-1}.$$

Exercises

6.1 (i) Show that starting from the sequences $r(t)$ and $c(t)$ defined by (6.1.4), the infinite array $[d_{n,k}]_{n,k\geq 0}$ defined by (6.1.8) is an exponential Riordan array.

Hence, $r(t)$ and $c(t)$ form the r- and c-sequence characterization of an exponential Riordan array. This characterization has been used to study the orthogonal polynomials determined by the exponential Riordan arrays. Note that sometimes these two sequences are denoted by the notations $A_e(t)$ and $Z_e(t)$, respectively, to serve as the analogs of the A- and Z- sequences of the classical Riordan arrays.

(ii) Use the r- and c-sequences of the exponential Riordan array $[g(t), f(t)]$ to characterize the inverse array $[1/g(\bar{f}(t)), \bar{f}(t)]$.

(iii) Suppose $r(t)=1+t^2$, $c(t)=t$ and $g_0=1$. Determine the corresponding exponential Riordan array $[g(t), f(t)]$.

6.2 (i) Use the result of (6.1.8) to find the recurrence relation of Stirling numbers of the second kind $\left\{ {n \atop k} \right\}=\left[\frac{t^n}{n!}\right]\frac{(e^t-1)^k}{k!}$.

(ii) Let the generalized Stirling numbers $S(n,k;\alpha,\beta,\gamma)$ be defined by

$$\frac{1}{k!}(1+\alpha t)^{\gamma/\alpha}\left(\frac{(1+\alpha t)^{\beta/\alpha}-1}{\beta}\right)^k=\sum_{n\geq 0}S(n,k;\alpha,\beta,\gamma)\frac{t^n}{n!} \tag{6.6.1}$$

for $(\alpha,\beta,\gamma)\in\mathbb{R}^3$ and $\beta\neq 0$, where $\lim_{\beta\to 0} S(n,k;1,\beta,0)=(-1)^{n-k}\left[{n\atop k}\right]$, the Stirling numbers of the first kind defined by $(-1)^{n-k}\left[{n\atop k}\right]=\left[\frac{t^n}{n!}\right]\frac{(\ln(1+t))^k}{k!}$, and $\lim_{\alpha\to 0} S(n,k;\alpha,1,0)=\left\{{n\atop k}\right\}$, the Stirling numbers of the second kind defined in (i). Use the result of (6.1.8) to find the recurrence relation of the Stirling numbers $S(n,k;\alpha,\alpha,\gamma)$.

(iii) Let $S(n,k;\alpha,\beta,\gamma)$ be the *generalized Stirling numbers* defined by (6.6.1). Prove the numbers $S(n,k;\alpha,\beta,\gamma)$ satisfy the following recursive relation:

$$S(n+1,k;\alpha,\beta,\gamma)=S(n,k-1;\alpha,\beta,\gamma)+(\gamma+k\beta-n\alpha)S(n,k;\alpha,\beta,\gamma) \tag{6.6.2}$$

for $n\geq 1$.

6.3 Assume that $f(a+t)$ has a formal power series expansion in t with $a\in\mathbb{R}$. Let $\phi(t)=\sum_{n=0}^{\infty}\alpha_n t^n$ be a given formal power series, and let $\bar{f}$ denote the compositional inverse of f so that $(\bar{f}\circ f)(t)=(f\circ\bar{f})(t)=t$. Let $\beta_n=[t^n](f\circ\phi)$ and $\phi(0)=a$. Show that there exists the pair of reciprocal relations

$$\beta_n=\sum_{\sigma(n)} f^{(k)}(a)\frac{\alpha_1^{k_1}\cdots\alpha_n^{k_n}}{k_1!\cdots k_n!}, \tag{6.6.3}$$

$$\alpha_n=\sum_{\sigma(n)} \bar{f}^{(k)}(f(a))\frac{\beta_1^{k_1}\cdots\beta_n^{k_n}}{k_1!\cdots k_n!}, \tag{6.6.4}$$

where the summation is extended over the set $\sigma(n)$ of all partitions of n, that is over all non-negative integral solutions $(k_1,k_2,\ldots,k_n)$ of the equations $k_1+2k_2+\cdots+nk_n=n$, $k_1+k_2+\cdots+k_n=k$, $k=1,2,\ldots,n$.

Replacing α_n by $x_n/n!$ and β_n by $y_n/n!$, we see that (6.6.3) and (6.6.4) may be expressed in terms of the exponential Bell polynomials, namely,

$$y_n=\sum_{k=1}^{n} f^{(k)}(a)B_{n,k}(x_1,x_2,\ldots,x_{n-k+1}), \tag{6.6.5}$$

$$x_n=\sum_{k=1}^{n} \bar{f}^{(k)}(a)B_{n,k}(y_1,y_2,\ldots,y_{n-k+1}), \tag{6.6.6}$$

where $B_{n,k}(\ldots)$ is defined by (cf. Sect. 6.2)

$$B_{n,k}(x_1,x_2,\ldots,x_{n-k+1})=\sum_{\sigma(n,k)}\frac{n!}{k_1!k_2!\cdots}\left(\frac{x_1}{1!}\right)^{k_1}\left(\frac{x_2}{2!}\right)^{k_2}\cdots$$

and $\sigma(n,k)$ is the set of the solutions of the partition equations for a given k $(1\leq k\leq n)$, i.e., the set of all partitions of n into k parts. Of course, the set $\sigma(n)$ is the union of all subsets $\sigma(n,k)$, $k=1,2,\ldots,n$.

6.4 (i) Verify the Faà di Bruno's relations

$$\beta_n = \sum_{\sigma(n)} \frac{\alpha_1^{k_1} \cdots \alpha_n^{k_n}}{k_1! \cdots k_n!},$$

$$\alpha_n = \sum_{\sigma(n)} (-1)^{k-1}(k-1)! \frac{\beta_1^{k_1} \cdots \beta_n^{k_n}}{k_1! \cdots k_n!},$$

which have the associated relations

$$\exp\left(\sum_{n=1}^{\infty} \alpha_n t^n\right) = 1 + \sum_{n=1}^{\infty} \beta_n t^n,$$

$$\ln\left(1 + \sum_{n=1}^{\infty} \beta_n t^n\right) = \sum_{n=1}^{\infty} \alpha_n t^n.$$

(ii) Verify the Faà di Bruno's relations

$$\beta_n = \sum_{\sigma(n)} (\alpha)_k \frac{\alpha_1^{k_1} \cdots \alpha_n^{k_n}}{k_1! \cdots k_n!},$$

$$\alpha_n = \sum_{\sigma(n)} (1/\alpha)_k \frac{\beta_1^{k_1} \cdots \beta_n^{k_n}}{k_1! \cdots k_n!},$$

where $(\alpha)_k = \alpha(\alpha-1)\cdots(\alpha-k+1)$ and $(\alpha)_0 = 1$, which have the associated relations

$$\left(1 + \sum_{n=1}^{\infty} \alpha_n t^n\right)^{\alpha} = 1 + \sum_{n=1}^{\infty} \beta_n t^n,$$

$$\left(1 + \sum_{n=1}^{\infty} \beta_n t^n\right)^{1/\alpha} = 1 + \sum_{n=1}^{\infty} \alpha_n t^n.$$

(iii) Let C_n be the Catalan numbers. Prove that

$$C_n := \frac{1}{n+1}\binom{2n}{n} = \frac{1}{n+1} \sum_{\sigma(n)} \frac{2^{2n-k}}{1^{k_1} k_1! 2^{k_2} k_2! \cdots n^{k_n} k_n!} \tag{6.6.7}$$

and

$$4^n = 2n \sum_{\sigma(n)} (-1)^{k-1}(k-1)! \prod_{j=1}^{n} \frac{\binom{2j}{j}^{k_j}}{k_j!}. \tag{6.6.8}$$

Furthermore,

$$C_n = \frac{1}{n+1} \sum_{[n/2]\le k\le n} \left(-\frac{1}{2}\right)_k \frac{(-4)^{n-k}}{(2k-n)!(n-k)!}. \tag{6.6.9}$$

6.5 (i) Show that the Riordan array $(d(xt), h(xt)/x) = (d_{n,k}x^{n-k})_{n,k\ge 0}$, where $d_{n,k} = [t^n]d(t)(h(t))^k$.

If $(d(t), h(t)) = (1/(1-t), t/(1-t))$, then $(1/(1-xt), t/(1-xt)) = \left(\binom{n}{k}x^{n-k}\right)_{n\ge k\ge 0}$ is called the Pascal matrix function denoted by $P[x]$.

(ii) Prove that $P[x]$ satisfies the homomorphism property

$$P[x+y] = P[x]P[y].$$

(iii) Let $B(x) = (B_0(x), B_1(x), \ldots)^T$ and $B = (B_0, B_1, \ldots)^T$ be a *Bernoulli polynomial sequence* and a *Bernoulli number sequence*, respectively. It is known that $B_n = B_n(0)$ and $B(x) = P[x]B$. Show $B = P[-x]B(x)$, which implies

$$B_n = \sum_{k=0}^{n} \binom{n}{k}(-x)^{n-k}B_k(x). \tag{6.6.10}$$

6.6 Denote by $(d_{n,k})_{0\le k\le n}$ the Bell-type Riordan array $(d(t), h(t))$ with respect to the sequence $(c_n)_{n\in\mathbb{N}}$. Prove the identity

$$\sum_{n=0}^{s} \frac{c_{s+1}}{c_n c_{s-n}} \frac{c_k c_m}{c_{k+m+1}} d_{n,k} d_{s-n,m} = d_{s+1,k+m+1} \tag{6.6.11}$$

for $s \ge n \ge k \ge 0$ and $s-n \ge m \ge 0$. Particularly, if $(d(t), h(t)) = (1/(1-t), t/(1-t))$, show this formula reduces to the Chu-Vandermonde type identity (cf. (3.3) in [H. W. Gould, Combinatorial Identities, Revised Edition, Morgantown, W. Va., 1972.]):

$$\sum_{n=k}^{s-m} \binom{n}{k}\binom{s-n}{m} = \binom{s+1}{k+m+1}. \tag{6.6.12}$$

6.7 Let $(d(t), h(t)) = (d_{n,k})_{n,k\ge 0}$ be a Riordan array, and let $p_n(x) = \sum_{n\ge k\ge 0} d_{n,k}x^k/k!$. Prove that $(p_n(x))_{n\ge 0}$ is a Sheffer polynomial sequence if and only if there is a unique A-sequence, $A = (a_j)_{j\ge 0}$ with $a_0 \ne 0$, and a unique Z-sequence, $Z = (z_j)_{j\ge 0}$, such that

$$p_n(x) = \sum_{j\ge 0} a_j \sum_{k=j}^{n} d_{n-1,k-1} \frac{x^{k-j}}{(k-j)!}, \tag{6.6.13}$$

$$p_n(0) = \sum_{j\ge 0} z_j \frac{\mathrm{d}^j}{\mathrm{d}x^j} p_{n-1}(x)\Big|_{x=0} \tag{6.6.14}$$

for all $n, k \ge 0$.

6.8 Use exponential Riordan arrays to prove the b-th order Gould's identity

$$\sum_{k=b}^{n} \frac{\gamma - \beta b}{\gamma - \beta k} \binom{\gamma - \beta k}{k - b} \binom{\alpha + k\beta}{n - k} = \binom{\alpha + \gamma}{n - b},$$

which includes Gould's identity (11), in [H.W. Gould, Some generalizations of Vandermonde's convolution. *Amer. Math. Monthly*, 63 (1956), 84–91], as the case of $b = 0$.

6.9 Elements of the stabilizer subgroup of the double Riordan group by a column vector $f = (f_0, f_1, f_2, ...)^T$ with generating function $f(t)$, where $f_0 \neq 0$, are the double Riordan arrays $(d(t); h_1(t), h_2(t))$ such that

$$(d(t); h_1(t), h_2(t)) f = f.$$

We denote this subgroup by

$$\mathcal{DS}_f = \{(d; h_1, h_2) \in \mathcal{DR} : (d; h_1, h_2) f = f\},$$

where $\mathcal{DR}$ is the double Riordan group. If $f = 1/(1 - kt), k \geq 1$, show that a double Riordan array $(d; h_1, h_2)$ is a stabilizer associated with f if and only if

$$d(t) = \frac{1 - k^2 t h_2(t)}{1 - k^2 t^2} \tag{6.6.15}$$

with $d_0 = 1$ and $d_2 \neq k^2$, where $d(t) = \sum_{k \geq 0} d_{2k} t^{2k}$.

6.10 Let $C(t)$ be the generating function of the Catalan numbers. (i) Find $h(t)$ such that the double Riordan array $(C^2(t^2); t, h(t))$ is a stabilizer associated with $f = 1/(1 - t)$. (ii) Show that there does not exist a function $h(t)$ such that $(C(t^2); t, h(t))$ or $(1/(1 - t^2); t, h(t))$ is a stabilizer associated with $f = 1/(1 - t)$. (iii) Find functions $h(t)$ such that the double Riordan arrays $(d(t); t, h(t))$ are stabilizers associated with $f = 1/(1 - 2t)$ for $d(t) = 1/(1 - t^2)$ and $d(t) = C(t^2)$, respectively.

References

1. M. Abramowitz, I.A. Stegun, *Handbook of Mathematical Functions with Formulas, Graphs, and Mathematical Tables*. Reprint of the 1972 edition (Dover Publications, Inc., New York, 1992)
2. G.E. Andrews, R. Askey, R. Roy, *Special Functions* (Cambridge University Press, Cambridge, 1999)
3. E. Barcucci, A. Del Lungo, E. Pergola, R. Pinzani, ECO: a methodology for the enumeration of combinatorial objects. J. Differ. Equ. Appl. **5**, 435–490 (1999)

4. P. Barry, Exponential Riordan arrays and permutation enumeration. J. Integer Seq. **13**(9), Article 10.9.1, 16 pp. (2010)
5. P. Barry, Combinatorial polynomials as moments, Hankel transforms, and exponential Riordan arrays. J. Integer Seq. **14**(6), Article 11.6.7, 14 pp. (2011)
6. P. Barry, General Eulerian polynomials as moments using exponential Riordan arrays. J. Integer Seq. **16**(9), Article 13.9.6, 15 pp. (2013)
7. P. Barry, Constructing exponential Riordan arrays from their A and Z sequences. J. Integer Seq. **17**(2), Art. 14.2.6, 19 pp. (2014)
8. P. Barry, *Riordan Arrays: A Primer* (Logic Press, Raleigh, 2016)
9. R.P. Boas Jr., R.C. Buck, Polynomials defined by generating relations. Am. Math. Mon. **63**, 626–632 (1956)
10. J.W. Brown, Generalized Appell connection sequences. II, J. Math. Anal. Appl. **50**, 458–464 (1975)
11. J.W. Brown, J.L. Goldberg, Generalized Appell connection sequences. J. Math. Anal. Appl. **46**, 242–248 (1974)
12. J.W. Brown, M. Kuczma, Self-inverse Sheffer sequences. SIAM J. Math. Anal. **7**(5), 723–728 (1976)
13. G.S. Call, D.J. Velleman, Pascal's matrices. Am. Math. Mon. **100**(4), 372–376 (1993)
14. G.-S. Cheon, A note on the Bernoulli and Euler polynomials. Appl. Math. Lett. **16**(3), 365–368 (2003)
15. G.-S. Cheon, S.-G. Hwang, S.-G. Lee, Several polynomials associated with the harmonic numbers. Discrete Appl. Math. **155**(18), 2573–2584 (2007)
16. G.-S. Cheon, J.-H. Jung, L.W. Shapiro, Generalized Bessel numbers and some combinatorial settings. Discrete Math. **313**(20), 2127–2138 (2013)
17. G.-S. Cheon, J.-S. Kim, Stirling matrix via Pascal matrix. Linear Algebra Appl. **329**(1–3), 49–59 (2001)
18. F.R.K. Chung, R.L. Graham, V.E. Hoggatt, M. Kleiman, The number of Baxter permutations. J. Combin. Theory Ser. A **24**, 382–394 (1978)
19. L. Comtet, *Advanced Combinatorics* (D. Reidel Publishing Co., Dordrecht, 1974)
20. C. Corsani, D. Merlini, R. Sprugnoli, Left-inversion of combinatorial sums. Discrete Math. **180**(1–3), 107–122 (1998)
21. F. Costabile, Expansions of real functions in Bernoulli polynomials and applications. Conf. Semin. Mat. Univ. Bari **273**, 13 pp. (1999)
22. F. Costabile, F. Dell'Accio, M.I. Gualtieri, A new approach to Bernoulli polynomials. Rend. Mat. Appl. (7) **26**(1), 1–12 (2006)
23. F.A. Costabile, E. Longo, A determinantal approach to Appell polynomials. J. Comput. Appl. Math. **234**(5), 1528–1542 (2010)
24. E. Deutsch, L. Ferrari, S. Rinaldi, Production matrices. Adv. Appl. Math. **34**(1), 101–122 (2005)
25. E. Deutsch, L. Ferrari, S. Rinaldi, Production matrices and Riordan arrays. Ann. Comb. **13**(1), 65–85 (2009)
26. E. Deutsch, L. Shapiro, Exponential Riordan Arrays, Lecture Notes (Nankai University, 2004), available electronically: http://www.combinatorics.net/ppt2004/Louis%20W.%20Shapiro/shapiro.pdf
27. D.E. Davenport, L.W. Shapiro, L.C. Woodson, The double Riordan group. Electron. J. Combin. **18**(2), P33 (2012)
28. L. Ferrari, E. Pergola, R. Pinzani, S. Rinaldi, An algebraic characterization of the set of succession rules. Selected papers in honor of Maurice Nivat, Theoret. Comput. Sci. **281**(1-2), 351–367 (2002)
29. H.W. Gould, T.-X. He, Characterization of (c)-Riordan arrays, Gegenbauer-Humbert-type polynomial sequences, and (c)-Bell polynomials. J. Math. Res. Appl. **33**(5), 505–527 (2013)
30. T.-X. He, A symbolic operator approach to power series transformation-expansion formulas. J. Integer Seq. **11**(2), Article 08.2.7, 19 pp. (2008)

31. T.-X. He, Riordan arrays associated with Laurent series and generalized Sheffer-type groups. Linear Algebra Appl. **435**(6), 1241–1256 (2011)
32. T.-X. He, Characterizations of orthogonal generalized Gegenbauer-Humbert polynomials and orthogonal Sheffer-type polynomials. J. Comput. Anal. Appl. **13**(4), 701–723 (2011)
33. T.-X. He, The characterization of Riordan arrays and Sheffer-type polynomial sequences. 24th MCCCC, J. Combin. Math. Combin. Comput. **82**, 249–268 (2012)
34. T.-X. He, A unified approach to generalized Stirling functions. J. Math. Res. Appl. **32**(6), 631–646 (2012)
35. T.-X. He, Expression and computation of generalized Stirling numbers. J. Combin. Math. Combin. Comput. **86**, 239–268 (2013)
36. T.-X. He, Matrix characterizations of Riordan arrays. Linear Algebra Appl. **465**, 15–42 (2015)
37. T.-X. He, Applications of Riordan matrix functions to Bernoulli and Euler polynomials. Linear Algebra Appl. **507**, 208–228 (2016)
38. T.-X. He, Sequence characterizations of double Riordan arrays and their compressions. Linear Algebra Appl. **549**, 176–202 (2018)
39. T.-X. He, L.C. Hsu, X. Ma, On an extension of Riordan arrays with an application to obtaining convolution-type identities. Eur. J. Combin. **42**, 112–134 (2014)
40. T.-X. He, L.C. Hsu, P.J.-S. Shiue, The Sheffer group and the Riordan group. Discrete Appl. Math. **155**(15), 1895–1909 (2007)
41. T.-X. He, J.H.-C. Liao, P.J.-S. Shiue, The Pascal matrix function and its applications to Bernoulli numbers and Bernoulli polynomials and Euler numbers and Euler polynomials. J. Comb. Number Theory **6**(3), 189–207 (2014)
42. Y. Horibe, Notes on Fibonacci trees and their optimality. Fibonacci Quart. **21**(2), 118–128 (1983)
43. L.C. Hsu, Generalized Stirling number pairs associated with inverse relations. Fibonacci Quart. **25**(4), 346–351 (1987)
44. Ch. Jordan, *Calculus of Finite Differences* (Chelsea, New York, 1965)
45. D.E. Knuth, *The Art of Computer Programming*, vol. 3. Sorting and Searching, 2nd edn. (Addison-Wesley, Reading, MA, 1998)
46. M. Koutras, Noncentral Stirling numbers and some applications. Discrete Math. **42**(1), 73–89 (1982)
47. A. Kuznetkov, I. Pak, A. Postnikov, Trees associated with the Motzkin numbers. J. Combin. Theory Series A **76**, 145–147 (1996)
48. D. Merlini, R. Sprugnoli, M.C. Verri, The Cauchy numbers. Discrete Math. **306**(16), 1906–1920 (2006)
49. D. Merlini, M.C. Verri, Generating trees and proper Riordan Arrays. Discrete Math. **218**(16), 167–183 (2000)
50. H. Pan, Z.-W. Sun, New identities involving Bernoulli and Euler polynomials. J. Combin. Theory Ser. A **113**(1), 156–175 (2006)
51. P. Peart, W.-J. Woan, Generating functions via Hankel and Stieltjes matrices. J. Integer Seq. **3**(2), Article 00.2.1 (2000)
52. E.D. Rainville, *Special Functions*. Reprint of 1960 first edition (Chelsea Publishing Co., Bronx, N.Y., 1971)
53. J. Riordan, Moment recurrence relations for binomial Poisson and hypergeometric frequency distribution. Ann. Math. Statist. **8**, 103–111 (1937)
54. J. Riordan, *Combinatorial Identities*. Reprint of the original (Krieger Publishing Co., Huntington, N.Y, Robert E, 1968)
55. D.G. Rogers, Pascal triangles, Catalan numbers and renewal arrays. Discrete Math. **22**(3), 301–310 (1978)
56. S. Roman, The theory of the umbral calculus. I, J. Math. Anal. Appl. **87**(1), 58–115 (1982)
57. S. Roman, *The Umbral Calculus* (Academic Press Inc., New York, 1984)
58. S. Roman, G.-C. Rota, The Umbral calculus. Adv. Math. **27**(2), 95–188 (1978)
59. G.-C. Rota, D. Kahaner, A. Odlyzko, On the foundations of combinatorial theory. VIII. Finite operator calculus. J. Math. Anal. Appl. **42**, 684–760 (1973)

60. L.W. Shapiro, Some open questions about random walks, involutions, limiting distributions and generating functions. Adv. Appl. Math. **27**(2–3), 585–596 (2001)
61. L.W. Shapiro, Bijections and the Riordan group. Theoret. Comput. Sci. **307**(2), 403–413 (2003)
62. L.W. Shapiro, S. Getu, W.J. Woan, L.C. Woodson, The Riordan group. Discrete Appl. Math. **34**(1–3), 229–239 (1991)
63. I.M. Sheffer, Some properties of polynomial sets of type zero. Duke Math. J. **5**, 590–622 (1939)
64. I.M. Sheffer, Note on Appell polynomials. Bull. Am. Math. Soc. **51**, 739–744 (1945)
65. R. Sprugnoli, Riordan arrays and combinatorial sums. Discrete Math. **132**(1–3), 267–290 (1994)
66. R. Sprugnoli, Riordan arrays and the Abel-Gould identity. Discrete Math. **142**(1–3), 213–233 (1995)
67. H.M. Srivastava, Á. Pintér, Remarks on some relationships between the Bernoulli and Euler polynomials. Appl. Math. Lett. **17**(4), 375–380 (2004)
68. R.P. Stanley, *Catalan Numbers* (Cambridge University Press, New York, 2015)
69. R. Vein, P. Dale, *Determinants and Their Applications in Mathematical Physics*. Applied Mathematical Sciences, vol. 134 (Springer, New York, 1999)
70. W. Wang, Generalized higher order Bernoulli number pairs and generalized Stirling number pairs. J. Math. Anal. Appl. **364**(1), 255–274 (2010)
71. W. Wang, A determinantal approach to Sheffer sequences. Linear Algebra Appl. **463**, 228–254 (2014)
72. W. Wang, T. Wang, A note on the relationships between the generalized Bernoulli and Euler polynomials. Ars Combin. **88**, 397–405 (2008)
73. W. Wang, T. Wang, Generalized Riordan arrays. Discrete Math. **308**(24), 6466–6500 (2008)
74. J. West, Generating trees and forbidden subsequences, in *Proceedings of the 6th Conference on Formal Power Series and Algebraic Combinatorics* (New Brunswick, NJ, 1994), Discrete Math. **157**(1–3), 363–374 (1996)
75. S.-L. Yang, Recurrence relations for the Sheffer sequences. Linear Algebra Appl. **437**(12), 2986–2996 (2012)
76. Y. Yang, Determinant representations of Appell polynomial sequences. Oper. Matrices **2**(4), 517–524 (2008)
77. X. Zhao, T. Wang, Some identities related to reciprocal functions. Discrete Math. **265**(1–3), 323–335 (2003)

Chapter 7
Extensions of the Riordan Group

Abstract We consider two groups, $\mathcal{F}_0 = \{g \in \mathbb{F}[[z]] \mid g(0) \neq 0\}$ under multiplication, and $\mathcal{F}_1 = z\mathcal{F}_0$ under composition where $\mathbb{F}$ is the real field $\mathbb{R}$ or complex field $\mathbb{C}$. As observed in Sect. 3.3, it is known that the Riordan group $\mathcal{R}$ is isomorphic to the semidirect product $\mathcal{F}_0 \rtimes \mathcal{F}_1$. It may be viewed as a group extension of $\mathcal{F}_1$ by $\mathcal{F}_0$. In this chapter, we develop the *group of three-dimensional Riordan arrays* [7] from an extension of the Riordan group $\mathcal{R}$ by $\mathcal{F}_0$. This concept extends to the group of multi-dimensional Riordan arrays. Moreover, we discuss the *multivariate Riordan group* [6, 17] defined by the semidirect product $\mathcal{F}_{\mathbf{0}}^d \rtimes \mathcal{F}_{\mathbf{1}}^d$ for an integer $d \geq 1$. The two groups $\mathcal{F}_{\mathbf{0}}^d$ and $\mathcal{F}_{\mathbf{1}}^d$ are obtained, respectively, from $\mathcal{F}_0$ and $\mathcal{F}_1$ by extending the ring $\mathbb{F}[[z]]$ of a single variable to the ring $\mathbb{F}[[z_1, \ldots, z_d]]$ of d variables. We will see some similarity with the Riordan group in a single variable, but its matrix representation, called a *multivariate Riordan array*, will be quite different to usual Riordan arrays.

7.1 Three-Dimensional Riordan Group

Recall that for any two groups H and K, a group G is called a *group extension* of K by H if there exists a normal subgroup N of G such that $H \cong N$ and $K \cong G/N$. Thus a group extension is a general means of describing a group in terms of a particular normal subgroup and quotient group. We note that $\mathcal{F}_0 \cong \mathcal{A}$ and $\mathcal{F}_1 \cong \mathcal{L} \cong \mathcal{R}/\mathcal{A}$. In this sense, the Riordan group $\mathcal{R} = \{(g, f) \mid g \in \mathcal{F}_0, f \in \mathcal{F}_1\}$ may be viewed as an extension of the group $\mathcal{F}_1$ by the group $\mathcal{F}_0$.

A short exact sequence of the groups

$$1 \to \mathcal{F}_0 \stackrel{\alpha}{\to} \mathcal{R} \stackrel{\beta}{\to} \mathcal{F}_1 \to 1 \tag{7.1.1}$$

is another way of presenting an extension. Note that the map α is a monomorphism defined as $\alpha(g) = (g, z)$ and β is an epimorphism defined as $\beta((g, f)) = \overline{f}$ so that $\mathrm{Im}\,(\alpha) = \mathcal{A} = \ker\,(\beta)$. In addition, if we define a map $\gamma : \mathcal{F}_1 \to \mathcal{R}$ by $\gamma(f) = (1, \overline{f})$ then γ is a homomorphism such that $\beta \circ \gamma$ is the identity map. Thus the exact

© The Author(s), under exclusive license to Springer Nature Switzerland AG 2022
L. Shapiro et al., *The Riordan Group and Applications*, Springer Monographs in Mathematics, https://doi.org/10.1007/978-3-030-94151-2_7

sequence (7.1.1) is split so that $\mathcal{R}$ is isomorphic to the semidirect product of $\mathcal{F}_0$ and $\mathcal{F}_1$ by the splitting lemma.

The homomorphism $\varphi_1 : \mathcal{F}_1 \to \mathrm{Aut}(\mathcal{F}_0)$, $f \mapsto \varphi_1(f) := \varphi_{1_f}$ with $\varphi_{1_f}(g) = g \circ \overline{f} = g(\overline{f})$ leads to the binary operation $*_{\varphi_1}$ on the semidirect product $\mathcal{F}_0 \rtimes_{\varphi_1} \mathcal{F}_1$ with respect to φ_1:

$$(g_1, f_1) *_{\varphi_1} (g_2, f_2) = \left(g_1 \cdot \varphi_{1_{f_1}}(g_2), f_1 \circ f_2\right) = \left(g_1 g_2(\overline{f_1}), f_1(f_2)\right). \quad (7.1.2)$$

In [11], Goins and Nkwanta observed that the linear representation $\pi : \mathcal{F}_0 \rtimes_{\varphi_1} \mathcal{F}_1 \to GL(V)$ uniquely defines a Riordan matrix by showing that $\pi(g, \overline{f}) = (g, f)$, where $GL(V)$ is the general linear group over the vector space $V := \mathbb{F}[[z]]$.

We now go through ideas analogous to an extension of $\mathcal{F}_1$ by $\mathcal{F}_0$ to address an interesting question:

"*what group is an extension of the Riordan group $\mathcal{R}$ by $\mathcal{F}_0$?*"

As a result, we see how the three-dimensional Riordan arrays arise in describing extensions of the Riordan group. We refer to the extension as the group of three-dimensional Riordan arrays. More generally, this concept leads us to the *group of multi-dimensional Riordan arrays* [7].

Lemma 7.1 *Let $\varphi_2 : \mathcal{F}_0 \rtimes_{\varphi_1} \mathcal{F}_1 \to \mathrm{Aut}(\mathcal{F}_0)$, $(g, f) \mapsto \varphi_2((g, f)) := \varphi_{2_{(g,f)}}$ be a map defined by $\varphi_{2_{(g,f)}}(h) = h \circ \overline{f}$ for $h \in \mathcal{F}_0$. Then φ_2 is a homomorphism.*

Proof For any $h \in \mathcal{F}_0$ we have $(h \circ \overline{f})(0) = h(\overline{f}(0)) = h(0) \neq 0$. Thus the map φ_2 is well-defined. Using (7.1.2) together with $\overline{f_2} \circ \overline{f_1} = \overline{f_1 \circ f_2}$, it can be easily shown that φ_2 is a homomorphism. □

It is also easy to see with Lemma 7.1 that the binary operation $*_{\varphi_2}$ on the semidirect product $\mathcal{F}_0 \rtimes_{\varphi_2} (\mathcal{F}_0 \rtimes_{\varphi_1} \mathcal{F}_1)$ with respect to φ_2 is given by

$$(h_1, (g_1, f_1)) *_{\varphi_2} (h_2, (g_2, f_2)) = \left(h_1 h_2(\overline{f_1}), \left(g_1 g_2(\overline{f_1}), f_1(f_2)\right)\right). \quad (7.1.3)$$

From now on, we set $G = \mathcal{F}_0 \rtimes_{\varphi_2} (\mathcal{F}_0 \rtimes_{\varphi_1} \mathcal{F}_1)$ and $V^{\mathbb{N}} = \prod_{n=1}^{\infty} \mathbb{F}[[z]]$, the infinite Cartesian product of $\mathbb{F}[[z]]$. Note that $V^{\mathbb{N}}$ is a $\mathbb{F}$-vector space with the Schauder basis $\{z^i \mathbf{e}_j\}_{i,j \in \mathbb{N}}$, where $\mathbf{e}_j = (0, \ldots, 0, 1, 0, \ldots)$ with jth position 1.

Theorem 7.1 *Let $\mu : G \times V^{\mathbb{N}} \to V^{\mathbb{N}}$ be a map defined by*

$$\mu(x, \xi) = x *_{\varphi_2} \xi := (x *_{\varphi_2} \xi_0, x *_{\varphi_2} \xi_1, \ldots) \quad (7.1.4)$$

where $x = (h, (g, f))$, $\xi = (\xi_0, \xi_1, \ldots)$, and $x *_{\varphi_2} \xi_j = gh^j \xi_j(\overline{f})$ for $j = 0, 1, \ldots$. Then μ is linear and a group action of G on $V^{\mathbb{N}}$.

Proof Obviously, the map μ is well-defined. Since the identity of G is $e = (1, (1, z))$ and $\bar{z} = z$, we have $(1, (1, z)) *_{\varphi_2} \xi_j = \xi_j$ for $j = 0, 1, \ldots$. Thus, $e *_{\varphi_2} \xi = \xi$. Now set $x_1 = (h_1, (g_1, f_1)), x_2 = (h_2, (g_2, f_2)) \in G$. Using $\overline{f_2} \circ \overline{f_1} = \overline{f_1 \circ f_2}$ and (7.1.3), for $j = 0, 1, \ldots$ we obtain

$$\begin{aligned} x_1 *_{\varphi_2} (x_2 *_{\varphi_2} \xi_j) &= g_1 h_1^j \left(g_2 h_2^j \xi_j(\overline{f_2}) \right)(\overline{f_1}) = g_1 g_2(\overline{f_1}) h_1^j h_2^j(\overline{f_1}) \xi_j(\overline{f_1 \circ f_2}) \\ &= (x_1 *_{\varphi_2} x_2) *_{\varphi_2} \xi_j. \end{aligned}$$

In addition, for any $\xi, \hat{\xi} \in V^{\mathbb{N}}$, $\alpha \in \mathbb{F}$ and $j = 0, 1, \ldots$ we obtain

$$\begin{aligned} x_1 *_{\varphi_2} (\xi_j + \hat{\xi}_j) &= g_1 h_1^j \xi_j(\overline{f_1}) + g_1 h_1^j \hat{\xi}_j(\overline{f_1}) = (x_1 *_{\varphi_2} \xi_j) + (x_1 *_{\varphi_2} \hat{\xi}_j), \\ x_1 *_{\varphi_2} \alpha \xi_j &= g_1 h_1^j \alpha \xi_j(\overline{f_1}) = \alpha(x_1 *_{\varphi_2} \xi_j). \end{aligned}$$

Hence μ is linear and a group action of G on $V^{\mathbb{N}}$. □

We now construct a matrix representation of the semidirect product

$$G = \mathcal{F}_0 \rtimes_{\varphi_2} (\mathcal{F}_0 \rtimes_{\varphi_1} \mathcal{F}_1)$$

acting on the $\mathbb{F}$-vector space $V^{\mathbb{N}}$. By linearity of the group action μ in (7.1.4), we obtain the linear representation $\sigma : G \to \operatorname{Aut}(V^{\mathbb{N}})$, which sends $x = (h, (g, f)) \in G$ to the automorphism

$$\sigma_x : (\xi_0, \xi_1, \ldots, \xi_k, \ldots) \mapsto (gh^0 \xi_0(\overline{f}), gh^1 \xi_1(\overline{f}), \ldots, gh^k \xi_k(\overline{f}), \ldots).$$

Consider the $\mathbb{F}$-vector space of infinite matrices denoted by $\operatorname{Mat}_{\mathbb{N}}(\mathbb{F})$. It has the Schauder basis $\{E_{i,j}\}_{i,j \in \mathbb{N}}$ with $E_{i,j}$ representing the infinite matrix equal to 1 at the (i, j)-position and 0 elsewhere. Then there exists an isomorphism $\iota : V^{\mathbb{N}} \to \operatorname{Mat}_{\mathbb{N}}(\mathbb{F})$ given by $\iota(\xi_0, \xi_1, \ldots) = [a_{i,j}]_{i,j \geq 0}$ where $a_{i,j} = [z^i]\xi_j$. Since ι is linear and invertible, conjugation by ι sends a homomorphism σ to the new homomorphism $\hat{\sigma} : G \to \operatorname{Aut}(\operatorname{Mat}_{\mathbb{N}}(\mathbb{F}))$ given by $x \mapsto \iota \circ \sigma(x) \circ \iota^{-1}$ so that it satisfies $\iota(\sigma_x(\xi)) = \hat{\sigma}_x(\iota(\xi))$. This identifies isomorphic representations (see [2]). Since $\iota(z^j \mathbf{e}_k) = E_{j,k}$ there exists $m_{i,j,k} = [z^i] gh^k \overline{f}^j$ such that

$$\hat{\sigma}_x(E_{j,k}) = \iota(\sigma_x(z^j \mathbf{e}_k)) = \iota((gh^k \overline{f}^j) \mathbf{e}_k) = \sum_{i \geq 0} m_{i,j,k} E_{i,k}.$$

Let $A = [a_{j,k}] \in \operatorname{Mat}_{\mathbb{N}}(\mathbb{F})$. By linearity we then obtain

$$\begin{aligned} \hat{\sigma}_x(A) = \hat{\sigma}_x \left(\sum_{j \geq 0} \sum_{k \geq 0} a_{j,k} E_{j,k} \right) &= \sum_{j \geq 0} \sum_{k \geq 0} a_{j,k} \left(\sum_{i \geq 0} m_{i,j,k} E_{i,k} \right) \\ &= \sum_{k \geq 0} \sum_{i \geq 0} \left(\sum_{j \geq 0} m_{i,j,k} a_{j,k} \right) E_{i,k}. \end{aligned}$$

Thus it follows that if $\hat{\sigma}_x(A) = [b_{i,k}]_{i,k\in\mathbb{N}}$, then $b_{i,k}$ can be expressed as

$$b_{i,k} = \sum_{j\geq 0} m_{i,j,k}a_{j,k}. \tag{7.1.5}$$

From this point of view, the homomorphism $\hat{\sigma} : G \to \mathrm{Aut}(\mathrm{Mat}_{\mathbb{N}}(\mathbb{F}))$ can be represented as a three-dimensional array:

$$\hat{\sigma}(x) = \hat{\sigma}_x = [m_{i,j,k}]_{i,j,k\in\mathbb{N}}, \quad m_{i,j,k} = [z^i]gh^k\overline{f}^j, \tag{7.1.6}$$

where $x = (h, (g, \overline{f}))$. Note that the subarray corresponding to $k = 0$ is the Riordan array $(g, \overline{f})$. Since $\overline{f} \in \mathcal{F}_1$, more generally, we call the array $\mathbb{M} = [m_{i,j,k}]_{i,j,k\in\mathbb{N}}$ with

$$m_{i,j,k} = [z^i]gf^jh^k$$

a *three-dimensional (or 3-D) Riordan array* [7] and we write $\mathbb{M}$ as (g, f, h) or $\mathcal{R}(g, f, h)$. The set of all three-dimensional Riordan arrays over a field $\mathbb{F}$ is denoted by

$$\mathcal{R}^{\langle 3\rangle} = \{(g, f, h) \mid g, h \in \mathcal{F}_0,\ f \in \mathcal{F}_1\}.$$

Theorem 7.2 *The set $\mathcal{R}^{\langle 3\rangle}$ forms a group under the binary operation*

$$(g_1, f_1, h_1) * (g_2, f_2, h_2) = (g_1g_2(f_1), f_2(f_1), h_1h_2(f_1)). \tag{7.1.7}$$

We call $\mathcal{R}^{\langle 3\rangle}$ the three-dimensional (or 3-D) Riordan group.

Proof Since $g_1(0)g_2(f_1(0)) = g_1(0)g_2(0) \neq 0$, $h_1(0)h_2(f_1(0)) = h_1(0)\,h_2(0) \neq 0$ and $f_2(f_1) \in \mathcal{F}_1$, the binary operation $* : \mathcal{R}^{\langle 3\rangle} \times \mathcal{R}^{\langle 3\rangle} \to \mathcal{R}^{\langle 3\rangle}$ defined in (7.1.7) is well-defined. One can easily show that $*$ is associative. Since

$$(1, z, 1) * (g, f, h) = (g, f, h) * (1, z, 1) = (g, f, h),$$

it follows that $\mathbb{I} = (1, z, 1)$ is the identity. In addition, if

$$(g_1, f_1, h_1) * (g_2, f_2, h_2) = (1, z, 1)$$

then $g_2 = \frac{1}{g_1(\bar{f}_1)}$, $h_2 = \frac{1}{h_1(\bar{f}_1)}$, and $f_2 = \bar{f}_1$. Thus

$$(g_1, f_1, h_1)^{-1} = \left(\frac{1}{g_1(\bar{f}_1)}, \bar{f}_1, \frac{1}{h_1(\bar{f}_1)}\right) \in \mathcal{R}^{\langle 3\rangle}.$$

□

Another way of describing the three-dimensional Riordan group $\mathcal{R}^{\langle 3\rangle}$ is via group extension [7].

Theorem 7.3 *The group $\mathcal{R}^{\langle 3\rangle}$ is an extension of the Riordan group $\mathcal{R}$ by $\mathcal{F}_0$.*

Proof Let us define two maps $\alpha : \mathcal{F}_0 \to \mathcal{R}^{\langle 3\rangle}$ and $\beta : \mathcal{R}^{\langle 3\rangle} \to \mathcal{R}$ by $\alpha(h) = (1, z, h)$ and $\beta((g, f, h)) = (g, f)$, respectively. Clearly, α is a monomorphism and β is an epimorphism in which $\mathrm{Im}\,(\alpha) = \mathcal{A}_2 = \ker\,(\beta)$, where $\mathcal{A}_2 = \{(1, z, h) | h \in \mathcal{F}_0\}$ is a normal subgroup of $\mathcal{R}^{\langle 3\rangle}$. Thus we obtain a short exact sequence

$$1 \to \mathcal{F}_0 \overset{\alpha}{\to} \mathcal{R}^{\langle 3\rangle} \overset{\beta}{\to} \mathcal{R} \to 1. \tag{7.1.8}$$

This implies that the group $\mathcal{R}^{\langle 3\rangle}$ is a group extension of $\mathcal{R}$ by $\mathcal{F}_0$ since $\mathcal{F}_0 \cong \mathcal{A}_2$. Further, the map $\gamma : \mathcal{R} \to \mathcal{R}^{\langle 3\rangle}$ defined by $\gamma((g, f)) = (g, f, 1)$ is a homomorphism such that $\beta \circ \gamma$ is the identity map. Thus the exact sequence (7.1.8) is split so that $\mathcal{R}^{\langle 3\rangle} = \mathcal{A}_2 \rtimes \mathcal{R}$. □

Subgroups

In Sect. 3.2, we studied special subgroups of the Riordan group $\mathcal{R}$. It is also possible to form special subgroups of 3-D Riordan group $\mathcal{R}^{\langle 3\rangle}$ by taking subsets of $\mathcal{R}^{\langle 3\rangle}$. It is straightforward to verify that each of the following special sets is a subgroup of $\mathcal{R}^{\langle 3\rangle}$. We leave it to the reader to verify this claim (see Exercise 7.1).

(a) The 3-D Appell subgroup of 1st kind, $\mathcal{A}_1 = \left\{(g, z, 1) \mid g \in \mathcal{F}_0\right\}$.
(b) The 3-D Appell subgroup of 2nd kind, $\mathcal{A}_2 = \left\{(1, z, h) \mid h \in \mathcal{F}_0\right\}$.
(c) The 3-D Lagrange subgroup, $\mathcal{L} = \left\{(1, f, 1) \mid f \in \mathcal{F}_1\right\}$.
(d) The 3-D Riordan subgroup, $\mathcal{R} = \left\{(g, f, 1) \mid g \in \mathcal{F}_0, f \in \mathcal{F}_1\right\}$.
(e) The 3-D multiplication subgroup, $\mathcal{M} = \left\{(g, z, h) \mid g, h \in \mathcal{F}_0\right\}$.
(f) The 3-D Bell subgroup, $\mathcal{B} = \left\{(f/z, f, 1) \mid f \in \mathcal{F}_1\right\}$.
(g) The 3-D convolution subgroup, $C = \left\{(gh, zg, h) \mid g, h \in \mathcal{F}_0\right\}$.
(h) The 3-D checkerboard subgroup, $\mathcal{K} = \left\{(g, f, h) \mid g, h \text{ an even, } f \text{ an odd }\right\}$.

7.2 Three-Dimensional Riordan Arrays

A three-dimensional array is a natural generalization of a two-dimensional array, i.e., a matrix. Let $\mathbb{A} = [a_{i,j,k}]$ be a three-dimensional array of size $n_1 \times n_2 \times n_3$. In the special case where $n_1, n_2, n_3 \to \infty$, $\mathbb{A}$ is called an infinite three-dimensional array.

Matrix algebra operations for addition, subtraction, multiplication by a scalar, and multiplication of two d-dimensional matrices have been studied by Solo [23]. The (j,k)-*multiplication* $\mathbb{A}\mathbb{B} = [c_{i_1,\ldots,i_d}]$ of finite multi-dimensional matrices $\mathbb{A} = [a_{i_1,\ldots,i_d}]$ and $\mathbb{B} = [b_{i_1,\ldots,i_d}]$ is defined by

$$c_{i_1,\ldots,i_d} = \sum_{x=1}^{n} a_{i_1,\ldots,i_{j-1},x,i_{j+1},\ldots,i_d} b_{i_1,\ldots,i_{k-1},x,i_{k+1},\ldots,i_d}$$

where n is the number of elements in the jth dimension of $\mathbb{A}$ and the number of elements in the kth dimension of $\mathbb{B}$ as well. Thus $(2,1)$-*multiplication* $\mathbb{A}\mathbb{B} = (c_{i,j,k})_{i,j,k\geq 0}$ of infinite 3-D arrays $\mathbb{A} = [a_{i,j,k}]_{i,j,k\in\mathbb{N}}$ and $\mathbb{B} = [b_{i,j,k}]_{i,j,k\in\mathbb{N}}$ is given by

$$c_{i,j,k} = \sum_{x\geq 0} a_{i,x,k} b_{x,j,k} \tag{7.2.9}$$

where the sum is assumed to have a finite number of non-zero terms.

Theorem 7.4 *The binary operation in (7.1.7) of the group* $\mathcal{R}^{\langle 3\rangle}$ *agrees with the* $(2,1)$-*multiplication.*

Proof Let $\mathbb{A} = (g_1, f_1, h_1) = [a_{i,j,k}]$, $\mathbb{B} = (g_2, f_2, h_2) = [b_{i,j,k}]$ in $\mathcal{R}^{\langle 3\rangle}$ and its product $\mathbb{A}\mathbb{B} = [c_{i,j,k}]$. Since $a_{i,x,k} = 0$ for all $x > i$, the $(2,1)$-multiplication (7.2.9) is well-defined. Thus, summing (7.2.9) over all i we get

$$\sum_{i\geq 0} c_{i,j,k} z^i = \sum_{i\geq 0}\left(\sum_{x\geq 0} a_{i,x,k} b_{x,j,k}\right) z^i = \sum_{x\geq 0}\left(\sum_{i\geq 0} a_{i,x,k} z^i\right) b_{x,j,k}.$$

Using $a_{i,x,k} = [z^i] g_1 f_1^x h_1^k$ and $b_{x,j,k} = [z^x] g_2 f_2^j h_2^k$ yields

$$\begin{aligned}\sum_{i\geq 0} c_{i,j,k} z^i &= \sum_{x\geq 0} \left(g_1 f_1^x h_1^k\right) b_{x,j,k} = g_1 h_1^k \sum_{x\geq 0} b_{x,j,k} f_1^x \\ &= g_1 h_1^k g_2(f_1)(f_2(f_1))^j h_2^k(f_1) \\ &= g_1 g_2(f_1)(f_2(f_1))^j (h_1 h_2(f_1))^k.\end{aligned}$$

Thus

$$c_{i,j,k} = [z^i] g_1 g_2(f_1)(f_2(f_1))^j (h_1 h_2(f_1))^k,$$

which agrees with the binary operation given by (7.1.7). □

Given a 3-D matrix $\mathbb{A} = [a_{i,j,k}]$ with i, j, k indexed by $\mathbb{N} = \{0, 1, \ldots\}$, it is often necessary to refer to a particular plane or line. The *planes* of $\mathbb{A}$ are the submatrices obtained by fixing one of i, j, k, and the *lines* of $\mathbb{A}$ are the vectors obtained by fixing two of i, j, k [5]. The plane obtained by fixing k is said to be *kth layer matrix* of $\mathbb{A}$ and will be denoted by $L_k(\mathbb{A})$. The plane obtained by fixing j is said to be *jth column matrix* of $\mathbb{A}$ and will be denoted by $C_j(\mathbb{A})$. If only i is varying then the entries obtained by fixing j, k are called *x-line* (or *i-line*) of $\mathbb{A}$. The *y (or j)-line* and *z (or k)-line* are similarly defined.

Now let $\mathbb{A} = [a_{i,j,k}] = (g, f, h)$ be a three-dimensional Riordan array where $a_{i,j,k}$ represents the entry of $\mathbb{A}$ in the ith row, jth column, and kth layer. Since

$$a_{i,j,k} = [z^i]gf^jh^k \quad (g, h \in \mathcal{F}_0,\ f \in \mathcal{F}_1)$$

we see that the kth layer matrix is a proper Riordan array given by

$$L_k(\mathbb{A}) = (gh^k, f) = \begin{bmatrix} a_{0,0,k} & 0 & 0 & \cdots \\ a_{1,0,k} & a_{1,1,k} & 0 & \cdots \\ a_{2,0,k} & a_{2,1,k} & a_{2,2,k} & \cdots \\ \vdots & \vdots & \vdots & \ddots \end{bmatrix}, \quad k = 0, 1, \ldots, \qquad (7.2.10)$$

and the jth column matrix is an *improper*[1] Riordan array given by

$$C_j(\mathbb{A}) = (gf^j, h) = \begin{bmatrix} O_{j\times 1} & O_{j\times 1} & O_{j\times 1} & \cdots \\ a_{j,j,0} & a_{j,j,1} & a_{j,j,2} & \cdots \\ a_{j+1,j,0} & a_{j+1,j,1} & a_{j+1,j,2} & \cdots \\ a_{j+2,j,0} & a_{j+2,j,1} & a_{j+2,j,2} & \cdots \\ \vdots & \vdots & \vdots & \ddots \end{bmatrix}, \quad j = 0, 1, \ldots \qquad (7.2.11)$$

where $O_{j\times 1}$ is the $j \times 1$ zero matrix. Thus, the (i, j, k) entry of $\mathbb{A}$ is the (i, j) entry of the kth layer matrix $L_k(\mathbb{A})$.

In addition, gf^jh^k is the generating function for entries $a_{0,j,k}, a_{1,j,k}, a_{2,j,k}, \ldots$ of the i-line in both jth column matrix $C_j(\mathbb{A})$ and kth layer matrix $L_k(\mathbb{A})$, i.e.

$$gf^jh^k = \sum_{i\geq 0} a_{i,j,k}z^i \in \mathcal{F}_j, \quad j = 0, 1, \ldots.$$

The array $\mathbb{A}$ can be expressed in terms of their i-line generating functions as follows:

$$\mathbb{A} = (g, f, h) = [a_{i,j,k}] = \left[\begin{array}{c|c|c|c|c} g & gf & gf^2 & gf^3 & \cdots \\ \hline gh & gfh & gf^2h & gf^3h & \cdots \\ \hline gh^2 & gfh^2 & gf^2h^2 & gf^3h^2 & \cdots \\ \hline gh^3 & gfh^3 & gf^2h^3 & gf^3h^3 & \cdots \\ \hline \vdots & \vdots & \vdots & \vdots & \ddots \end{array}\right].$$

Note that the columns and the rows of $\mathbb{A}$ represent jth column matrices $C_j(\mathbb{A}) = (gf^j, h)$, $j = 0, 1, \ldots$, and kth layer matrices $L_k(\mathbb{A}) = (gh^k, f)$, $k = 0, 1, \ldots$, respectively. Thus $\mathbb{A}$ can be also represented in terms of either its layer matrices

[1] A Riordan matrix (G, F) is improper provided that $G \notin \mathcal{F}_0$ or $F \notin \mathcal{F}_1$.

or its column matrices as follows:

$$\mathbb{A} = \left[L_0(\mathbb{A}) \mid L_1(\mathbb{A}) \mid \cdots\right]^T \quad \text{or} \quad \mathbb{A} = \left[C_0(\mathbb{A}) \mid C_1(\mathbb{A}) \mid \cdots\right].$$

Example 7.1 (*3-D Pascal matrix*) One of the interesting examples in which three-dimensional approach works concerns the 3-D Pascal matrix defined by $\mathbb{P} = \left(\frac{1}{1-z}, \frac{z}{1-z}, \frac{1}{1-z}\right) = [a_{i,j,k}]$. Using formal series

$$\frac{1}{(1-z)^{m+1}} = \sum_{n\geq 0}\binom{m+n}{n}z^n, \quad m \in \mathbb{Z}, \tag{7.2.12}$$

we obtain

$$\begin{aligned} a_{i,j,k} &= [z^i]\left(\frac{1}{1-z}\right)\left(\frac{z}{1-z}\right)^j\left(\frac{1}{1-z}\right)^k \\ &= [z^{i-j}]\frac{1}{(1-z)^{j+k+1}} = \binom{i+k}{j+k}. \end{aligned} \tag{7.2.13}$$

In addition, for $k, j = 0, 1, \ldots$

$$L_k(\mathbb{P}) = \left(\frac{1}{(1-z)^{k+1}}, \frac{z}{1-z}\right) \quad \text{and} \quad C_j(\mathbb{P}) = \left(\frac{z^j}{(1-z)^{j+1}}, \frac{1}{1-z}\right).$$

As shown in Fig. 7.1, these matrices can be obtained from (7.2.10) and (7.2.11) via (7.2.13), or i-line generating functions, e.g.

$$gfh^2 = \frac{z}{(1-z)^4} = \sum_{i\geq 0}\binom{i+2}{1+2}z^i = z + 4z^2 + 10z^3 + 20z^4 + 35z^5 + \cdots.$$

3-D Riordan matrix multiplication

Given two 3-D Riordan arrays $\mathbb{A} = (g_1, f_1, h_1)$ and $\mathbb{B} = (g_2, f_2, h_2)$, where $g_i, h_i \in \mathcal{F}_0$ and $f_i \in \mathcal{F}_1$ $(i = 1, 2)$, how can we carry out matrix multiplication of $\mathbb{AB}$? If we partition $\mathbb{A}$ and $\mathbb{B}$ into layer matrices

$$L_0(\mathbb{A}), L_1(\mathbb{A}), \ldots \quad \text{and} \quad L_0(\mathbb{B}), L_1(\mathbb{B}), \ldots,$$

respectively, then we may be able to compute the product $\mathbb{AB}$ efficiently. Indeed, since the layer matrices are proper Riordan arrays we see that

$$\begin{aligned} L_k(\mathbb{A})L_k(\mathbb{B}) &= (g_1h_1^k, f_1)(g_2h_2^k, f_2) = \left(g_1g_2(f_1)(h_1h_2(f_1))^k,\ f_2(f_1)\right) \\ &= L_k(\mathbb{AB}). \end{aligned}$$

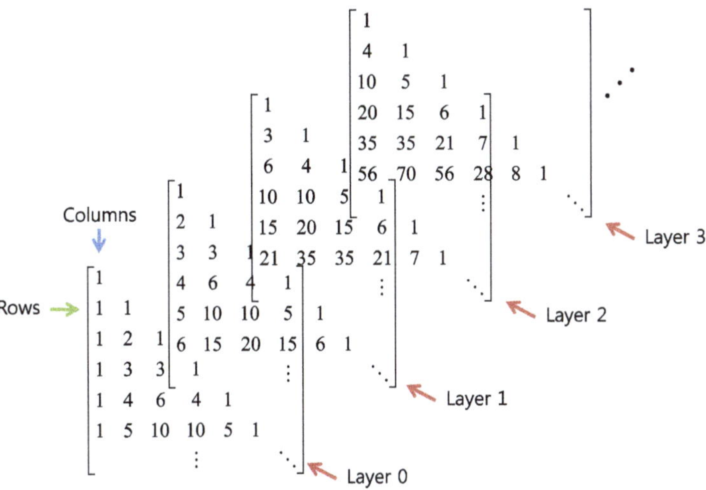

Fig. 7.1 3-D Pascal matrix $\mathbb{P} = \left(\frac{1}{1-z}, \frac{z}{1-z}, \frac{1}{1-z}\right)$

Thus, the product $\mathbb{AB}$ is the array whose kth layer matrix is $L_k(\mathbb{A})L_k(\mathbb{B})$, $k = 0, 1, 2, \ldots$, i.e.

$$\begin{aligned}(g_1, f_1, h_1)(g_2, f_2, h_2) &= \mathbb{AB} = \left[L_0(\mathbb{A})L_0(\mathbb{B}) \mid L_1(\mathbb{A})L_1(\mathbb{B}) \mid \cdots\right]^T \\ &= \Big(g_1g_2(f_1), f_2(f_1), h_1h_2(f_1)\Big).\end{aligned}$$

Note that the (i, j, k) entry of $\mathbb{AB}$ for a fixed k is the (i, j)-entry of $L_k(\mathbb{AB})$, and the entry is determined by multiplying the ith row vector of $L_k(\mathbb{A})$ times the jth column vector of $L_k(\mathbb{B})$.

Recall that the notion of a Riordan array can be used in a constructive way to find generating functions of many combinatorial sums. Indeed, suppose we multiply the Riordan array (g, f) by a column vector $A = (a_0, a_1, \ldots)^T$ with its generating function $A(z)$ and we get a column vector $B = (b_0, b_1, \ldots)^T$:

$$(g, f)\begin{bmatrix} a_0 \\ a_1 \\ a_2 \\ \vdots \end{bmatrix} = \begin{bmatrix} b_0 \\ b_1 \\ b_2 \\ \vdots \end{bmatrix}.$$

As shown in Theorem 3.1, the resulting column vector has the generating function $g(z)A(f(z))$. If we identify the sequences A and B with its generating functions $A(z)$ and $B(z)$, respectively, this rule can be represented as

$$\Big(g(z), f(z)\Big)A(z) = g(z)A(f(z)),$$

which is called the *fundamental theorem of Riordan arrays* (FTRA).

Theorem 7.5 *Let $\mathbb{A} = (g_1, f_1, h_1)$ and $\mathbb{B} = (g_2, f_2, h_2)$ be 3-D Riordan arrays, where $g_i, h_i \in \mathcal{F}_0$, $f_i \in \mathcal{F}_1$ $(i = 1, 2)$. Then*

$$C_j(\mathbb{AB}) = \left[L_0(\mathbb{A})\mathbf{b}_{0_j}, L_1(\mathbb{A})\mathbf{b}_{1_j}, L_2(\mathbb{A})\mathbf{b}_{2_j}, \ldots\right], \quad j = 0, 1, \ldots, \tag{7.2.14}$$

where $\mathbf{b}_{k_j}$ is the kth column vector of $C_j(\mathbb{B})$.

Proof Let $\mathbb{C} = \mathbb{AB} = [c_{i,j,k}]_{i,j,k\in\mathbb{N}}$. Consider layer matrices $L_0(\mathbb{A}), L_1(\mathbb{A}), \ldots$ of $\mathbb{A}$ and column matrices $C_0(\mathbb{B}), C_1(\mathbb{B}), \ldots$ of $\mathbb{B}$. Since

$$c_{i,j,k} = \sum_{x\geq 0} a_{i,x,k} b_{x,j,k} = (a_{i,0,k}, a_{i,1,k}, a_{i,2,k}, \ldots) \begin{bmatrix} b_{0,j,k} \\ b_{1,j,k} \\ b_{2,j,k} \\ \vdots \end{bmatrix},$$

the entry $c_{i,j,k}$ can be represented as the product of the ith row vector $\mathbf{a}_{i_k}^T$ of $L_k(\mathbb{A})$ and the kth column vector $\mathbf{b}_{k_j}$ of $C_j(\mathbb{B})$, i.e., $c_{i,j,k} = \mathbf{a}_{i_k}^T \mathbf{b}_{k_j}$.

Let $\mathbf{b}_{k_j}(z)$ be the kth column generating function of $C_j(\mathbb{B}) = \left(g_2 f_2^j, h_2\right)$. Since $L_k(\mathbb{A}) = \left(g_1 h_1^k, f_1\right)$ is a proper Riordan array and $\mathbf{b}_{k_j}(z) = g_2 f_2^j h_2^k$, it follows from the FTRA that

$$\begin{aligned} L_k(\mathbb{A})\mathbf{b}_{k_j}(z) &= \left(g_1 h_1^k, f_1\right) g_2 f_2^j h_2^k \\ &= g_1 h_1^k g_2(f_1) f_2^j(f_1) h_2^k(f_1) \\ &= g_1 g_2(f_1) f_2^j(f_1) \left(h_1 h_2(f_1)\right)^k, \quad k = 0, 1, \ldots. \end{aligned}$$

Since

$$C_j(\mathbb{AB}) = \left(g_1 g_2(f_1) f_2^j(f_1),\ h_1 h_2(f_1)\right), \tag{7.2.15}$$

we see that $L_k(\mathbb{A})\mathbf{b}_{k_j}(z)$ represents the kth column generating function of $C_j(\mathbb{AB})$. That is, the kth column vector of $C_j(\mathbb{AB})$ is obtained from multiplying the kth layer matrix $L_k(\mathbb{A})$ by the kth column vector $\mathbf{b}_{k_j}$ of $C_j(\mathbb{B})$. Thus, we obtain (7.2.14). □

Note that $\mathbb{A} = [L_0(\mathbb{A}) \mid L_1(\mathbb{A}) \mid \cdots]^T$ and $C_j(\mathbb{B}) = \left[\mathbf{b}_{0_j}, \mathbf{b}_{1_j}, \ldots\right]$. In view of Theorem 7.5, we can define the *formal product* $\bullet$ of $\mathbb{A}$ and $C_j(\mathbb{B})$ as the column matrix of $\mathbb{AB}$, i.e.

$$\mathbb{A} \bullet C_j(\mathbb{B}) = C_j(\mathbb{AB}) = \left[L_0(\mathbb{A})\mathbf{b}_{0_j}, L_1(\mathbb{A})\mathbf{b}_{1_j}, \ldots\right], \quad j = 0, 1, \ldots. \tag{7.2.16}$$

The result will be a Riordan array given by (7.2.15). Thus the product $\mathbb{AB}$ can be also determined by

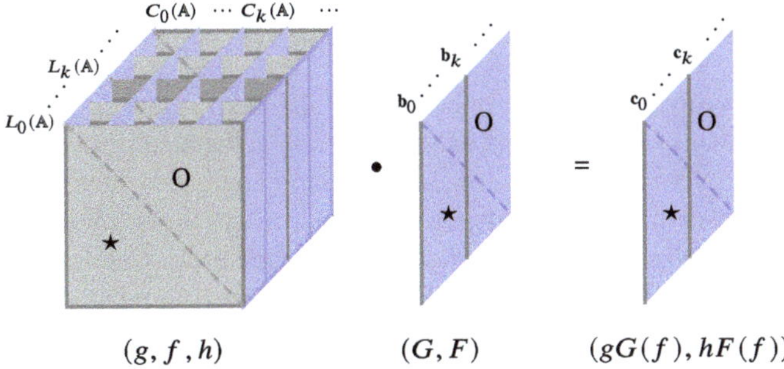

Fig. 7.2 Fundamental theorem of 3-D Riordan arrays

$$\mathbb{A}\mathbb{B} = \left[\mathbb{A} \bullet C_0(\mathbb{B}) \mid \mathbb{A} \bullet C_1(\mathbb{B}) \mid \cdots \right].$$

More generally, in view of (7.2.16) we can define the $(3D, 2D)$-*product* of a 3-D Riordan array $\mathbb{A} = (g, f, h)$ and a usual(2-D) Riordan array $B = (G, F)$ as the formal product

$$\mathbb{A} \bullet B = \left[L_0(\mathbb{A})\mathbf{b}_0, L_1(\mathbb{A})\mathbf{b}_1, \ldots \right] \tag{7.2.17}$$

where $\mathbf{b}_k$ is the kth column vector of B.

Theorem 7.6 (Fundamental theorem of 3-D Riordan arrays)
Let $\mathbb{A} = (g, f, h) \in \mathcal{R}^{(3)}$ and $B = (G, F) \in \mathcal{R}$. Then

$$(g, f, h) \bullet (G, F) = \Big(gG(f), hF(f)\Big). \tag{7.2.18}$$

Proof For $k = 0, 1, \ldots$, consider a kth column generating function $L_k(\mathbb{A})\mathbf{b}_k(z)$ of $\mathbb{A}B$, where $\mathbf{b}_k$ is the kth column vector of B. Since $L_k(\mathbb{A}) = \left(gh^k, f\right)$ and $\mathbf{b}_k(z) = GF^k$, it follows from the FTRA that

$$\begin{aligned} L_k(\mathbb{A})\mathbf{b}_k(z) &= \left(gh^k, f\right) GF^k = gh^k G(f) F^k(f) \\ &= gG(f)\,(hF(f))^k := \mathbf{c}_k(z), \end{aligned}$$

which is the kth column generating function of $\Big(gG(f), hF(f)\Big)$ for $k = 0, 1, \ldots$. □

Figure 7.2 pictorially shows the fundamental theorem of 3-D Riordan arrays.

Applications

The Riordan group energizes applications of the matrix approach to various combinatorial problems. As we studied in Sect. 5.1, one of the method derives generating functions for combinatorial sums involving counting numbers. Moreover, Sprugnoli [24] and Merlini et al. [18] examine the Riordan arrays as the combinatorial problem concerning the walks on the line, which includes the Pascal, Catalan, Motzkin, and other triangular arrays originated by lattice walks with colored steps. We now discuss how 3-D Riordan arrays can be applied to combinatorial sums and lattice path counting in three-dimensional lattice space.

Combinatorial sums Generating functions give us several methods to compute combinatorial sums of the form $s(n) = \sum_{k\geq 0} a_{n,k}$, such as the Vandermonde identity $\sum_{k=0}^{n}\binom{x}{k}\binom{y}{n-k} = \binom{x+y}{n}$, or to establish combinatorial identities between the sums. We will see how 3-D Riordan arrays can be applied to combinatorial double or triple sums. The key idea is to use (2, 1)-multiplication of two *shifted* three-dimensional Riordan arrays [7] defined as follows.

Let $\mathbb{A} = [a_{i,j,k}] = (g, f, h)$ be a 3-D Riordan array with $g, h \in \mathcal{F}_0$ and $f \in \mathcal{F}_1$. The *shifted 3-D Riordan array*, $\hat{\mathbb{A}} := (g, f, zh) = [\hat{a}_{i,j,k}]$ associated with $\mathbb{A}$ is defined by $\hat{a}_{i,j,k} = [z^i]gf^j(zh)^k$. Note that

$$\hat{a}_{i,j,k} = [z^i]gf^j(zh)^k = [z^{i-k}]gf^jh^k = a_{i-k,j,k}. \tag{7.2.19}$$

The kth layer matrix and jth column matrix of $\hat{\mathbb{A}}$ are defined similarly to those of $\mathbb{A}$ as follows:

$$L_k(\hat{\mathbb{A}}) = (g \cdot (zh)^k, f) \quad \text{and} \quad C_j(\hat{\mathbb{A}}) = (gf^j, zh).$$

Let $\hat{\mathbb{A}}$ be as above, and for $\hat{\mathbb{B}} = [\hat{b}_{i,j,k}] = (G, F, zH)$ let $\hat{\mathbb{A}}\hat{\mathbb{B}} = (\hat{c}_{i,j,k})$. By (2, 1)-multiplication, we have

$$\hat{c}_{i,j,k} = c_{i-k,j,k} = \sum_{x\geq 0} a_{i-k,x,k}b_{x,j,k} = \sum_{x\geq 0} \hat{a}_{i,x,k}\hat{b}_{x+k,j,k}. \tag{7.2.20}$$

Using an argument similar to the proof of Theorem 7.4 makes it easy to prove (see Exercise 7.3) that

$$\hat{\mathbb{A}}\hat{\mathbb{B}} = (g, f, zh)(G, F, zH) = \Big(gG(f), F(f), zhH(f)\Big). \tag{7.2.21}$$

We are ready to show how shifted 3-D Riordan arrays can serve as a tool in certain situations for deriving combinatorial identities. We begin with an example related to the Catalan numbers, which are almost as ubiquitous as the binomial coefficients in various counting problems. Recall that the Catalan sequence $(C_n)_{n\geq 0}$ is defined by convolution, $C_{n+1} = \sum_{k=0}^{n} C_kC_{n-k}$, $C_0 = 1$. Using the Catalan generating function $C = \sum_{n\geq 0} C_nz^n$, the convolution translates into $C = zC^2 + 1$. Solving this quadratic equation with $C_0 = 1$, we obtain

$$C = \frac{1-\sqrt{1-4z}}{2z} = \sum_{n\geq 0} \frac{1}{n+1}\binom{2n}{n} = 1 + z + 2z^2 + 5z^3 + 14z^4 + \cdots .$$

It is also known [1] that for $k = 1, 2, \ldots$

$$[z^n]C^k = \frac{k}{2n+k}\binom{2n+k}{n}. \tag{7.2.22}$$

Example 7.2 *(3-D Catalan matrix)* Let $\hat{\mathbb{C}} = [\hat{a}_{i,j,k}] = (C, zC, zC)$, a shifted 3-D Riordan array. From (7.2.22) we obtain

$$\hat{a}_{i,j,k} = [z^{i-j-k}]C^{j+k+1} = \frac{j+k+1}{i+1}\binom{2i-j-k}{i}.$$

Now consider the shifted Pascal matrix $\hat{\mathbb{P}} = [\hat{b}_{i,j,k}] = \left(\frac{1}{1-z}, \frac{z}{1-z}, \frac{z}{1-z}\right)$ with

$$\hat{b}_{i,j,k} = [z^{i-j-k}]\left(\frac{1}{1-z}\right)^{j+k+1} = \binom{i}{j+k}. \tag{7.2.23}$$

Since $C = zC^2 + 1$ it follows from (7.2.21) and (7.2.22) that

$$\hat{\mathbb{C}}\,\hat{\mathbb{P}} = [\hat{c}_{i,j,k}] = \left(\frac{C}{1-zC}, \frac{zC}{1-zC}, \frac{zC}{1-zC}\right) = \left(C^2, zC^2, zC^2\right),$$

and

$$\hat{c}_{i,j,k} = \frac{j+k+1}{i+1}\binom{2i+2}{i-j-k}.$$

Using the (2,1)-multiplication of $\hat{\mathbb{C}}\hat{\mathbb{P}}$ in (7.2.20) we obtain

$$\binom{2i+2}{i-j-k} = \sum_{x=j}^{i-k} \frac{x+k+1}{j+k+1}\binom{2i-x-k}{i}\binom{x+k}{j+k} \tag{7.2.24}$$

$$= \sum_{x=j}^{i-k} \binom{2i-x-k}{i-x-k}\binom{x+k+1}{j+k+1}. \tag{7.2.25}$$

Set $i - j - k = m$ and consider the substitution $x - j \mapsto n$ in (7.2.24). For integers $i, m \geq 0$ with $i \geq m-1$, this gives

$$\sum_{n=0}^{m} \binom{i+m-n}{m-n}\binom{i-m+n+1}{n} = \binom{2i+2}{m}, \quad m \geq 1. \tag{7.2.26}$$

More generally, we have the following theorem that will be referred to as the *3-D Riordan array method.*

Theorem 7.7 *Let* $(g, f_1, f_2) = [d_{n,j,k}]$ *be a shifted* 3-D *Riordan array, where* $g \in \mathcal{F}_0$ *and* $f_1, f_2 \in \mathcal{F}_1$. *Then for any two sequences* $(a_n)_{n\geq 0}$ *and* $(b_n)_{n\geq 0}$ *we have*

$$\sum_{k=0}^{n}\sum_{j=0}^{n-k} d_{n,j,k}a_j b_k = [z^n]gA(f_1)B(f_2), \tag{7.2.27}$$

where $A = \sum_{n\geq 0} a_n z^n$ *and* $B = \sum_{n\geq 0} b_n z^n$.

Proof Let $\mathbb{D} = (g, f_1, f_2) = [d_{n,j,k}]$ where $g \in \mathcal{F}_0$ and $f_1, f_2 \in \mathcal{F}_1$. For a fixed k, consider the kth layer matrix $\left(g \cdot (f_2/z)^k, f_1\right) = [c_{n,j}]$ of $(g, f_1, f_2/z)$. Then

$$c_{n,j} = [z^n]g \cdot (f_2/z)^k f_1^j = [z^{n+k}]gf_1^j f_2^k = d_{n+k,j,k}.$$

Thus $d_{n,j,k} = c_{n-k,j}$. Now let $A(z) = \sum_{n\geq 0} a_n z^n$. Using the FTRA yields that

$$\begin{aligned}\sum_{j=0}^{n-k} c_{n-k,j}a_j &= [z^{n-k}]\left(g \cdot (f_2/z)^k, f_1\right)A(z)\\ &= [z^n]gf_2^k A(f_1),\end{aligned}$$

which is equal to the (n, k)-entry of the Riordan array $\left(gA(f_1), f_2\right)$. Again, for a generating function $B(z) = \sum_{n\geq 0} b_n z^n$ we thus obtain

$$\begin{aligned}\sum_{k=0}^{n}\left(\sum_{j=0}^{n-k} d_{n,j,k}a_j\right)b_k &= \sum_{k=0}^{n}\left(\sum_{j=0}^{n-k} c_{n-k,j}a_j\right)b_k\\ &= \sum_{k=0}^{n}\left([z^n]gf_2^k A(f_1)\right)b_k\\ &= [z^n]\left(gA(f_1), f_2\right)B(z)\\ &= [z^n]gA(f_1)B(f_2).\end{aligned}$$

□

Example 7.3 We want to compute the sum

$$\sum_{k=0}^{n}\sum_{j=0}^{n-k}\binom{n}{j+k}\binom{n}{j}\binom{n}{k}.$$

Let us try to use the 3-D Riordan array method in Theorem 7.7. In Example 7.2 we saw that $\binom{n}{j+k}$ is the (n, j, k)-entry of the shifted Pascal matrix $\left(\frac{1}{1-z}, \frac{z}{1-z}, \frac{z}{1-z}\right)$. In addition, we note that $\binom{n}{j} = [z^j](1+z)^n$ and $\binom{n}{k} = [z^k](1+z)^n$. Take $A(z) = (1+z)^n$ and $B(z) = (1+z)^n$. By Theorem 7.7 we thus have

$$\begin{aligned}\sum_{k=0}^{n}\sum_{j=0}^{n-k}\binom{n}{j+k}\binom{n}{j}\binom{n}{k} &= [z^n]\frac{1}{1-z}\left(1+\frac{z}{1-z}\right)^n\left(1+\frac{z}{1-z}\right)^n \\ &= [z^n]\left(\frac{1}{1-z}\right)^{1+2n} \\ &= \binom{3n}{n} \text{ by (7.2.12).}\end{aligned}$$

Lattice path counting A *lattice path* in $\mathbb{Z}^d$ with steps in the step set S is a sequence $P_0, P_1, \ldots, P_k$ of points P_i in $\mathbb{Z}^d$ such that each consecutive difference $P_i - P_{i-1}$ lies in S. The lattice path may take any restrictions that affect where the path may go. The most common lattice path counting problems are arising in the lattice plane $\mathbb{Z}^2$, but there have also been studies [16] of 3-D or higher dimensional lattice space. For the survey of lattice path enumeration, see [15].

We see that several authors have used Riordan arrays to construct matrices that are useful to study diverse aspects of lattice path counting problems coming from interesting integer sequences, such as the Catalan numbers, the Delannoy numbers, Motzkin numbers, and the Schröder numbers, etc. (see [3, 8, 9, 15, 19, 21, 22, 26, 27]).

For example, suppose that we want to count the lattice paths from $(0, 0)$ to (n, k) using the horizontal steps $H = (1, 0)$ and the diagonal steps $D = (1, 1)$ in the coordinate plane. We note that HD-paths from (0,0) to (n, k) cannot go both above the line $y = x$ and below the x-axis in the xy-lattice plane. If we assign each element $p_{n,k}$ to the number of these HD-paths from $(0, 0)$ to (n, k), then we get the familiar Pascal matrix $[p_{n,k}]$ with $p_{n,k} = \binom{n}{k}$. It can be expressed as the Riordan array, see Fig. 7.3.

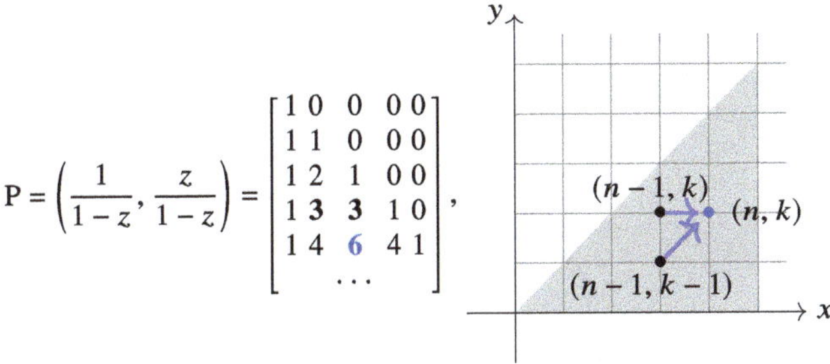

Fig. 7.3 Pascal matrix and possible last two steps H, D

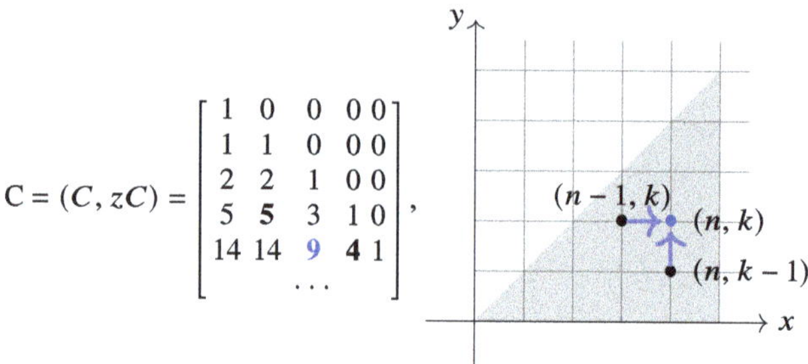

Fig. 7.4 Catalan matrix and possible last two steps H, V

To give another example, recall that *subdiagonal rectangular lattice paths* from $(0, 0)$ to (n, k) are rectangular lattice paths using the horizontal steps $H = (1, 0)$ and the vertical steps $V = (0, 1)$ that are restricted to lie on or below the line $y = x$ in the coordinate plane. As is well known [1, 4], the nth Catalan number $C_n = \frac{1}{n+1}\binom{2n}{n}$ equals the number of subdiagonal rectangular lattice paths from $(0, 0)$ to (n, n). More generally, if we count the number of subdiagonal rectangular lattice paths from $(0, 0)$ to $(n, n-k)$, say $c_{n,k}$, we get the Catalan triangle matrix $[c_{n,k}]$ with $c_{n,k} = \frac{k+1}{n+1}\binom{2n-k}{n}$. It can be also expressed as the Riordan array [8], where $C = \frac{1-\sqrt{1-4z}}{2z}$ is the generating function for the Catalan numbers, see Fig. 7.4.

The matrices P and C will be referred to as *path matrices* associated with the step sets $S = \{(1, 0), (1, 1)\}$ and $S = \{(1, 0), (0, 1)\}$, respectively.

We now consider a lattice path L from the origin O to the point $P(n, k, \ell)$ with $n \geq k \geq \ell \geq 0$ using the step set $S = S_1 \cup S_2$ in three-dimensional lattice space $\mathbb{Z}^3$, where

(i) S_1 is a subset of $\{(a, b, 0) \in \mathbb{Z}^3 \mid a \geq 0\}$;
(ii) S_2 is a subset of $\{(c, 0, d) \in \mathbb{Z}^3 \mid c \geq 0, c \geq d\}$.

For our purposes, we start by defining a specific type of lattice paths, called *xyz-lattice paths* in $\mathbb{Z}^3$. Assume that L consists of two lattice paths L_1 and L_2 in the lattice planes of $\mathbb{Z}^3$ such that, for any $i \in \mathbb{N}$ with $k \leq i \leq n - \ell$

(i) L_1 is a lattice path from $(0, 0, 0)$ to $(i, k, 0)$ using the step set S_1 in the xy-plane of $\mathbb{Z}^3$;
(ii) L_2 is a lattice path from $(i, k, 0)$ to (n, k, ℓ) using the step set S_2 in the plane $y = k$ of $\mathbb{Z}^3$,

where steps used in L_1 cannot go above the line $y = x$ and below the x-axis in the xy-plane, and steps used in L_2 cannot go above the plane $z = x$ and below the xy-plane.

We will refer to $L\colon O \xrightarrow{L_1} Q_i \xrightarrow{L_2} P$ as the *xyz-lattice path* from O to P where L_1 and L_2 touch the line $y = k$ at $Q_i = (i, k, 0)$ in the xy-plane, see Fig. 7.5.

For simplicity, we also use the notation $\sigma_S(A \to B)$ to denote the number of lattice paths from A to B using the step set S in $\mathbb{Z}^d$. By the definition of the xyz-lattice path

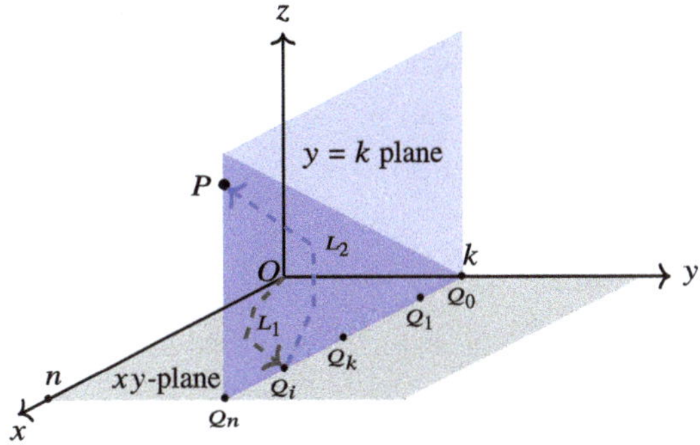

Fig. 7.5 xyz-lattice path $L : O \xrightarrow{L_1} Q_i \xrightarrow{L_2} P$

from O to $P(n, k, \ell)$ we see that

$$\sigma_{S_1 \cup S_2}(O \to P) = \sum_{i=k}^{n-\ell} \sigma_{S_1}(O \to Q_i)\sigma_{S_2}(Q_i \to P). \tag{7.2.28}$$

In working with lattice path counting problems in the plane $y = k$, it is generally more convenient to find the solutions in the xz-plane ($y = 0$) obtained by orthogonal projection so that a point (u, k, v) in the plane $y = k$ is projected to the point (u, v) in the xz-plane. From this idea we see that $\sigma_{S_2}(Q_i \to P)$ in the plane $y = k$ equals the number of lattice paths from $(i, 0)$ to (n, ℓ), $k \le i \le n - \ell$, using the step set S_2 of a subset $\{(c, d) \in \mathbb{Z}^2 \mid c \ge 0, c \ge d\}$ in the xz-plane, where paths cannot go above the line $z = x$ and below the x-axis.

Now look at $\sigma_{S_2}(Q_i \to P)$ in the xz-plane where $Q_i = (i, 0)$ and $P = (n, \ell)$. Since $k \le i \le n - \ell$, it follows that

$$\frac{\overline{PQ_n}}{\overline{Q_iQ_n}} = \frac{\ell}{n - i} \le 1. \tag{7.2.29}$$

Since every step in S_2 is of the form (c, d) where $c \ge 0$ and $c \ge d$, the step cannot go above the line $z = x$. Thus it follows from (7.2.29) that

$$\sigma_{S_2}(Q_i \to P) = \sigma_{S_2}\Big(O \to (n - i, \ell)\Big). \tag{7.2.30}$$

Denote π_{S_i} the path matrix associated with the step set S_i for $i = 1, 2$.

Theorem 7.8 *Let $D(n, k, \ell)$ denote the number of xyz-lattice paths from* $(0, 0, 0)$ *to* (n, k, ℓ) *using the step set $S_1 \cup S_2$, where $D(0, 0, 0) = 1$ and $n \ge k + \ell$ for $n, k, \ell \in \mathbb{N}$. If $\pi_{S_1} = (g, f)$ and $\pi_{S_2} = (G, F)$ are proper Riordan arrays for some S_1 and S_2, then*

$$D(n,k,\ell)=[z^n]gGf^kF^\ell. \tag{7.2.31}$$

Proof Let $\pi_{S_1}=(g,f)=[a_{n,k}]$ and $\pi_{S_2}=(G,F)=[b_{n,k}]$ for some S_1 and S_2. Consider a xyz-lattice path

$$L: O \xrightarrow{L_1} Q_i \xrightarrow{L_2} P,$$

where $Q_i=(i,k,0)$ and $P=(n,k,\ell)$. Then it follows from (7.2.28) and (7.2.30) that

$$\begin{aligned}D(n,k,\ell)&=\sum_{i=k}^{n-\ell}\sigma_{S_1}\Big(O\to(i,k,0)\Big)\sigma_{S_2}\Big(O\to(n-i,\ell)\Big)\\&=\sum_{i=k}^{n-\ell}a_{i,k}b_{n-i,\ell}.\end{aligned} \tag{7.2.32}$$

Now, consider the shifted 3-D Riordan arrays $(g,f,z)=[\hat{a}_{n,k,\ell}]$ and $(G,z,F)=[\hat{b}_{n,k,\ell}]$. Then

$$\hat{a}_{n,k,\ell}=[z^n]gf^kz^\ell=a_{n-\ell,k}\quad\text{and}\quad \hat{b}_{n,k,\ell}=[z^n]Gz^kF^\ell=b_{n-k,\ell}. \tag{7.2.33}$$

Since $(G,z,zh)(g,f,z)=(gG,f,zh)$ where $F=zh$, it follows from (7.2.20) and (7.2.33) that

$$[z^n]gGf^kF^\ell=\sum_{i\ge0}\hat{b}_{n,i,\ell}\hat{a}_{i+\ell,k,\ell}=\sum_{i\ge0}a_{i,k}b_{n-i,\ell}. \tag{7.2.34}$$

We note that $a_{i,k}=0$ for $0\le i\le k-1$ and $b_{n-i,\ell}=0$ for $i\ge n-\ell+1$. By comparing (7.2.32) and (7.2.34), we obtain

$$D(n,k,\ell)=[z^n]gGf^kF^\ell,$$

as desired. □

Corollary 7.1 *Let $D(n,k,\ell)$ be the same number in Theorem 7.8. If $\pi_{S_1}=(g,f)$ and $\pi_{S_2}=(1,F)$ for some S_1 and S_2, then $D(n,k,\ell)=[z^n]gf^kF^\ell$.*

7.3 The Riordan Group in Several Variables

There are various generalizations of Riordan arrays in one or several variables [7, 10, 13, 28]. Extending several known results, we discuss the *multivariate Riordan group* to be defined by formal power series in several variables (see [6, 17]). We will see some similarity with the Riordan group in a single variable, but its matrix representation, called a *multivariate Riordan array*, will be quite different to usual

Riordan arrays.

Formal power series in several variables

We have seen in Chap. 2 that formal power series and generating functions are a useful tool in various branches of mathematics. For our purposes, formal power series include series with a finite number of variables and with coefficients in an arbitrary ring. We briefly describe the basic notion of formal power series in several variables which can be found in [20, 25].

Let $\mathbb{F}$ be the real $\mathbb{R}$ or complex field $\mathbb{C}$, and $z_1, \ldots, z_d$ be indeterminate. We consider formal series of the form

$$F(\mathbf{z}) := F(z_1, \ldots, z_d) = \sum_{\mathbf{i}} a_{\mathbf{i}} \mathbf{z}^{\mathbf{i}} \tag{7.3.35}$$

where the sum runs over all $\mathbf{i} = (i_1, \ldots, i_d) \in \mathbb{N}^d$, each $a_{\mathbf{i}}$ is an element of $\mathbb{F}$, and $\mathbf{z}^{\mathbf{i}}$ is a short-hand notation for the monomial $z_1^{i_1} \cdots z_d^{i_d}$. A common notation for $a_{\mathbf{i}}$ is $[\mathbf{z}^{\mathbf{i}}]F(\mathbf{z})$, read as the $\mathbf{z}^{\mathbf{i}}$-coefficient of $F(\mathbf{z})$. We also refer to $F(\mathbf{z})$ as the *generating function* for the d-way array $\{a_{\mathbf{i}}\}$. For example, if $a_{\mathbf{i}}$ denotes the multinomial coefficient with $|\mathbf{i}| = i_1 + \cdots + i_d$, then the generating function is

$$F(\mathbf{z}) = \frac{1}{1 - z_1 - \cdots - z_d} = \sum_{\mathbf{i}} \binom{|\mathbf{i}|}{i_1 \cdots i_d} z_1^{i_1} \cdots z_d^{i_d}.$$

The set of formal expressions of the form $F(\mathbf{z}) = \sum_{\mathbf{i}} f_{\mathbf{i}} \mathbf{z}^{\mathbf{i}}$ with $f_{\mathbf{i}} \in \mathbb{F}$ is denoted by $\mathbb{F}[[z_1, \ldots, z_d]]$. Addition is defined by

$$(F + G)(\mathbf{z}) = F(\mathbf{z}) + G(\mathbf{z}) = \sum_{\mathbf{i}} (f_{\mathbf{i}} + g_{\mathbf{i}}) \mathbf{z}^{\mathbf{i}}$$

and multiplication is defined by convolution:

$$(F \cdot G)(\mathbf{z}) = F(\mathbf{z}) G(\mathbf{z}) = \sum_{\mathbf{i}} \left(\sum_{\mathbf{j}} f_{\mathbf{j}} g_{\mathbf{i}-\mathbf{j}} \right) \mathbf{z}^{\mathbf{i}}, \tag{7.3.36}$$

where it is assumed that the inner sum is always finite, so there is no question of convergence. The *additive identity* in $\mathbb{F}[[z_1, \ldots, z_d]]$ is the zero series and the *multiplicative identity* is the series with a 1 in position $\mathbf{i} = (0, \ldots, 0)$ and zero elsewhere. It is easy to show that $F(\mathbf{z})$ has the multiplicative inverse $\frac{1}{F(\mathbf{z})}$ if and only if $F(\mathbf{0}) = f_0 \neq 0$. Thus, $\mathbb{F}[[z_1, \ldots, z_d]]$ is a *local ring*, i.e., there is a unique maximal ideal.

For $G \in \mathbb{F}[[z_1, \ldots, z_d]]$ and d generating functions $F_1, \ldots, F_d$ in any number of variables, all with vanishing constant terms, one may define the *formal composition* [20] as

$$G \circ \mathbf{F} = G \circ (F_1, \ldots, F_d) := \lim_{n\to\infty} \sum_{|\mathbf{i}|\le n} g_{\mathbf{i}} \mathbf{F}^{\mathbf{i}}.$$

The degree of any monomial in $\mathbf{F}^{\mathbf{i}} := F_1^{i_1} \cdots F_d^{i_d}$ is at least $|\mathbf{i}| = \sum_{j=1}^{d} i_j$ by the assumption that $F_j(\mathbf{0}) = 0$ for all j. Moreover, the formal composition can be extended to any finite number of generating functions $G_i \in \mathbb{F}[[z_1, \ldots, z_d]]$ as follows:

$$\mathbf{G} \circ \mathbf{F} = \mathbf{G}(\mathbf{F}) := (G_1 \circ \mathbf{F}, \ldots, G_\ell \circ \mathbf{F}), \tag{7.3.37}$$

where $\mathbf{G} = (G_1, \ldots, G_\ell)$ and $\mathbf{F} = (F_1, \ldots, F_d)$ with $\mathbf{F}(\mathbf{0}) = \mathbf{0}$, i.e., $F_j(\mathbf{0}) = 0$ for all j.

The *Jacobian matrix* of $\mathbf{F}$ with respect to $z_1, \ldots, z_d$ is defined to be the $d \times d$ matrix whose (i, j)-entry is $\frac{\partial F_i}{\partial z_j}$, denoted by

$$J_{\mathbf{F}}(\mathbf{z}) := \frac{\partial \mathbf{F}}{\partial \mathbf{z}} = \frac{\partial(F_1, \ldots, F_d)}{\partial(z_1, \ldots, z_d)} = \left[\frac{\partial F_i}{\partial z_j}\right]_{1\le i,j\le d}. \tag{7.3.38}$$

A standard application of partial derivatives gives the chain rule

$$\frac{\partial \mathbf{G}(\mathbf{F})}{\partial \mathbf{z}} = \frac{\partial \mathbf{G}}{\partial \mathbf{z}}(\mathbf{F})\frac{\partial \mathbf{F}}{\partial \mathbf{z}}.$$

Since $\mathbf{F}(\mathbf{0}) = \mathbf{0}$ it follows that

$$\det\Big(J_{\mathbf{G}\circ\mathbf{F}}(\mathbf{0})\Big) = \det\Big(J_{\mathbf{G}}(\mathbf{F}(\mathbf{0}))J_{\mathbf{F}}(\mathbf{0})\Big) = \det\Big(J_{\mathbf{G}}(\mathbf{0})\Big)\det\Big(J_{\mathbf{F}}(\mathbf{0})\Big). \tag{7.3.39}$$

It is also known [14] that $\mathbf{F}$ has a compositional inverse, i.e., $\mathbf{G} \circ \mathbf{F} = \mathbf{F} \circ \mathbf{G} = \mathbf{z}$ if and only if $F_i(\mathbf{0}) = 0$ for all i and $\det\Big(J_{\mathbf{F}}(\mathbf{0})\Big) = |J_{\mathbf{F}}(\mathbf{0})| \neq 0$. The compositional inverse of $\mathbf{F}$ is denoted by $\overline{\mathbf{F}}$.

Multivariate Riordan group

In Sect. 3.3, we have seen how the Riordan group $\mathcal{R}$ can be represented as the semidirect product of $\mathcal{F}_0$ and $\mathcal{F}_1$ where $\mathcal{F}_0 = \{g \in \mathbb{F}[[z]] \mid g(0) \neq 0\}$ and $\mathcal{F}_1 = \{f \in \mathbb{F}[[z]] \mid f(0) = 0, f'(0) \neq 0\}$. It is also possible to represent the *Riordan group by multiple variables* using a similar argument.

We begin with the two sets for generating functions of several variables defined by

$$\mathcal{F}_{\mathbf{0}}^d = \Big\{G \in \mathbb{F}[[z_1, \ldots, z_d]] \mid G(\mathbf{0}) \neq 0\Big\};$$
$$\mathcal{F}_{\mathbf{1}}^d = \Big\{\mathbf{F} = (F_1, \ldots, F_d),\ F_i \in \mathbb{F}[[z_1, \ldots, z_d]] \mid \mathbf{F}(\mathbf{0}) = \mathbf{0},\ |J_{\mathbf{F}}(\mathbf{0})| \neq 0\Big\}.$$

It is easy to show that $\mathcal{F}_{\mathbf{0}}^d$ and $\mathcal{F}_{\mathbf{1}}^d$ form groups with respect to convolution $\cdot$ in (7.3.36) and composition $\circ$ in (7.3.37), respectively. We now show that the semidirect product

$$\mathcal{F}_{\mathbf{0}}^d \rtimes \mathcal{F}_{\mathbf{1}}^d = \left\{(G, \mathbf{F}) \mid G \in \mathcal{F}_{\mathbf{0}}^d,\ \mathbf{F} \in \mathcal{F}_{\mathbf{1}}^d\right\}$$

forms a group for an appropriately chosen binary operation.

Theorem 7.9 *Let $\mathcal{R}^d$ be the semidirect product $\mathcal{F}_{\mathbf{0}}^d \rtimes \mathcal{F}_{\mathbf{1}}^d$. Then $\mathcal{R}^d$ is a group with binary operation*

$$(G, \mathbf{F}) * (H, \mathbf{L}) = \left(GH(\mathbf{F}), \mathbf{L}(\mathbf{F})\right). \tag{7.3.40}$$

Proof Clearly, $GH(\mathbf{F}) \in \mathcal{F}_{\mathbf{0}}^d$. Since $\mathbf{L}(\mathbf{F}(\mathbf{0})) = \mathbf{0}$ and $|J_{\mathbf{L}\circ\mathbf{F}}(\mathbf{0})| \neq 0$ from (7.3.39), we have $\mathbf{L}(\mathbf{F}) \in \mathcal{F}_{\mathbf{1}}^d$. Thus $\mathcal{R}^d$ is closed under the operation $*$. It is easy to see that $*$ is associative. In addition, $(1, \mathbf{z})$ is the identity of $\mathcal{R}^d$, and for any $(G, \mathbf{F}) \in \mathcal{R}^d$ we have

$$(G, \mathbf{F})^{-1} = \left(\frac{1}{G\left(\overline{\mathbf{F}}\right)}, \overline{\mathbf{F}}\right) \in \mathcal{R}^d.$$

Thus $\mathcal{R}^d$ is a group under binary operation $*$ in (7.3.40). □

We call $\mathcal{R}^d$ the *multivariate Riordan group*. Certain subgroups of the Riordan group can be generalized immediately from one variable to several variables. We leave to the reader to verify that each of the following is a subgroup of $\mathcal{R}^d$ (see Exercises 7.10).

(a) The multivariate Appell subgroup, $\mathcal{A}^d = \left\{(G, \mathbf{z}) \mid G \in \mathcal{F}_{\mathbf{0}}^d\right\}$.
(b) The multivariate Lagrange subgroup, $\mathcal{L}^d = \left\{(1, \mathbf{F}) \mid F \in \mathcal{F}_{\mathbf{1}}^d\right\}$.
(c) The multivariate Bell subgroup, $\mathcal{B}^d = \left\{(G, \mathbf{z}G) \mid G \in \mathcal{F}_{\mathbf{0}}^d\right\}$ where $\mathbf{z}G = (z_1G, \ldots, z_dG)$.
(d) The multivariate checkerboard subgroup,

$$\mathcal{C}^d = \left\{(G_e, \mathbf{F}_o) \mid G_e \in \mathcal{F}_{\mathbf{0}}^d, \mathbf{F}_o = (F_1, \ldots, F_d) \in \mathcal{F}_{\mathbf{1}}^d\right\},$$

where $G_e = \sum_{\mathbf{i}} g_{\mathbf{i}}\mathbf{z}^{\mathbf{i}}$ and $F_i = \sum_{\mathbf{i}} f_{\mathbf{i}}\mathbf{z}^{\mathbf{i}}$ with the sums running over all $\mathbf{i}$ such that $|\mathbf{i}|$ is even and odd, respectively.

Note that $(G, \mathbf{F}) = (G, \mathbf{z}) * (1, \mathbf{F})$. Thus we see that the multivariate Riordan group is the semidirect product of the multivariate Appell subgroup and the multivariate Lagrange group, i.e.

$$\mathcal{R}^d = \mathcal{A}^d \rtimes \mathcal{L}^d.$$

Moreover, it follows from (7.3.39) that we can derive a new subgroup of $\mathcal{R}^d$ called the *special linear subgroup* as shown in Theorem 7.10.

(e) The special linear subgroup,

$$\mathcal{S}^d = \left\{(G, \mathbf{F}) \in \mathcal{R}^d \mid |J_{\mathbf{F}}(\mathbf{0})| = 1\right\}.$$

Theorem 7.10 *The set $\mathcal{S}^d$ is a normal subgroup of $\mathcal{R}^d$.*

Proof Since $|J_{\mathbf{z}}(\mathbf{0})| = |I_d| = 1$, clearly $(1, \mathbf{z}) \in \mathcal{S}^d$ so that $\mathcal{S}^d$ is non-empty. Now let $(G, \mathbf{F}), (H, \mathbf{L}) \in \mathcal{S}^d$. Then it follows from (7.3.39) that the operation is closed in $\mathcal{S}^d$. In addition, since $|J_{\overline{\mathbf{F}}}(\mathbf{0})| = |J_{\mathbf{F}}(\mathbf{0})^{-1}| = 1$, we have $(G, \mathbf{F})^{-1} \in \mathcal{S}^d$. Thus $\mathcal{S}^d$ is a subgroup of $\mathcal{R}^d$.

Moreover, for any elements $(H, \mathbf{L}) \in \mathcal{R}^d$ and $(G, \mathbf{F}) \in \mathcal{S}^d$, we have

$$(H, \mathbf{L})^{-1}(G, \mathbf{F})(H, \mathbf{L}) = \left(\frac{G\left(\overline{\mathbf{L}}\right) H\left(\overline{\mathbf{F}}\left(\overline{\mathbf{L}}\right)\right)}{H\left(\overline{\mathbf{L}}\right)}, \mathbf{L}\left(\overline{\mathbf{F}}\left(\overline{\mathbf{L}}\right)\right)\right).$$

Since $|J_{\mathbf{L}\circ\overline{\mathbf{F}}(\overline{\mathbf{L}})}(\mathbf{0})| = |J_{\mathbf{L}}(\mathbf{0})||J_{\overline{\mathbf{F}}}(\mathbf{0})||J_{\overline{\mathbf{L}}}(\mathbf{0})| = 1$ by (7.3.39), it follows that

$$(H, \mathbf{L})^{-1}(G, \mathbf{F})(H, \mathbf{L}) \in \mathcal{S}^d.$$

Thus $\mathcal{S}^d$ is a normal subgroup of $\mathcal{R}^d$. □

Example 7.4 Consider power series $G = \frac{1}{1-x-y}$ and $\mathbf{F} = (F_1, F_2) = \left(\frac{x}{1-x-y}, \frac{y}{1-x-y}\right)$. Since

$$|J_{\mathbf{F}}(\mathbf{0})| = \det \begin{bmatrix} \frac{\partial}{\partial x}\left(\frac{x}{1-x-y}\right) & \frac{\partial}{\partial y}\left(\frac{x}{1-x-y}\right) \\ \frac{\partial}{\partial x}\left(\frac{y}{1-x-y}\right) & \frac{\partial}{\partial y}\left(\frac{y}{1-x-y}\right) \end{bmatrix}_{(x,y)=(0,0)} = \det \begin{bmatrix} 1 & 0 \\ 0 & 1 \end{bmatrix} = 1,$$

and $\mathbf{F} = (xG, yG)$, we have $(G, \mathbf{F}) \in \mathcal{S}^2 \cap \mathcal{B}^2$.

Multivariate Riordan arrays

As with the case of one variable, an element in the multivariate Riordan group can be represented as an infinite matrix determined by a numeration of monomials $z_1^{i_1} z_2^{i_2} \cdots z_d^{i_d}$ in the ring $\mathbb{F}[[z_1, \ldots, z_d]]$. The numeration is assumed to respect the total degree, that is, a monomial must precede other monomials with larger total degrees. Monomials with the same total degree can be arranged arbitrarily in the numeration. For example, we can use the graded reverse lexicographic order.

Let $\mathbf{i} = (i_1, i_2, \ldots, i_d)$ and $\mathbf{j} = (j_1, j_2, \ldots, j_d)$ be monomial exponents in $\mathbb{N}^d$. We say that $\mathbf{i}$ precedes $\mathbf{j}$, denoted by $\mathbf{i} \prec \mathbf{j}$, in the *graded reverse lexicographic order* if

(i) $|\mathbf{i}| < |\mathbf{j}|$, or
(ii) $|\mathbf{i}| = |\mathbf{j}|$ and the right-most non-zero entry of $\mathbf{j} - \mathbf{i}$ is negative.

By this monomial ordering, it is possible to list the monomial exponents in a linear order $\mathbf{i}_0 \prec \mathbf{i}_1 \prec \mathbf{i}_2 \prec \ldots$ starting with $\mathbf{i}_0 = \mathbf{0} = (0, \ldots, 0)$, where $|\mathbf{i}_0| \leq |\mathbf{i}_1| \leq |\mathbf{i}_2| \leq \cdots$. For example, if $d = 2$ then

$$(0,0) \prec (0,1) \prec (1,0) \prec (0,2) \prec (1,1) \prec$$

$$(2,0) \prec (0,3) \prec (1,2) \prec (2,1) \prec (3,0) \prec \cdots$$

We now consider a matrix representation for elements in the multivariate Riordan group $\mathcal{R}^d$. Given an element $(G, \mathbf{F}) \in \mathcal{R}^d$ with $\mathbf{F} = (F_1, \ldots, F_d)$, let

$$a_{\mathbf{i},\mathbf{j}} = \left[\mathbf{z}^{\mathbf{i}}\right] G\mathbf{F}^{\mathbf{j}} = \left[z_1^{i_1} \cdots z_d^{i_d}\right] G F_1^{j_1} \cdots F_d^{j_d}.$$

So we have

$$G\mathbf{F}^{\mathbf{j}} = \sum_{\mathbf{i}} a_{\mathbf{i},\mathbf{j}} \mathbf{z}^{\mathbf{i}}.$$

The element $(G, \mathbf{F})$ gives rise to an infinite matrix $M = \left[a_{\mathbf{i},\mathbf{j}}\right]$ whose rows and columns are indexed by the graded reverse lexicographic order in $\mathbb{N}^d$. We call the matrix M a *multivariate Riordan array* associated with $(G, \mathbf{F})$ and we write M as $\mathcal{M}(G, \mathbf{F})$ or $\mathcal{M}(G, F_1, \ldots, F_d)$. Note that if $d = 1$ then it is a usual Riordan array.

Given another element $(H, \mathbf{L})$ in $\mathcal{R}^d$, we write $H\mathbf{L}^{\mathbf{k}} = \sum_{\mathbf{j}} b_{\mathbf{j},\mathbf{k}} \mathbf{z}^{\mathbf{j}}$. Since

$$\sum_{\mathbf{i}} \left(\sum_{\mathbf{j}} a_{\mathbf{i},\mathbf{j}} b_{\mathbf{j},\mathbf{k}} \right) \mathbf{z}^{\mathbf{i}} = \sum_{\mathbf{j}} b_{\mathbf{j},\mathbf{k}} G\mathbf{F}^{\mathbf{j}} = G \cdot (H\mathbf{L}^{\mathbf{k}})(\mathbf{F}) = GH(\mathbf{F})\mathbf{L}^{\mathbf{k}}(\mathbf{F}),$$

the multiplication in $\mathcal{R}^d$ is compatible with the matrix product. In terms of our notation,

$$\mathcal{M}(G, \mathbf{F})\mathcal{M}(H, \mathbf{L}) = \mathcal{M}\left(GH(\mathbf{F}), \mathbf{L}(\mathbf{F})\right) = \mathcal{M}\left((G, \mathbf{F}) * (H, \mathbf{L})\right).$$

In particular, $GH(\mathbf{F})$ is the generating function of the product of $\mathcal{M}(G, \mathbf{F})$ and the first column of $\mathcal{M}(H, \mathbf{L})$. Thus we immediately obtain the following theorem. In the case of one variable, this particular situation is known as the fundamental theorem of Riordan arrays (see Theorem 3.1).

Theorem 7.11 (Fundamental theorem of multivariate Riordan arrays)
Let $\mathcal{M}(G, \mathbf{F})$ be a multivariate Riordan array and $H(\mathbf{z}) = \sum_{n \geq 0} h_{\mathbf{i}_n} \mathbf{z}^{\mathbf{i}_n}$ be a generating function. Then

$$\mathcal{M}(G, \mathbf{F})H = GH(\mathbf{F}).$$

More precisely, we look at the structure of the matrix $\mathcal{M}(G, \mathbf{F}) = \left[a_{\mathbf{i},\mathbf{j}}\right]$. Recall that its rows and columns are indexed by the graded reverse lexicographic order in

$\mathbb{N}^d$. Let α_k be the number of monomial exponents in $\mathbb{N}^d$ whose total degree is k, i.e.,

$$\alpha_k = |\{(i_1, \ldots, i_d) \in \mathbb{N}^d \mid i_1 + \cdots + i_d = k\}|.$$

Since α_k is equal to the number of non-negative solutions of $x_1 + \cdots + x_d = k$, we have $\alpha_k = \binom{k+d-1}{k}$. Let M_{ij} be the matrix of size $\binom{i+d-1}{i} \times \binom{j+d-1}{j}$ with entries $a_{\mathbf{i},\mathbf{j}}$ arranged by the graded reverse lexicographic order, where the total degrees of $\mathbf{z}^{\mathbf{i}}$ and $\mathbf{z}^{\mathbf{j}}$ are i and j, respectively. If $|\mathbf{i}| < |\mathbf{j}|$ then $a_{\mathbf{i},\mathbf{j}} = [\mathbf{z}^{\mathbf{i}}]\, G\mathbf{F}^{\mathbf{j}} = 0$ since each F_i does not have a constant term. Therefore, the multivariate Riordan array with d variables is an infinite, lower block triangular matrix of the form

$$\mathcal{M}(G, \mathbf{F}) = [M_{ij}] = \begin{bmatrix} M_{00} & 0 & 0 & 0 & \cdots \\ M_{10} & M_{11} & 0 & 0 & \cdots \\ M_{20} & M_{21} & M_{22} & 0 & \cdots \\ M_{30} & M_{31} & M_{32} & M_{33} & \cdots \\ \vdots & \vdots & \vdots & \vdots & \ddots \end{bmatrix}. \tag{7.3.41}$$

As is well-known, the Taylor series $f(x) = \sum_{n\geq 0} \frac{f^{(n)}(a)}{n!}(x-a)^n$ is a series expansion of an analytic function f about a point a. The Taylor series can be generalized to a multivariate function $f(\mathbf{z})$ with an arbitrary number of variables $z_1, \ldots, z_d$ which is continuously differentiable at the point $\mathbf{a} \in \mathbb{F}^d$:

$$f(\mathbf{z}) = \sum_{|\mathbf{i}|\geq 0} \frac{D^{\mathbf{i}} f(\mathbf{a})}{\mathbf{i}!}(\mathbf{z}-\mathbf{a})^{\mathbf{i}}, \quad D^{\mathbf{i}} f = \frac{\partial^{|\mathbf{i}|} f}{\partial z_1^{i_1} \cdots \partial z_d^{i_d}}$$

where $\mathbf{i}! = i_1! \cdots i_d!$. For example, if $d = 2$ then the Taylor series of $f(\mathbf{z})$ at $\mathbf{0}$ gives

$$\begin{aligned} f(\mathbf{z}) = f(\mathbf{0}) &+ \frac{\partial f(\mathbf{0})}{\partial z_1} z_1 + \frac{\partial f(\mathbf{0})}{\partial z_2} z_2 \\ &+ \frac{\partial^2 f(\mathbf{0})}{\partial z_1^2}\frac{z_1^2}{2!} + \frac{\partial^2 f(\mathbf{0})}{\partial z_1 \partial z_2} z_1 z_2 + \frac{\partial^2 f(\mathbf{0})}{\partial z_2^2}\frac{z_2^2}{2!} \\ &+ \frac{\partial^3 f(\mathbf{0})}{\partial z_1^3}\frac{z_1^3}{3!} + \frac{\partial^3 f(\mathbf{0})}{\partial z_1^2 \partial z_2}\frac{z_1^2 z_2}{2!} + \frac{\partial^3 f(\mathbf{0})}{\partial z_1 \partial z_2^2}\frac{z_1 z_2^2}{2!} + \frac{\partial^3 f(\mathbf{0})}{\partial z_2^3}\frac{z_2^3}{3!} + \cdots \end{aligned}$$

Theorem 7.12 *Let M_{11} be the $(1, 1)$-block of the matrix $\mathcal{M}(G, \mathbf{F})$ in (7.3.41) and $\alpha = G(\mathbf{0})$. Then*

$$M_{11} = P\left(\alpha J_{\mathbf{F}}(\mathbf{0})^T\right) P^T,$$

where P is the $d \times d$ backward identity matrix defined by

$$P = \begin{bmatrix} 0 & \cdots & 0 & 1 \\ \vdots & \iddots & 1 & 0 \\ 0 & \iddots & \iddots & \vdots \\ 1 & 0 & \cdots & 0 \end{bmatrix}.$$

Proof Let $\mathbf{e}_i$ denote the ith row vector of the $d \times d$ identity matrix. Consider the $d \times d$ matrix $D = [a_{\mathbf{e}_i,\mathbf{e}_j}]$ defined by $a_{\mathbf{e}_i,\mathbf{e}_j} = [\mathbf{z}^{\mathbf{e}_i}]G\mathbf{F}^{\mathbf{e}_j}$. Since $\mathbf{F}(\mathbf{0}) = \mathbf{0}$ it follows from the Taylor series of GF_j at $\mathbf{0}$ that

$$a_{\mathbf{e}_i,\mathbf{e}_j} = [\mathbf{z}^{\mathbf{e}_i}]G\mathbf{F}^{\mathbf{e}_j} = [z_i]GF_j = \frac{\partial(GF_j)(\mathbf{0})}{\partial z_i} = G(\mathbf{0})\frac{\partial F_j(\mathbf{0})}{\partial z_i}.$$

By the definition of the Jacobian matrix in (7.3.38), we see that if $G(\mathbf{0}) = \alpha$ then $D = \alpha J_{\mathbf{F}}(\mathbf{0})^T$. Since the rows and columns of M_{11} are ordered by $\mathbf{e}_d, \mathbf{e}_{d-1}, \ldots, \mathbf{e}_1$ from the graded reverse lexicographic order, it follows from that $M_{11} = PDP^T$. Thus $M_{11} = P\left(\alpha J_{\mathbf{F}}(\mathbf{0})^T\right)P^T$. □

Example 7.5 Consider the multivariate Riordan array with two variables x, y given by

$$\mathcal{M}(G, F_1, F_2) = [M_{ij}] = \left(\frac{1}{1-x-y-xy}, x+y, x+2y\right).$$

Let $\mathbf{F} = (x+y, x+2y)$. Since $\mathbf{F}(\mathbf{0}) = (0, 0)$ and

$$|J_{\mathbf{F}}(\mathbf{0})| = \det\begin{bmatrix} \frac{\partial}{\partial x}(x+y) & \frac{\partial}{\partial y}(x+y) \\ \frac{\partial}{\partial x}(x+2y) & \frac{\partial}{\partial y}(x+2y) \end{bmatrix}_{(x,y)=(0,0)} = \begin{vmatrix} 1 & 1 \\ 1 & 2 \end{vmatrix} = 1 \neq 0,$$

we have $\mathbf{F} \in \mathcal{F}_{\mathbf{1}}^2$. By using the graded reverse lexicographic order, both rows and columns of $\mathcal{M}(G, \mathbf{F})$ are indexed by the monomial exponents

$$(0,0), (0,1), (1,0), (0,2), (1,1), (2,0), (0,3), (1,2), (2,1), (3,0), \cdots.$$

Thus each block $M_{ij} = [a_{\mathbf{i},\mathbf{j}}]$ is the $(i+1) \times (j+1)$ matrix with entries

$$a_{\mathbf{i},\mathbf{j}} = [x^{i_1}y^{i_2}]GF_1^{j_1}F_2^{j_2},$$

where $|\mathbf{i}| = i_1 + i_2 = i$ and $|\mathbf{j}| = j_1 + j_2 = j$. For instance, using the Taylor series

$$\begin{aligned} GF_1^0F_2^1 &= \frac{x+2y}{1-x-y-xy} = x + 2y + x^2 + 3xy + 2y^2 + x^3 + 5x^2y + 7xy^2 + 2y^3 + \cdots \\ GF_1^1F_2^0 &= \frac{x+y}{1-x-y-xy} = x + y + x^2 + 2xy + y^2 + x^3 + 4x^2y + 4xy^2 + y^3 + \cdots \end{aligned}$$

we obtain

$$M_{21}=\begin{bmatrix}[x^0y^2]GF_1^0F_2^1 & [x^0y^2]GF_1^1F_2^0\\ [x^1y^1]GF_1^0F_2^1 & [x^1y^1]GF_1^1F_2^0\\ [x^2y^0]GF_1^0F_2^1 & [x^2y^0]GF_1^1F_2^0\end{bmatrix}=\begin{bmatrix}2&1\\3&2\\1&1\end{bmatrix}.$$

Moreover,

$$\mathcal{M}(G,\mathbf{F})=\left[\begin{array}{c|cc|ccc|cccc}1&0&0&0&0&0&0&0&0&0\\ \hline 1&2&1&0&0&0&0&0&0&0\\ 1&1&1&0&0&0&0&0&0&0\\ \hline 1&2&1&4&2&1&0&0&0&0\\ 3&3&2&4&3&2&0&0&0&0\\ 1&1&1&1&1&1&0&0&0&0\\ \hline 1&2&1&4&2&1&8&4&2&1\\ 5&7&4&8&5&3&12&8&5&3\\ 5&5&4&5&4&3&6&5&4&3\\ 1&1&1&1&1&1&1&1&1&1\\ &&&&\cdots&&&&&\end{array}\right].$$

As noted in [6], the matrix M_{nn} involves only the constant term α of G and the linear part of $\mathbf{F}$ as power series in $\mathbf{z}$. Indeed, the matrix $\alpha^{-1}M_{nn}$ is the image of $\alpha^{-1}M_{11}$ under the nth symmetric power representation $GL(V)\to GL(\mathrm{sym}^n V)$, where V is the $\mathbb{F}$-vector space generated by $\mathbf{z}$. For instance, if $\alpha=1$ and $M_{11}=\begin{bmatrix}a&c\\b&d\end{bmatrix}$ then

$$M_{22}=\begin{bmatrix}a^2 & ac & c^2\\ 2ab & ad+bc & 2cd\\ b^2 & bd & d^2\end{bmatrix}.$$

In the case of one variable, A-sequences and Z-sequences have been used to characterize Riordan arrays. The idea is transparent from our viewpoint, even for several variables. Given $(G,\mathbf{F})$ in the multivariate Riordan group $\mathcal{R}^d$, $\mathbf{F}$ can be written as $\mathbf{F}=\mathbf{z}\mathbf{A}$, where $\mathbf{A}$ is a $d\times d$ matrix with power series as entries. Let α be the constant term of G. Then G can be written as $G=\alpha(1-\mathbf{z}\mathbf{B})^{-1}$, where $\mathbf{B}$ is a $d\times 1$ matrix with power series as entries. The fact that $\mathbf{A}$, $\mathbf{B}$ and α determine $(G,\mathbf{F})$ was observed in the case of one variable [12], where the matrices $\mathbf{A}$ and $\mathbf{B}$ in terms of $\mathbf{F}$ are the generating functions for the A-sequence and Z-sequence of the multivariate Riordan array $\mathcal{M}(G,\mathbf{F})$, respectively. Note that the matrices $\mathbf{A}$ and $\mathbf{B}$ in several variables are not unique.

Exercises

7.1 Show that each of the sets $\mathcal{A}_1,\mathcal{A}_2,\mathcal{L},\mathcal{R},\mathcal{M},\mathcal{B},\mathcal{C},\mathcal{K}$ is a subgroup of $\mathcal{R}^{\langle 3\rangle}$ (see (a)–(h) in Sect. 7.1).

7.2 Show that $\mathcal{A}_1, \mathcal{A}_2$ and $\mathcal{M}$ are normal subgroups of $\mathcal{R}^{\langle 3\rangle}$.

7.3 Show that for $g, h, G, H \in \mathcal{F}_0$ and $f, F \in \mathcal{F}_1$,

$$(g, f, zh)(G, F, zH) = (gG(f), F(f), zhH(f)).$$

7.4 Let $\hat{\mathcal{R}}^{\langle 3\rangle}$ be the set of shifted 3-D Riordan arrays, i.e.,

$$\hat{\mathcal{R}}^{\langle 3\rangle} = \Big\{(g, f, zh) \mid g, h \in \mathcal{F}_0, f \in \mathcal{F}_1\Big\}.$$

Prove that $\hat{\mathcal{R}}^{\langle 3\rangle}$ forms a group under the binary operation defined in Exercise 7.3.

7.5 Prove that $\mathcal{R}^{\langle 3\rangle}$ is isomorphic to $\hat{\mathcal{R}}^{\langle 3\rangle}$.

7.6 Apply the fundamental theorem of 3-D Riordan array to prove that

$$\sum_{k=0}^{n}\sum_{i=0}^{\lfloor\frac{k+\ell}{2}\rfloor}(-1)^i\frac{1}{n-i+\ell}\binom{k}{\ell}\binom{k-i+\ell}{i}\binom{2(n-i+\ell)}{n-i+\ell} = \frac{3\ell+2}{n+2\ell+2}\binom{2n+\ell+1}{n+2\ell+1}.$$

7.7 Apply the identity (7.2.26) for the case $i = m-1$ to prove that

$$\frac{1-zC}{\sqrt{1-4z}} = 1 + \sum_{m\geq 1}\binom{2m-1}{m-1}z^m.$$

7.8 Use the 3-D Riordan array method to derive the identities:

(a) $\sum_{\ell=0}^{1}\sum_{k=0}^{n-\ell}\binom{n}{k+\ell}\binom{n}{k} = \binom{2n+1}{n}$;
(b) $\sum_{\ell=0}^{n}\sum_{k=0}^{n-\ell}\binom{n}{k+\ell} = (n+2)2^{n-1}$;
(c) $\sum_{\ell=0}^{n}\sum_{k=0}^{n-\ell}\frac{k+\ell+1}{n+1}\binom{2n-(k+\ell)}{n} = \frac{3}{n+3}\binom{2n+2}{n}$.

7.9 Consider the 3-D shifted Riordan array $\left(\frac{C}{1-z}, zC, \frac{z}{1-z}\right)$ where C is the Catalan generating function.

(a) Show that $[z^{n-k-\ell}]\frac{C^{k+1}}{(1-z)^{\ell+1}} = \sum_{i=k}^{n-\ell}\frac{k+1}{i+1}\binom{2i-k}{i}\binom{n-i}{\ell}$ for $n \geq k+\ell$.
(b) Give a combinatorial interpretation for the number in (a) in terms of xyz-lattice paths.

7.10 Show that each of the sets $\mathcal{A}^d, \mathcal{L}^d, \mathcal{B}^d, \mathcal{C}^d$ is a subgroup of $\mathcal{R}^d$ (see (a)-(d) in Sect. 7.3).

7.11 Show that $\mathcal{A}^d$ is a normal subgroup of $\mathcal{R}^d$.

7.12 Let $\varphi(\mathbf{z}) = a_1z_1 + \cdots + a_dz_d$, $a_i \in \mathbb{F}$. Prove that if (G, zH) is an involution in $\mathcal{R}$ where $G, H \in \mathcal{F}_0$ then $\Big(G(\varphi), \mathbf{z}H(\varphi)\Big)$ is an involution in $\mathcal{R}^d$.

7.13 (Multivariate Pascal matrix) Let $\mathbb{P}_{\mathbf{z}} = \mathcal{M}\left(\frac{1}{1-z_1-\cdots-z_d}, \frac{\mathbf{z}}{1-z_1-\cdots-z_d}\right)$.

(a) For $d = 2$, give the first 6×6 submatrix of $\mathbb{P}_{\mathbf{z}}$.
(b) Find the inverse matrix $\mathbb{P}_{\mathbf{z}}^{-1}$ in terms of generating functions.
(c) Show that $\mathbb{P}_{\mathbf{z}}$ is a lower triangular matrix.

References

1. M. Aigner, *A Course in Enumeration*, GTM 238 (Springer, 2007)
2. A.K. Ahlawat, S. Prakash, A new approach for representation of multi-dimensional matrix multiplication operations. Int. J. Comput. Appl. **29**(8), 0975–8887 (2011)
3. P. Barry, The central coefficients of a family of Pascal-like triangles and colored lattice paths. J. Integer Sequences **22** (2019), Article 19.1.3
4. R.A. Brualdi, *Introductory Combinatorics*, 4th edn. (Pearson, 2010)
5. R.A. Brualdi, H.J. Ryser, *Combinatorial Matrix Theory, Encyclopedia of Mathematics and its Applications*, vol. 39 (Cambridge University Press, 1991)
6. G.-S. Cheon, I.-C. Huang, S. Kim, Multivariate Riordan groups and their representations. Linear Algebra Appl. **514**, 198–207 (2017)
7. G.-S. Cheon, S.-T. Jin, The group of multi-dimensional Riordan arrays. Linear Algebra Appl. **524**, 263–277 (2017)
8. G.-S. Cheon, H. Kim, L.W. Shapiro, Combinatorics of Riordan arrays with identical A and Z sequences. Discrete Math. **312**, 2040–2049 (2012)
9. G.-S. Cheon, H. Kim, L.W. Shapiro, A generalization of Lucas polynomial sequence. Discrete Appl. Math. **157**, 920–927 (2009)
10. D.E. Davenport, L.W. Shapiro, L.C. Woodson, The Double Riordan group. Electron. J. Comb. **18**(2), #P331 (2012)
11. E.H. Goins, A. Nkwanta, Riordan matrix representations of Euler's constant γ and Euler's number e. Int. J. Comb. Article ID 8324150 (2016)
12. T.-X. He, R. Sprugnoli, Sequence characterization of Riordan arrays. Discrete Math. **309**(12), 3962–3974 (2009)
13. T.-X. He, L. Hsu, P.J.-S. Shiue, The Sheffer group and the Riordan group. Discrete Appl. Math. **155**, 1895–1909 (2007)
14. I-C. Huang, Residue methods in combinatorial analysis, Local Cohomology and its Applications. Lecture Notes Pure Appl. Math. **226**, 255–342 (2001)
15. K. Humphrey, A history and a survey of lattice path enumeration. J. Stat. Plann. Infer. **140**, 2237–2254 (2010)
16. S. Kaparthi, H. Raghav Rao, Higher dimensional restricted lattice paths with diagonal steps. Discrete Appl. Math. **31**, 279–289 (1991)
17. S. Kim, Algebric structure of multivariate Riordan matrices, MS thesis, Sungkyunkwan University (2016)
18. D. Merlini, R. Sprugnoli, M.C. Verri, Algebraic and combinatorial properties of simple, coloured walks. *Trees in Algebra and Programming, CAAP* (Edinburgh, 1994), Lecture Notes in Computer Science, vol. 787 (Springer, Berlin, 1994), pp. 218–233
19. D. Merlini, D.G. Rogers, R. Sprugnoli, M.C. Verri, On some alternative characterizations of Riordan arrays. Can. J. Math. **49**, 301–320 (1997)
20. R. Pemantle, M.C. Wilson, *Analytic Combinatorics in Several Variables*, vol. 140 (Cambridge University Press, 2013)
21. J.L. Ramírez, V.F. Sirvent, Generalized Schröder matrix and its combinatorial interpretation. Linear Multilinear Algebra **66**, 418–433 (2018)
22. L.W. Shapiro, S. Getu, W.-J. Woan, L. Woodson, The Riordan group. Discrete Appl. Math. **34**, 229–239 (1991)

23. A.M.G. Solo, Multidimensional matrix mathematics: multidimensional matrix equality, addition, subtraction, and multiplication, Part 2 of 6, in *Proceedings of the World Congress on Engineering 2010*, vol. III (2010), pp. 1829–1833
24. R. Sprugnoli, Riordan arrays and combinatorial sums. Discrete Math. **132**, 267–290 (1994)
25. R. Stanley, *Enumerative Combinatorics, Vol 1*, Cambridge Studies in advanced Mathematics, vol. 49 (Cambridge University Press, 2017)
26. L. Yang, S.-L. Yang, T.-X. He, Generalized Schröder matrices arising from enumeration of lattice paths. Czechoslovak Math. J. **70**(145), 411–433 (2020)
27. S.-L. Yang, S.-N. Zheng, S.-P. Yuan, T.-X. He, Schröder matrix as inverse of Delannoy matrix. Linear Algebra Appl. **439**, 3605–3614 (2013)
28. W. Wang, T. Wang, Generalized Riordan arrays. Discrete Math. **308**(24), 6466–6500 (2008)

Chapter 8
q-Analogs of Riordan Arrays

Abstract The Riordan group consisting of proper Riordan arrays shows up naturally in a variety of combinatorial settings. In this chapter, we define the q-analog of a Riordan array called a q-Riordan array and denoted as $(g, f)_q$. It is defined by using a pair of Eulerian generating functions g, f of the form

$$\sum_{n\geq 0} a_n \frac{z^n}{[n]_q!}$$

where $[n]_q! = 1(1+q)(1+q+q^2)\cdots(1+q+\cdots+q^{n-1})$. A q-analog appears naturally in several contexts of combinatorics, quantum group theory, and so on. We establish some algebraic properties for q-Riordan arrays. Noticing that $[n]_q!$ reduces to $n!$ upon setting $q=1$, we see that a q-Riordan array reduces to an exponential Riordan array for the case of $q=1$. This suggests that q-Riordan arrays might be useful not only for studying enumeration problem [2, 10, 11] but also for defining q-analogs of some orthogonal polynomials [3, 4]. Indeed, it is shown that q-Riordan arrays associated to the counting functions may be applied to the enumeration problem on set partitions by block inversions. This notion also leads us to find q-analogs of the composition formula and the exponential formula, respectively.

8.1 Combinatorial q-Analogs

A *q-analog* of a theorem, identity, or expression in the areas of combinatorics and special functions is a generalization involving a new parameter $q \in \mathbb{C}$ that returns the original theorem, identity, or expression in the limit as $q \to 1$.

For example, the equality

$$\lim_{q\to 1} \frac{q^n - 1}{q-1} = n$$

suggests that the q-analog of a non-negative integer n written $[n]_q$ can be defined as

© The Author(s), under exclusive license to Springer Nature Switzerland AG 2022

L. Shapiro et al., *The Riordan Group and Applications*, Springer Monographs in Mathematics, https://doi.org/10.1007/978-3-030-94151-2_8

$$[n]_q = \frac{q^n - 1}{q - 1} = \sum_{k=1}^{n} q^{k-1}, \quad [0]_q = 0.$$

It is also known as the *q-bracket* or *q-number* of n. The q-analog of the factorial known as the *q-factorial* is defined by

$$[n]_q! = \prod_{k=1}^{n} [k]_q = 1(1+q)\cdots(1+q+\cdots q^{n-1}), \quad [0]_q! = 1.$$

This q-analog appears naturally in several contexts (for example, see [6–8]). Notably, while $n!$ counts the number of permutations of length n, $[n]_q!$ counts permutations while keeping track of the number of inversions. That is, if $\mathrm{inv}(w)$ denotes the number of inversions of the permutation w and S_n denotes the set of permutations of length n, we have

$$\sum_{w \in S_n} q^{\mathrm{inv}(w)} = [n]_q!.$$

From the q-factorials, the *q-binomial coefficients* known as *Gaussian coefficients* are defined by

$$\begin{bmatrix} n \\ k \end{bmatrix}_q = \frac{[n]_q!}{[n-k]_q![k]_q!}.$$

The q-binomial coefficients $\begin{bmatrix} n \\ k \end{bmatrix}_q$ count subspaces of a finite vector space. Indeed, if q denotes the number of elements in a finite field then the number of k-dimensional subspaces of the n-dimensional vector space over the q-element field equals $\begin{bmatrix} n \\ k \end{bmatrix}_q$.

In classical calculus, the derivative of a function f is defined by

$$\frac{df}{dz} = \lim_{h \to 0} \frac{f(z+h) - f(z)}{h}.$$

Equivalently, we could define the derivative to be

$$\frac{df}{dz} = \lim_{q \to 1} \frac{f(qz) - f(z)}{qz - z}.$$

One might wonder, what happens when we do not take the limit, and just use the expression inside the limit as the definition of our derivative. This leads us into the exciting world of quantum calculus, also known as *q-calculus*.

A *q-differential* of a function f is defined to be $d_q f = f(qz) - f(z)$. Now the *q-derivative* of f, $D_q f$ is defined by

$$D_q f := \frac{d_q f}{d_q z} = \frac{f(qz) - f(z)}{qz - z}, \quad q \in \mathbb{C} \setminus \{1\}.$$

For example,

$$D_q(z^n) = \frac{(qz)^n - z^n}{qz - z} = \frac{q^n - 1}{q - 1} z^{n-1} = [n]_q z^{n-1}.$$

The formulae for the q-derivative of a sum, a product and a quotient of functions are, respectively, given by

(i) $D_q(f + g) = D_q f + D_q g$
(ii) $D_q(fg) = f(qx) D_q g + g(x) D_q f$
(iii) $D_q(f/g) = (g D_q f - f D_q g)/g(qz)g(z),\ g(qz)g(z) \neq 0$

A function $f(z)$ that possesses derivatives of all orders is analytic at $z = a$ if $f(z)$ can be expressed as a power series about $z = a$. *Taylor's formula* is the expansion of an analytic function of $f(z)$ about $z = a$ as

$$f(z) = \sum_{n=0}^{\infty} f^{(n)}(a) \frac{(z - a)^n}{n!}.$$

It often allows us to extend the definition of the function to a larger and more interesting domain. The following q-analog of Taylor's formula was introduced by Jackson in 1909.

(*q-Taylor formula*) Let $f(x)$ be capable of expansion at $x = a$ as a convergent power series. If $q \neq$ root of unity, then

$$f(z) = \sum_{n=0}^{\infty} \left(D_q^n f\right)(a) \frac{(z - a)^{(n)}}{[n]_q!}. \tag{8.1.1}$$

Here, we have $(z - a)^{(0)} = 1$ and $(z - a)^{(n)} = \prod_{i=0}^{n-1}(z - aq^i)$.

8.2 Eulerian Generating Functions

One of the fundamental concepts in combinatorics is that of enumeration, and one of the basic techniques for dealing with problems of enumeration is that of generating functions. Thus generating functions provide an algebraic machinery for solving combinatorial problems. Heuristically, a generating function f is a representation of a counting function $N : \mathbb{N} \to \mathbb{N}$ as an element $f(N)$ of some algebra. There are several types of generating functions which have actually arisen in specific enumeration problems:

– ordinary generating functions;

- exponential generating functions;
- Eulerian generating functions;
- doubly-exponential generating functions;
- chromatic generating functions;
- Lambert generating functions, and so on.

It is well known [7] that many properties of exponential generating functions have analogs for *Eulerian generating functions* of the form

$$\sum_{n\geq 0} a_n \frac{z^n}{[n]_q!},$$

where $[n]_q! = 1(1+q)\cdots(1+q+\cdots+q^{n-1})$ and a_n is a polynomial in q. It can be thought of as the q-analog of the exponential generating function for the sequence $(a_n)_{n\geq 0}$. In particular,

$$e_q(z) = \sum_{n=0}^{\infty} \frac{z^n}{[n]_q!}$$

is called the *q-exponential*. It arises in several combinatorial applications such as finite vector spaces, partitions and counting permutations by inversions.

For $k \in \mathbb{N}$, the *kth symbolic power* $f^{[k]}(z)$ of $f(z)$ with $f(0) = 0$ is inductively given by

$$D_q f^{[k]}(z) = [k]_q f^{[k-1]}(z) D_q f(z) \quad \text{for } k \geq 1 \quad \text{and} \quad f^{[0]}(z) = 1,$$

where $f^{[k]}(0) = 0$ for $k \geq 1$. In particular, $f^{[1]}(z) = f(z)$. If $g(z) = \sum_{n\geq 0} g_n \frac{z^n}{[n]_q!}$ then the *q-composition* $\circ_q$ of g with f is defined as

$$(g \circ_q f)(z) = g[f(z)] = \sum_{n\geq 0} g_n \frac{f^{[n]}(z)}{[n]_q!}.$$

Since $z^{[n]} = z^n$, we have $g[z] = g(z)$.

8.3 q-Riordan Arrays

For $n \in \mathbb{N}$ let $\mathcal{E}_q(n)$ be the set of the Eulerian generating functions of the form

$$\sum_{k\geq n} a_k \frac{z^k}{[k]_q!} = a_n \frac{z^n}{[n]_q!} + a_{n+1} \frac{z^{n+1}}{[n+1]_q!} + \cdots, \quad a_n = 1.$$

To simplify expressions, the coefficients of $z^k/[k]_q!$ of the Eulerian generating functions g, f, h, ℓ etc. are always denoted by g_k, f_k, h_k, ℓ_k etc.

Remark 8.1 With a pair of functions $g \in \mathcal{E}_q(0)$ and $f \in \mathcal{E}_q(1)$, a *q-Riordan array* [5] written $(g, f)_q = (\ell_{n,k})_{n,k\in\mathbb{N}}$ is defined by

$$\sum_{n\geq k} \ell_{n,k} \frac{z^n}{[n]_q!} = g(z) \frac{f^{[k]}(z)}{[k]_q!}, \tag{8.3.1}$$

i.e., its k-column has the Eulerian generating function $g(z)f^{[k]}(z)/[k]_q! \in \mathcal{E}_q(k)$.

Since $\ell_{n,k} = 0$ for $n < k$ and $\ell_{n,n} = 1$ for $n \in \mathbb{N}$, every q-Riordan array is an infinite lower triangular matrix whose diagonal elements are all 1s. If $q = 0$ or $q = 1$ then $(g, f)_q$ reduces to the usual Riordan array and the exponential Riordan array, respectively.

The product on the right-hand side of (8.3.1) is the convolution of two functions so that

$$\ell_{n,k} = \sum_{j=k}^{n} \begin{bmatrix} n \\ j \end{bmatrix}_q g_{n-j} a_{j,k}, \tag{8.3.2}$$

where

$$\frac{f^{[k]}(z)}{[k]_q!} = \sum_{j\geq k} a_{j,k} \frac{z^j}{[j]_q!}.$$

As observed by Gessel [7], the coefficients $a_{j,k}$ for $k \geq 1$ may be recursively expressed as

$$a_{j,k} = \sum_{i=k-1}^{j-1} \begin{bmatrix} j-1 \\ i \end{bmatrix}_q f_{j-i} a_{i,k-1}, \tag{8.3.3}$$

where $a_{i,0} = 0$ for $i \geq 1$ and $a_{0,0} = 1$.

Example 8.1 Let $L = (e_q(z), z)_q = (\ell_{n,k})$ where $e_q(z) = \sum_{n=0}^{\infty} z^n/[n]_q!$. Since $z^{[k]} = z^k$, it follows from (8.3.2) that $\ell_{n,k} = \begin{bmatrix} n \\ k \end{bmatrix}_q$. Thus L is a q-Riordan array displayed in the matrix form:

$$L = \begin{bmatrix} 1 & 0 & 0 & 0 & 0 \dots \\ 1 & 1 & 0 & 0 & 0 \dots \\ 1 & 1+q & 1 & 0 & 0 \dots \\ 1 & 1+q+q^2 & 1+q+q^2 & 1 & 0 \dots \\ 1 & 1+q+q^2+q^3 & (1+q+q^2)(1+q^2) & 1+q+q^2+q^3 & 1 \dots \\ \vdots & \vdots & \vdots & \vdots & \vdots \ddots \end{bmatrix}.$$

The following theorem is useful for deriving the q-analogs of the Composition Formula and the Exponential Formula.

Theorem 8.1 *Let* $h, \ell \in \mathcal{E}_q(0)$. *Then*

$$(g, f)_q(h_0, h_1, \ldots)^T = (\ell_0, \ell_1, \ldots)^T \text{ if and only if } gh[f] = \ell.$$

Proof Expressing the multiplication in terms of the generating functions yields

$$\ell(z) = \sum_{k\geq 0} h_k g(z) \frac{f^{[k]}}{[k]_q!} = g(z) \sum_{k\geq 0} h_k \frac{f^{[k]}}{[k]_q!} = g(z)h[f(z)]$$

as required. □

Theorem 8.1 will be called the *fundamental theorem of* q*-Riordan arrays*, and will be written as

$$(g, f)_q h = gh[f].$$

This simple observation reduces many computations to merely q-composition of functions.

Algebraic structure of q-Riordan arrays

A set S with a binary operation $S \times S \to S$; $(a, b) \mapsto a * b$ is called a *groupoid* if it contains an identity 1, i.e., $a * 1 = 1 * a = a$ for all $a \in S$. For $a \in S$, an element $b \in S$ is called a left (resp. right) inverse of a if $b * a = 1$ (resp. $a * b = 1$). A groupoid S is called a *left (resp. right)* loop if there is a unique solution $x \in S$ of the equation $a * x = b$ (resp. $x * a = b$) for all $a, b \in S$. If S is a left and a right loop, then S is called a *loop* [9], i.e., a quasigroup with an identity element.

An Eulerian generating function g is called a *left (resp. right) q-compositional inverse* of $f \in \mathcal{E}_q(n)$ if $(g \circ_q f)(z) = z$ (resp. $(f \circ_q g)(z) = z$). The notations $\overline{f}_L$ and $\overline{f}_R$ denote the left and right q-compositional inverses of f, respectively.

Theorem 8.2 *Let* $f(z) = \sum_{k\geq 1} f_k z^k/[k]_q!$. *Then there exist unique left and right q-compositional inverses of f if and only if* $f_1 \neq 0$.

Proof Assume that $\overline{f}_L = \sum_{k\geq 1} \overline{f}_k z^k/[k]_q!$ satisfies $\overline{f}_L[f(z)] = z$. Since

$$\begin{aligned}\overline{f}_L[f(z)] &= \sum_{k\geq 1} \overline{f}_k \frac{f^{[k]}(z)}{[k]_q!} = \sum_{n\geq 1} \left(\sum_{k=1}^{n} \overline{f}_k \widehat{f}_{n,k} \right) \frac{z^n}{[n]_q!} \\ &= \overline{f}_1 f_1 z + \sum_{n\geq 2} \left(\overline{f}_n f_1^n + \sum_{k=1}^{n-1} \overline{f}_k \widehat{f}_{n,k} \right) \frac{z^n}{[n]_q!} = z,\end{aligned}$$

we have

$$\overline{f}_1 f_1 = 1, \quad \overline{f}_n f_1^n + \sum_{k=1}^{n-1} \overline{f}_k \widehat{f}_{n,k} = 0 \quad n \geq 2. \tag{8.3.4}$$

From the first equation of (8.3.4), $\overline{f}_1$ exists uniquely if and only if $f_1 \neq 0$. Then by the second equation, $\overline{f}_k$ exists uniquely for every $k \geq 2$. Hence $\overline{f}_L$ exists if and only if $f_1 \neq 0$.

In a similar way, an analogous result holds for the right q-compositional inverse $\overline{f}_R$ of f. □

Now let $\mathcal{R}_q$ denote the set of q-Riordan arrays, and define [5] the *q-multiplication* $*_q$ on the set $\mathcal{R}_q$ to be

$$(g, f)_q *_q (h, \ell)_q = (gh[f], \ell[f])_q.$$

Theorem 8.3 *The set $\mathcal{R}_q$ is a loop under the q-multiplication $*_q$.*

Proof Clearly, $(\mathcal{R}_q, *_q)$ is a groupoid with the identity $(1, z)_q$. First we show that $\mathcal{R}_q$ is a left loop. For any $A, B \in \mathcal{R}_q$, consider the equation $A *_q X = B$. Let $A = (g, f)_q$, $B = (h, \ell)_q$ and $X = (x, y)_q$. Since

$$(h, \ell)_q = (g, f)_q *_q (x, y)_q = (gx[f], y[f])_q,$$

it follows that $h = gx[f]$ and $\ell = y[f]$. Let $\frac{f^{[k]}}{[k]_q!} = \sum_{n\geq k} \widehat{f}_{n,k} \frac{z^n}{[n]_q!}$. Since

$$x[f] = \sum_{k\geq 0} x_k \frac{f^{[k]}(z)}{[k]_q!} = \sum_{k\geq 0} x_k \left(\sum_{n\geq k} \widehat{f}_{n,k} \frac{z^n}{[n]_q!} \right) = \sum_{n\geq 0} \left(\sum_{k=0}^{n} x_k \widehat{f}_{n,k} \right) \frac{z^n}{[n]_q!},$$

it follows from $h = gx[f]$ that

$$h_n = \sum_{j=0}^{n} \begin{bmatrix} n \\ j \end{bmatrix} g_{n-j} \left(\sum_{k=0}^{j} x_k \widehat{f}_{j,k} \right), \quad n \geq 0.$$

Solving the above equation for x_n, $n = 0, 1, \ldots$, we obtain uniquely $x_0 = 1$ and for $n \geq 1$

$$x_n = h_n - \sum_{j=0}^{n-1} \left\{ \begin{bmatrix} n \\ j \end{bmatrix}_q g_{n-j} \left(\sum_{k=0}^{j} x_k \widehat{f}_{j,k} \right) + x_j \widehat{f}_{n,j} \right\}. \tag{8.3.5}$$

In a similar way, from $\ell = y[f]$ we obtain uniquely $y_1 = 1$ and for $n \geq 2$

$$y_n = \ell_n - \sum_{k=1}^{n-1} y_k \widehat{f}_{n,k}. \tag{8.3.6}$$

Since $x \in \mathcal{E}_q(0)$ and $y \in \mathcal{E}_q(1)$ by (8.3.5) and (8.3.6), $X = (x, y)_q \in \mathcal{R}_q$ and it is a unique solution of $A *_q X = B$. Hence $\mathcal{R}_q$ is a left loop.

In a similar way, one can show that $\mathcal{R}_q$ is a right loop. Hence the proof is completed. □

In particular, if $q = 0$ or $q = 1$ then $(\mathcal{R}_q, *_q)$ is the usual Riordan group and the exponential Riordan group respectively.

A non-empty subset H of a loop $(S, *)$ is a subloop if H is closed under $*$. The subloops of $\mathcal{R}_q$ under the q-multiplication $*_q$ are

(i) $\{(1, f)_q : f \in \mathcal{E}_q(1)\}$,
(ii) $\{(g, z)_q : g \in \mathcal{E}_q(0)\}$,
(iii) $\{(f', f)_q : f \in \mathcal{E}_q(1)\}$.

By definition, a q-Riordan matrix B is the left (resp. right) *q-inverse* of A if $B *_q A = (1, z)_q$ (resp. $A *_q B = (1, z)_q$).

Theorem 8.4 *For all $A \in \mathcal{R}_q$, there are unique left and right q-inverses of A.*

Proof Let $A = (g, f) \in \mathcal{R}_q$. Since

$$\left(\frac{1}{g[\overline{f}_R]}, \overline{f}_R\right)_q *_q (g, f)_q = (1, z)_q,$$

$(1/g[\overline{f}_R], \overline{f}_R)_q$ is the left q-inverse of $(g(z), f(z))_q$.

Since there is the function v such that $v[f] = 1/g$, we have

$$(g, f)_q *_q (v, \overline{f}_L)_q = (gv[f], \overline{f}_L[f])_q = (1, z)_q.$$

Thus $(v, \overline{f}_L)_q$ is the right q-inverse of $(g, f)_q$. □

We denote the left q-inverse and right q-inverse of A by A_L^{-1} and A_R^{-1}, respectively.

Example 8.2 Let $A = (e_q(z), e_q(z) - 1)_q$. Since $e_q[\mathrm{loq}(1 + z)] = 1 + z$ where

$$\mathrm{loq}(1 + z) = \sum_{n \geq 1} (-1)^{n-1} (n - 1)!_q z^n / [n]_q!,$$

we have

$$\begin{aligned}&\left(\frac{1}{1+z}, \mathrm{loq}(1 + z)\right)_q *_q (e_q(z), e_q(z) - 1)_q \\ &= \left(\frac{1}{1+z} e_q[\mathrm{loq}(1 + z)], e_q[\mathrm{loq}(1 + z)] - 1\right)_q = (1, z)_q.\end{aligned}$$

Thus $A_L^{-1} = (1/(1+z), \text{loq}(1+z))_q$. Since $e_q[\text{loq}(1+z)] - 1 = z$ we see that $e_q(z) - 1$ is the left q-compositional inverse of $\text{loq}(1+z)$, and $\text{loq}(1+z)$ is the right q-compositional inverse of $e_q(z) - 1$.

8.4 Combinatorial Applications of the q-Riordan Arrays

In this section, we prove [5] that if G and F are the Eulerian generating functions whose coefficients are associated to counting functions $\mathbb{N} \to \mathbb{C}[[q]]$, then the q-Riordan array $(G, F)_q$ can be applied to the enumeration problem of set partitions by block inversions.

By a *k-partition* of the n-set $[n] := \{1, 2, \ldots, n\}$ we mean a partition of the set $[n]$ into k non-empty disjoint sets. Each set is called a *block* of the partition. We denote the family of all k-partitions of $[n]$ by $\Pi_{n,k}$.

Consider a k-partition $\pi = \{B_1, \ldots, B_k\} \in \Pi_{n,k}$. Let $\ell(B_i)$ denote the least element in the block B_i. By an *inversion of a pair of blocks* (B_i, B_j) we shall mean a pair $(a, b) \in B_i \times B_j$ such that $a > b$ but $\ell(B_i) < \ell(B_j)$. Assume that $1 = \ell(B_1) < \ldots < \ell(B_k)$. Note that $1 \in B_1$. If the elements of blocks are arranged in increasing order then the block inversions of (B_i, B_j) coincide with inversions of the permutation $\sigma := B_i \cup B_j$ in which the braces are ignored and the elements are treated as a linear array preserving the order. The number of block inversions of (B_i, B_j) is denoted by $\text{inv}(B_i, B_j)$.

Example 8.3 Let $B_1 = \{1, 2, 5\}$ and $B_2 = \{3, 4, 6, 7\}$. Then the block inversions are $(5, 3), (5, 4) \in B_1 \times B_2$. Indeed, these are the same as inversions of the permutation $\sigma = (1\ 2\ 5\ 3\ 4\ 6\ 7)$.

The number of *inversions of a k-partition* $\pi = \{B_1, \ldots, B_k\} \in \Pi_{n,k}$ is now denoted by $\text{inv}(\pi)$ and is defined by

$$\text{inv}(\pi) = \sum_{1 \le i < j \le k} \text{inv}(B_i, B_j), \quad k \ge 2, \tag{8.4.1}$$

where $\text{inv}(\pi) = 0$ if $k = 1$.

In particular, if $\mathcal{B}_n$ is a collection of bi-partitions $\{A, B\}$ of $[n]$ in which $|B| = j$ and $\ell(A) < \ell(B)$ then it follows from [7, Lemma 5.1] that

$$\sum_{\{A,B\} \in \mathcal{B}_n} q^{\text{inv}(A,B)} = \begin{bmatrix} n-1 \\ j \end{bmatrix}_q. \tag{8.4.2}$$

We can immediately extend (8.4.2) to k factors. For fixed positive integers $b_1, \ldots, b_k$ with $b_1 + \cdots + b_k = n$, let

$$\widetilde{\Pi}_{n,k} = \left\{\alpha := \{B_1, \ldots, B_k\} \in \Pi_{n,k} : |B_i| = b_i\right\}.$$

Then

$$\sum_{\alpha\in\widetilde{\Pi}_{n,k}} q^{\mathrm{inv}(\alpha)} = \begin{bmatrix} n-1 \\ n-b_1 \end{bmatrix}_q \begin{bmatrix} n-1-b_1 \\ n-b_1-b_2 \end{bmatrix}_q \cdots \begin{bmatrix} n-1-b_1-\cdots-b_{k-2} \\ n-b_1-\cdots-b_{k-1} \end{bmatrix}_q. \tag{8.4.3}$$

To find a combinatorial interpretation of q-Riordan arrays, consider the Eulerian generating function $F(z)$ associated to a counting function $f:\mathbb{N}\to\mathbb{C}[[q]]$:

$$F(z)=\sum_{n\geq 0} f(n)\frac{z^n}{[n]_q!}.$$

First we observe that Eq. (8.3.3) may be rewritten as

$$f_{n,k} = \sum_{0<j_1<\ldots<j_{k-1}<n} \begin{bmatrix} n-1 \\ j_{k-1} \end{bmatrix}_q \begin{bmatrix} j_{k-1}-1 \\ j_{k-2} \end{bmatrix}_q \cdots \begin{bmatrix} j_2-1 \\ j_1 \end{bmatrix}_q f_{n-j_{k-1}} f_{j_{k-1}-j_{k-2}} \cdots f_{j_1} \tag{8.4.4}$$

Theorem 8.5 *Let $g, f:\mathbb{N}\to\mathbb{C}[[q]]$ be counting functions with $g(0)=1$, $f(0)=0$ and $f(1)=1$. If $h_k:\mathbb{N}\to\mathbb{C}[[q]]$ for a fixed k is defined by*

$$h_k(n) = \sum_{\pi=\{B_1,\ldots,B_{k+1}\}\in\Pi_{n+1,k+1}} g(|B_1|-1)f(|B_2|)\cdots f(|B_{k+1}|)q^{\mathrm{inv}(\pi)}, \tag{8.4.5}$$

then the array $(a_{n,k})_{n,k\in\mathbb{N}}$ where $a_{n,k}=h_k(n)$, may be expressed as the q-Riordan array given by $(G,F)_q$.

Proof For a fixed k, consider $(k+1)$-partitions of the set $[n+1]$. Let $\alpha=\{B_1,\ldots,B_{k+1}\}\in\widetilde{\Pi}_{n+1,k+1}$ where $|B_i|=b_i$. For brevity, let $j_{k-i+1}=n+1-\sum_{\ell=1}^{i} b_\ell$ for $i=1,2,\ldots,k$. From (8.4.3), we have

$$\sum_{\alpha\in\widetilde{\Pi}_{n+1,k+1}} q^{\mathrm{inv}(\alpha)} = \begin{bmatrix} n \\ j_k \end{bmatrix}_q \begin{bmatrix} j_k-1 \\ j_{k-1} \end{bmatrix}_q \cdots \begin{bmatrix} j_2-1 \\ j_1 \end{bmatrix}_q. \tag{8.4.6}$$

Since $j_1<j_2<\cdots<j_k$, it follows from (8.4.6) that (8.4.5) can be expressed as

$$\begin{aligned} h_k(n) &= \sum_{b_1+\cdots+b_{k+1}=n+1} \left(g(b_1-1)f(b_2)\cdots f(b_{k+1}) \sum_{\alpha\in\widetilde{\Pi}_{n+1,k+1}} q^{\mathrm{inv}(\alpha)} \right) \\ &= \sum_{0<j_1<\ldots<j_k<n+1} g(n-j_k)f(j_k-j_{k-1})\cdots f(j_1) \begin{bmatrix} n \\ j_k \end{bmatrix}_q \begin{bmatrix} j_k-1 \\ j_{k-1} \end{bmatrix}_q \\ &\quad \times\cdots\times \begin{bmatrix} j_2-1 \\ j_1 \end{bmatrix}_q. \end{aligned}$$

Letting $F^{[k]}(z)/[k]_q! = \sum_{n\geq k} f(n,k)z^n/[n]_q!$, from (8.3.2) and (8.4.4) we obtain

$$h_k(n) = \sum_{j=k}^{n} \begin{bmatrix} n \\ j \end{bmatrix}_q g(n-j) f(j,k),$$

which is the (n,k)-entry of the q-Riordan array $(G, F)_q$. Hence the result follows as required. □

By reversing the direction of the proof in Theorem 8.5, we have the following corollary.

Corollary 8.1 *Let $F(z)$ be an Eulerian generating function associated to the counting function $f : \mathbb{N} \to \mathbb{C}[[q]]$ with $f(0) = 0$ and $f(1) = 1$. If $(1, F)_q = (h_{n,k})_{n,k\in\mathbb{N}}$ then*

$$h_{n,k} = \sum_{\pi=\{B_1,\ldots,B_k\}\in\Pi_{n,k}} f(|B_1|)\cdots f(|B_k|) q^{\mathrm{inv}(\pi)} \quad (n \geq k \geq 1)$$

with $h_{0,0} = 1$ and $h_{n,0} = 0$ for $n \geq 1$.

Proof Let $(1, F)_q = (h_{n,k})_{n,k\in\mathbb{N}}$. Since the first column of $(h_{n,k})_{n,k\in\mathbb{N}}$ is $(1, 0, \ldots)^T$, we clearly obtain $h_{0,0} = 1$ and $h_{n,0} = 0$ for $n \geq 1$. Let $n, k \geq 1$. Since $\sum_{n\geq k} h_{n,k} z^n/[n]_q! = F^{[k]}(z)/[k]_q!$ by (8.3.1), it follows from (8.4.4) that

$$h_{n,k} = \sum_{0=j_0<j_1<\ldots<j_{k-1}<j_k=n} \begin{bmatrix} j_k - 1 \\ j_{k-1} \end{bmatrix}_q \begin{bmatrix} j_{k-1} - 1 \\ j_{k-2} \end{bmatrix}_q \cdots \begin{bmatrix} j_2 - 1 \\ j_1 \end{bmatrix}_q \times f(j_k - j_{k-1}) f(j_{k-1} - j_{k-2}) \cdots f(j_1 - j_0). \tag{8.4.7}$$

Set $b_i = j_{k-i+1} - j_{k-i}$ for $i = 1, 2, \ldots, k$. Since $\sum_{i=1}^{k} b_i = j_k - j_0 = n$ and $b_i > 0$, it follows from (8.4.6) and (8.4.7) that

$$\begin{aligned} h_{n,k} &= \sum_{b_1+\cdots+b_k=n} f(b_1)f(b_2)\cdots f(b_k) \begin{bmatrix} n-1 \\ n-b_1 \end{bmatrix}_q \begin{bmatrix} n-1-b_1 \\ n-b_1-b_2 \end{bmatrix}_q \\ &\quad \times \cdots \times \begin{bmatrix} n-1-b_1-\cdots-b_{k-2} \\ n-b_1-\cdots-b_{k-1} \end{bmatrix}_q \\ &= \sum_{b_1+\cdots+b_k=n} f(b_1)f(b_2)\cdots f(b_k) \left(\sum_{\alpha\in\widetilde{\Pi}_{n,k}} q^{\mathrm{inv}(\alpha)} \right) \\ &= \sum_{\pi\in\Pi_{n,k}} f(|B_1|)f(|B_2|)\cdots f(|B_k|) q^{\mathrm{inv}(\pi)} \quad (n, k \geq 1). \end{aligned}$$

This completes the proof. □

To simplify expressions, the coefficients of $z^n/[n]_q!$ of the Eulerian generating functions $F(z), G(z), H(z), L(z)$ associated to the counting functions $f, g, h, \ell : \mathbb{N} \to \mathbb{C}[[q]]$ are denoted by the $f(n), g(n), h(n), \ell(n)$, respectively.

By using the fundamental theorem of q-Riordan arrays, we obtain the following theorem.

Theorem 8.6 *Let* $g, f, \ell : \mathbb{N} \to \mathbb{C}[[q]]$ *be counting functions with* $g(0) = 1$, $f(0) = 0$ *and* $f(1) = 1$. *If* $h : \mathbb{N} \to \mathbb{C}[[q]]$ *is defined by*

$$h(n) = \sum_{k=0}^{n} \sum_{\pi=\{B_1,\ldots,B_{k+1}\}\in\Pi_{n+1,k+1}} g(|B_1|-1) f(|B_2|)\cdots f(|B_{k+1}|)\ell(k) q^{\mathrm{inv}(\pi)},$$

then $H = GL[F]$.

Proof By Theorem 8.1 and Theorem 8.5 we immediately obtain

$$H = (G, F)_q L = GL[F],$$

as required. □

In particular, if $g(0) = 1$ and $g(n) = 0$ for all $n \geq 1$ i.e., $G = 1$ then from Theorem 8.5 we obtain the q-analogs of the *Composition Formula* and of the *Exponential Formula* addressed in text books, respectively, by Aigner [1, p. 113] and Stanley [12, pp. 3–5].

Corollary 8.2 (The q-analog of the Composition Formula)
Let $f, \ell : \mathbb{N} \to \mathbb{C}[[q]]$ *be a counting functions with* $f(0) = 0$ *and* $f(1) = 1$. *If* $h : \mathbb{N} \to \mathbb{C}[[q]]$ *is defined by*

$$h(n) = \sum_{k=1}^{n} \sum_{\pi=\{B_1,\ldots,B_k\}\in\Pi_{n,k}} f(|B_1|)\cdots f(|B_k|)\ell(k) q^{\mathrm{inv}(\pi)},$$
$$h(0) = \ell(0),$$

then $H = L[F]$.

In particular, if $\ell(n) = 1$ for all $n \geq 0$ i.e., $L = e_q(z)$ then we obtain the following.

Corollary 8.3 (The q-analog of the Exponential Formula)
Let $f : \mathbb{N} \to \mathbb{C}[[q]]$ *be the counting function with* $f(0) = 0$ *and* $f(1) = 1$. *If* $h : \mathbb{N} \to \mathbb{C}[[q]]$ *is defined by*

$$h(n) = \sum_{k=1}^{n} \sum_{\pi=\{B_1,\ldots,B_k\}\in\Pi_{n,k}} f(|B_1|) f(|B_2|)\cdots f(|B_k|) q^{\mathrm{inv}(\pi)}, \quad (n \geq 1),$$
$$h(0) = 1,$$

then $H = e_q[F]$.

Example 8.4 *(A combinatorial interpretation for the q-Stirling numbers)*
Recall [8] that the q-Stirling numbers $\left\{ {n \atop k} \right\}_q$ of the second kind are defined by

$$\sum_{n \geq k} \left\{ {n \atop k} \right\}_q \frac{z^n}{[n]_q!} = \frac{\left(e_q(z) - 1\right)^{[k]}}{[k]_q!}.$$

By definition, the array of the q-Stirling numbers $\left\{ {n \atop k} \right\}_q$ coincides with the q-Riordan matrix given by

$$\left(1, e_q(z) - 1\right)_q = \left(\left\{ {n \atop k} \right\}_q \right)_{n,k \in \mathbb{N}}.$$

Since $e_q(z) - 1 = \sum_{n \geq 1} z^n / [n]_q!$, it follows from Corollary 8.1 that

$$\left\{ {n \atop k} \right\}_q = \sum_{\pi \in \Pi_{n,k}} q^{\text{inv}(\pi)}, \tag{8.4.8}$$

which gives a new combinatorial interpretation for $\left\{ {n \atop k} \right\}_q$.

If we consider 2-partitions of $\{1, 2, 3, 4\}$ then we obtain:

$\pi \in \Pi_{4,2}$	$\text{inv}(\pi)$	$\pi \in \Pi_{4,2}$	$\text{inv}(\pi)$
$\{\{1\}, \{2, 3, 4\}\}$	0	$\{\{1, 2, 4\}, \{3\}\}$	1
$\{\{1, 2\}, \{3, 4\}\}$	0	$\{\{1, 4\}, \{2, 3\}\}$	2
$\{\{1, 2, 3\}, \{4\}\}$	0	$\{\{1, 3, 4\}, \{2\}\}$	2
$\{\{1, 3\}, \{2, 4\}\}$	1		

Thus

$$\left\{ {4 \atop 2} \right\}_q = \sum_{\pi \in \Pi_{4,2}} q^{\text{inv}(\pi)} = 3 + 2q + 2q^2.$$

Exercises

8.1 Show that the q-factorial $[n]_q!$ counts permutations while keeping track of the number of inversions.

8.2 Prove the formulae for the q-derivative of a sum, a product, and a quotient of functions respectively, given by (i), (ii), (iii) in Section 8.1.

8.3 Prove the *q-Pascal identity* for $0 < |q| < 1$:

$$\binom{\alpha+1}{k}_q = \binom{\alpha}{k}_q q^k + \binom{\alpha}{k-1}_q = \binom{\alpha}{k}_q + \binom{\alpha}{k-1}_q q^{\alpha+1-k}.$$

8.4 Let V_n denote a vector space of dimension n over the Galois field $GF(q)$ where $q = p^n$. Prove that the number called the *Galois number* of subspaces of V_n is

$$G_n = \sum_{k=0}^{n} \binom{n}{k}_q.$$

8.5 Prove the followings:

(1) The loop $\mathcal{R}_q$ is a group if and only if $q = 0$ or $q = 1$.

(2) The subloop $\{(g, z)_q : g \in \mathcal{E}_q(0)\}$ forms a group.

8.6 The q-Stirling numbers $\left\{ {n \atop k} \right\}_q$ of the second kind are defined by

$$\sum_{n\geq k} \left\{ {n \atop k} \right\}_q \frac{z^n}{[n]_q!} = \frac{(e_q(z)-1)^{[k]}}{[k]_q!}.$$

Use (8.4.8) to find the first 5 rows of the q-Riordan matrix given by

$$(1, e_q(z)-1)_q = \left(\left\{ {n \atop k} \right\}_q \right)_{n,k\in\mathbb{N}}.$$

8.7 The q-Stirling numbers $\left[{n \atop k} \right]_q$ of the first kind are defined by

$$\sum_{n\geq k} \left[{n \atop k} \right]_q \frac{z^n}{[n]_q!} = \frac{(\mathrm{loq}(1+z))^{[k]}}{[k]_q!}.$$

(1) Let $\sigma = C_1C_2\cdots C_k$ be a *k-cyclic partition* of the n-set $[n]$ where every cycle C_i is written with its minimal element first and $1 = \min C_1 < \min C_2 < \cdots < \min C_k$. Then the set of such k-cyclic partitions is denoted by $\mathcal{S}_{n,k}$. Prove that the unsigned Stirling numbers $c_q(n, k) := (-1)^{n-k}\left[{n \atop k} \right]_q$ of the first kind are

$$c_q(n,k) = \sum_{\sigma\in\mathcal{S}_{n,k}} q^{\mathrm{cinv}(\sigma)}, \tag{8.4.9}$$

where $\mathrm{cinv}(\sigma)$ is the number of inversions of a k-cyclic partition σ of $[n]$.

(2) Use (8.4.9) to find first 5 rows of the (unsigned) q-Riordan matrix given by

$$(1, -\mathrm{loq}_q(1+z))_q)_q = \big(c_q(n,k)\big)_{n,k\in\mathbb{N}},$$

where $\mathrm{loq}(1+z) = \sum_{n\geq 1}(-1)^{n-1}[n-1]_q!\frac{z^n}{[n]_q!}$. since $\sum_{\sigma\in\mathcal{S}_{m,1}} q^{\mathrm{cinv}(\sigma)} = (m-1)!_q$.

8.8 Let $f(z)$ be the Eulerian generating function in Theorem 8.2. Prove that there exists unique right q-compositional inverses of f if and only if $f_1 \neq 0$.

References

1. M. Aigner, *A Course in Enumeration*. Graduate Texts in Mathematics (Springer, Berlin, 2007)
2. G.-S. Cheon, J.-H. Jung, Some combinatorial applications of the q-Riordan matrix. Linear Algebra Appl. **482**, 241–260 (2015)
3. G.-S. Cheon, J.-H. Jung, The q-Sheffer sequences of a new type and associated orthogonal polynomials. Linear Algebra Appl. **491**, 171–186 (2016)
4. G.-S. Cheon, J.-H. Jung, S.-R. Kim, New q-Laguerre polynomials having factorized permutation interpretations. J. Math. Anal. Appl. **470**, 118–134 (2019)
5. G.-S. Cheon, J.-H. Jung, Y. Lim, A q-analogue of the Riordan group. Linear Algebra Appl. **439**, 4119–4129 (2013)
6. I.M. Gessel, A noncommutative generalization and q-analog of the Lagrange inversion formula. Trans. Am. Math. Soc. **257**, 455–482 (1980)
7. I.M. Gessel, A q-analog of the exponential formula. Discrete Math. **306**, 1022–1031 (2006)
8. W.P. Johnson, Some applications of the q-exponential formula. Discrete Math. **157**, 207–225 (1996)
9. H. Kiechle, *Theory of K-loops*. Lecture Notes in Mathematics, vol. 1778 (Springer, Berlin, 2002)
10. M.K. Srinivasan, The Eulerian generating function of q-derangements. Discrete Math. **306**, 2134–2140 (2006)
11. R.P. Stanley, *Enumerative Combinatorics*, vol. 1 (Cambridge University Press, Cambridge, 1997)
12. R.P. Stanley, *Enumerative Combinatorics, Vol. 2*. Cambridge Studies in Advanced Mathematics, vol. 62 (Cambridge University Press, Cambridge, 1999)

Chapter 9
Orthogonal Polynomials

Abstract In this chapter, we give an overview of the links between Riordan arrays and orthogonal polynomials, and then we study some specialized areas including classical and semi-classical orthogonal polynomials defined by Riordan arrays, orthogonal polynomials that can be described as the moment sequences of Riordan arrays, applications of exponential Riordan arrays to the Toda lattice equations, and combinatorial polynomials that are moments of Riordan arrays. Orthogonal polynomials enjoy a special place both in pure mathematics and in applied mathematics. Stieltjes, in studying the moment sequences associated to families of orthogonal polynomials, defined the integral that now bears his name. Chebyshev and others, by putting the theory of orthogonal polynomials on a firm basis, provided a tool that has proved invaluable to mathematicians working in the area of approximation of functions, in the area of differential equations, and in many branches of mathematical physics. Traditional orthogonal polynomials are studied on the real line and on the circle. More sophisticated approaches study orthogonal polynomials defined on curves. We shall see that the orthogonal polynomials that can be defined by Riordan arrays are defined either over finite intervals on the real line, or intervals and some discrete points, or in the case of exponential Riordan arrays, over intervals of infinite extent, such as $[0, \infty)$ or even $(-\infty, \infty)$. Families of orthogonal polynomials over the real line are typically associated with a measure, which is often realized through a density or weight function. In the case of orthogonal polynomials defined by ordinary Riordan arrays, this weight function can be determined. However, in the case of orthogonal polynomials determined by exponential Riordan arrays, such weight functions are only known in special cases. The production matrix plays a vital role in this theory, as it is precisely when the production matrix is tri-diagonal in form that the corresponding Riordan array (either ordinary or exponential) defines a family of orthogonal polynomials. In the theory of orthogonal polynomials, a distinction is made between so-called "classical" orthogonal polynomials and those that are not "classical". Such a distinction can also be made for those orthogonal polynomials that can be defined by Riordan arrays.

© The Author(s), under exclusive license to Springer Nature Switzerland AG 2022

L. Shapiro et al., *The Riordan Group and Applications*, Springer Monographs in Mathematics, https://doi.org/10.1007/978-3-030-94151-2_9

9.1 Orthogonal Polynomials and Riordan Arrays

By an *orthogonal polynomial sequence* $(p_n(x))_{n\geq 0}$ we shall understand [10, 18, 37] an infinite sequence of polynomials $p_n(x)$, $n \geq 0$, of degree n, with real coefficients (often integer coefficients) that are mutually orthogonal on an interval $[x_0, x_1]$ (where $x_0 = -\infty$ is allowed, as well as $x_1 = \infty$), with respect to a weight function $w : [x_0, x_1] \to \mathbb{R}$:

$$\int_{x_0}^{x_1} p_n(x)p_m(x)w(x)dx = \delta_{nm}\sqrt{h_n h_m},$$

where

$$h_n = \int_{x_0}^{x_1} p_n^2(x)w(x)dx.$$

We assume that w is strictly positive on the interval (x_0, x_1). Every such sequence obeys a so-called "three-term recurrence" :

$$p_{n+1}(x) = (a_n x + b_n)p_n(x) - c_n p_{n-1}(x)$$

for coefficients a_n, b_n and c_n that depend on n but not x. We note that if

$$p_j(x) = k_j x^j + k_j' x^{j-1} + \ldots \qquad j = 0, 1, \ldots$$

then

$$a_n = \frac{k_{n+1}}{k_n}, \qquad b_n = a_n\left(\frac{k_{n+1}'}{k_{n+1}} - \frac{k_n'}{k_n}\right), \qquad c_n = a_n\left(\frac{k_{n-1}h_n}{k_n h_{n-1}}\right).$$

Since the degree of $p_n(x)$ is n, the coefficient array of the polynomials is a lower triangular (infinite) matrix. In the case of monic orthogonal polynomials the diagonal elements of this array will all be 1. In this case, we can write the three-term recurrence as

$$p_{n+1}(x) = (x - \alpha_n)p_n(x) - \beta_n p_{n-1}(x), \qquad p_0(x) = 1, \qquad p_1(x) = x - \alpha_0.$$

The *moments* associated to the orthogonal polynomial sequence are the numbers

$$\mu_n = \int_{x_0}^{x_1} x^n w(x)dx.$$

We can find $p_n(x)$, α_n and β_n from a knowledge of these moments. To do this, we let Δ_n be the Hankel determinant $|\mu_{i+j}|_{i,j\geq 0}^n$ and $\Delta_{n,x}$ be the same determinant, but with the last row equal to $1, x, x^2, \ldots$. Then

$$p_n(x) = \frac{\Delta_{n,x}}{\Delta_{n-1}}.$$

More generally, we let $H\begin{pmatrix} u_1 & \dots & u_k \\ v_1 & \dots & v_k \end{pmatrix}$ be the determinant with (i, j)th term $\mu_{u_i+v_j}$.

Let

$$\Delta_n = H\begin{pmatrix} 0 \; 1 \dots n \\ 0 \; 1 \dots n \end{pmatrix}, \qquad \Delta' = H\begin{pmatrix} 0 \; 1 \dots n-1 & n \\ 0 \; 1 \dots n-1 & n+1 \end{pmatrix}.$$

Then we have

$$\alpha_n = \frac{\Delta_n'}{\Delta_n} - \frac{\Delta_{n-1}'}{\Delta_{n-1}}, \qquad \beta_n = \frac{\Delta_{n-2}\Delta_n}{\Delta_{n-1}^2}.$$

A sequence $(p_n(x))_{n\geq 0}$ of polynomials is called (formally) orthogonal if $p_n(x)$ has degree n, $n = 0, 1, \dots$, and if there exists a linear functional L such that

$$L(p_n(x)p_m(x)) = \delta_{mn}c_n$$

for some sequence $(c_n)_{n\geq 0}$ of nonzero numbers.

Of importance to this study are the following results (the first is the well-known "Favard's Theorem"), which we essentially reproduce from [25].

Theorem 9.1 *[25] (Cf. [38], Théorème 9 on p. I-4, or [40], Theorem 50.1). Let $(p_n(x))_{n\geq 0}$ be a sequence of monic polynomials, the polynomial $p_n(x)$ having degree $n = 0, 1, \dots$ Then the sequence $(p_n(x))$ is (formally) orthogonal if and only if there exist sequences $(\alpha_n)_{n\geq 0}$ and $(\beta_n)_{n\geq 1}$ with $\beta_n \neq 0$ for all $n \geq 1$, such that the three-term recurrence*

$$p_{n+1} = (x - \alpha_n)p_n(x) - \beta_n(x), \quad \textit{for} \quad n \geq 1,$$

holds, with initial conditions $p_0(x) = 1$ and $p_1(x) = x - \alpha_0$.

Theorem 9.2 *[25] (Cf. [38], Proposition 1, (7), on p. V-5, or [40], Theorem 51.1). Let $(p_n(x))_{n\geq 0}$ be a sequence of monic polynomials, which is orthogonal with respect to some functional L. Let*

$$p_{n+1} = (x - \alpha_n)p_n(x) - \beta_n(x), \quad \textit{for} \quad n \geq 1,$$

be the corresponding three-term recurrence which is guaranteed by Favard's theorem. Then the generating function

$$g(x) = \sum_{k=0}^{\infty} \mu_k x^k$$

for the moments $\mu_k = L(x^k)$ satisfies

$$g(x) = \cfrac{\mu_0}{1 - \alpha_0 x - \cfrac{\beta_1 x^2}{1 - \alpha_1 x - \cfrac{\beta_2 x^2}{1 - \alpha_2 x - \cfrac{\beta_3 x^2}{1 - \alpha_3 x - \cdots}}}}.$$

Given a family of monic orthogonal polynomials

$$p_{n+1}(x) = (x - \alpha_n)p_n(x) - \beta_n p_{n-1}(x), \qquad p_0(x) = 1, \qquad p_1(x) = x - \alpha_0,$$

we can write

$$p_n(x) = \sum_{k=0}^{n} a_{n,k} x^k.$$

Then we have

$$\sum_{k=0}^{n+1} a_{n+1,k} x^k = (x - \alpha_n) \sum_{k=0}^{n} a_{n,k} x^k - \beta_n \sum_{k=0}^{n-1} a_{n-1,k} x^k$$

from which we deduce

$$a_{n+1,0} = -\alpha_n a_{n,0} - \beta_n a_{n-1,0} \tag{9.1.1}$$

and

$$a_{n+1,k} = a_{n,k-1} - \alpha_n a_{n,k} - \beta_n a_{n-1,k}. \tag{9.1.2}$$

The question immediately arises as to the conditions under which a Riordan array (g, f) can be the coefficient array of a family of orthogonal polynomials. It is already clear that the elements of (g, f) must satisfy a recurrence of the form above.

Proposition 9.1 *Every Riordan array of the form*

$$\left(\frac{1 - \gamma x - \delta x^2}{1 + \alpha x + \beta x^2}, \frac{x}{1 + \alpha x + \beta x^2} \right)$$

where $\beta \neq 0$ is the coefficient array of a family of monic orthogonal polynomials.

Proof We calculate the production matrix of the inverse matrix $(g, f)^{-1}$. We have

$$A(x) = \frac{x}{f(x)} = \frac{x}{x/(1 + \alpha x + \beta x^2)} = 1 + \alpha x + \beta x^2,$$

and

$$\begin{aligned}
Z(x) &= \frac{1}{f(x)}(1-g(x)) \\
&= \frac{1+\alpha x+\beta x^2}{x}\left(1-\frac{1-\gamma x-\delta x^2}{1+\alpha x+\beta x^2}\right) \\
&= \frac{1+\alpha x+\beta x^2}{x}\frac{(1+\alpha x+\beta x^2)-(1-\gamma x-\delta x^2)}{1+\alpha x+\beta x^2} \\
&= (\alpha+\gamma)+(\beta+\delta)x.
\end{aligned}$$

Thus the inverse of (g, f) has a production matrix that is tri-diagonal and that begins

$$P = \begin{bmatrix} \alpha+\gamma & 1 & 0 & 0 & 0 & 0 & \dots \\ \beta+\delta & \alpha & 1 & 0 & 0 & 0 & \dots \\ 0 & \beta & \alpha & 1 & 0 & 0 & \dots \\ 0 & 0 & \beta & \alpha & 1 & 0 & \dots \\ 0 & 0 & 0 & \beta & \alpha & 1 & \dots \\ 0 & 0 & 0 & 0 & \beta & a & \dots \\ \vdots & \vdots & \vdots & \vdots & \vdots & \vdots & \ddots \end{bmatrix}.$$

By the vertical property of Riordan arrays [9], we then get that

$$a_{n+1,k} = a_{n,k-1} - \alpha_n a_{n,k} - \beta_n a_{n-1,k}$$

for the elements of (g, f). Now (g, f) begins

$$\begin{bmatrix} 1 & 0 & 0 \\ -\alpha-\gamma & 1 & 0 \\ \alpha^2+\alpha\gamma-\beta-\delta & -2\alpha-\gamma & 1. \end{bmatrix}.$$

Thus the elements of $(g, f).(1, x, x^2, \dots)^t$ are polynomials $P_n(x)$ with

$$P_0(x) = 1, \quad P_1(x) = x-(\alpha+\gamma), \quad P_2(x) = x^2-(2\alpha+\gamma)x+\alpha(\alpha+\gamma)-\beta-\delta,$$

and

$$P(n+1, x) = (x-\alpha)P_n(x) - \beta P_{n-1}(x).$$

This completes the proof. □

We see that the family of orthogonal polynomials is defined by the α-sequence

$$\alpha_0 = \alpha+\gamma, \alpha, \alpha, \alpha, \dots$$

and the β-sequence

$$\beta_1 = \beta+\delta, \beta, \beta, \beta, \dots.$$

Proposition 9.2 *The elements in the leftmost column of*

$$M = \left(\frac{1-\gamma x-\delta x^2}{1+\alpha x+\beta x^2}, \frac{x}{1+\alpha x+\beta x^2}\right)^{-1}$$

are the moments corresponding to the family of orthogonal polynomials with coefficient array $\left(\frac{1-\gamma x-\delta x^2}{1+\alpha x+\beta x^2}, \frac{x}{1+\alpha x+\beta x^2}\right)$.

Proof We let

$$(g, f) = \left(\frac{1-\gamma x-\delta x^2}{1+\alpha x+\beta x^2}, \frac{x}{1+\alpha x+\beta x^2}\right).$$

Then

$$M = (g, f)^{-1} = \left(\frac{1}{g(\bar{f})}, \bar{f}\right).$$

Now $\bar{f}(x)$ is the solution to

$$\frac{u}{1+\alpha x+\beta x^2} = x,$$

thus

$$\bar{f}(x) = \frac{1-\beta x-\sqrt{1-2\beta x+(\beta^2-4\alpha)x^2}}{2\alpha x}.$$

Then

$$\frac{1}{g(\bar{f}(x))} = \frac{1+\alpha\bar{f}(x)+\beta(\bar{f}(x))^2}{1-\gamma\bar{f}(x)-\delta(\bar{f}(x))^2}.$$

Simplifying, we find that

$$\frac{1}{g(\bar{f}(x))} = \frac{2\beta}{(\beta+\gamma)\sqrt{1-2\alpha x+(\alpha^2-4\beta)x^2}-(\alpha(\beta-\delta)+2\beta\gamma)x+\beta-\delta}.$$

We now consider the continued fraction

$$\tilde{g}(x) = \cfrac{1}{1-(\alpha+\gamma)x-\cfrac{(\beta+\delta)x^2}{1-\alpha x-\cfrac{\beta x^2}{1-\alpha x-\cfrac{\beta x^2}{1-\alpha x-\cdots}}}}.$$

This is equivalent to

$$\tilde{g}(x) = \frac{1}{1-(\alpha+\gamma)x-(\beta+\delta)x^2h(x)},$$

where

$$h(x) = \frac{1}{1-\alpha x-\beta x^2 h(x)}.$$

Solving for $h(x)$ and subsequently for $\tilde{g}(x)$, we find that

$$\tilde{g}(x) = \frac{1}{g(\bar{f}(x))}$$

as expected. □

The Chebyshev polynomials of the second kind $U_n(x)$ can be defined by

$$U_n(x) = \sum_{k=0}^{\lfloor \frac{n}{2} \rfloor} \binom{n-k}{k}(-1)^k(2x)^{n-k}.$$

We then have

$$U_n(x) = [t^n]\frac{1}{1-2xt+t^2}.$$

Proposition 9.3 *The Riordan array* $\left(\frac{1}{1+\alpha x+\beta x^2}, \frac{x}{1+\alpha x+\beta x^2}\right)$ *is the coefficient array of the modified Chebyshev polynomials of the second kind given by*

$$P_n(x) = (\sqrt{\beta})^n U_n\left(\frac{x-\alpha}{2\sqrt{\beta}}\right), \quad n=0,1,2,\ldots$$

Proof We have

$$\frac{1}{1-2xt+t^2} = \sum_{n=0}^{\infty} U_n(x)t^n.$$

Thus

$$\frac{1}{1-2\frac{x-\alpha}{2\sqrt{\beta}}\sqrt{\beta}t+\beta t^2} = \sum_{n=0}^{\infty} U_n\left(\frac{x-\alpha}{2\sqrt{\beta}}\right)(\sqrt{\beta}t)^n.$$

Now

$$\begin{aligned}\frac{1}{1-2\frac{x-\alpha}{2\sqrt{\beta}}\sqrt{\beta}t+\beta t^2} &= \frac{1}{1-(x-\alpha)t+\beta t^2}\\ &= \left(\frac{1}{1+\alpha t+\beta t^2}, \frac{t}{1+\alpha t+\beta t^2}\right)\cdot\frac{1}{1-xt}.\end{aligned}$$

Thus

$$\left(\frac{1}{1+\alpha t+\beta t^2}, \frac{t}{1+\alpha t+\beta t^2}\right) \cdot \frac{1}{1-xt} = \sum_{n=0}^{\infty} (\sqrt{\beta})^n U_n\left(\frac{x-\alpha}{2\sqrt{\beta}}\right) t^n$$

as required. □

Corollary 9.1 *The Riordan array* $\left(\frac{1-\gamma x-\delta x^2}{1+\alpha x+\beta x^2}, \frac{x}{1+\alpha x+\beta x^2}\right)$ *is the coefficient array of the generalized Chebyshev polynomials of the second kind given by*

$$Q_n(x) = (\sqrt{\beta})^n U_n\left(\frac{x-\alpha}{2\sqrt{\beta}}\right) - \gamma(\sqrt{\beta})^{n-1} U_{n-1}\left(\frac{x-\alpha}{2\sqrt{\beta}}\right) - \delta(\sqrt{\beta})^{n-2} U_{n-2}\left(\frac{x-\alpha}{2\sqrt{\beta}}\right),$$

with $n = 0, 1, 2, \ldots$.

Proof We have

$$U_n(x) = [t^n]\frac{1}{1-2xt+t^2}.$$

By the method of coefficients we then have

$$[t^n]\frac{t}{1-2xt+t^2} = [t^{n-1}]\frac{1}{1-2xt+t^2} = U_{n-1}(x)$$

and similarly

$$[t^n]\frac{t^2}{1-2xt+t^2} = [t^{n-2}]\frac{1}{1-2xt+t^2} = U_{n-2}(x)$$

as expected. □

The following result justifies our calling the first column elements of the inverse matrix of the coefficient array of a family of orthogonal polynomials the moments associated to that family.

Proposition 9.4 *Let* $M = (\mu(x), \nu(x))$ *be a Riordan array with tri-diagonal production matrix* P_M*. Then*

$$[x^n]\mu(x) = \mathcal{L}(x^n),$$

where $\mathcal{L}$ *is the linear functional that defines the associated family of orthogonal polynomials.*

Proof Let $M = (m_{i,j})_{i,j\geq 0}$. We have [38]

$$x^n = \sum_{i=0}^{n} m_{n,i}\, p_i(x).$$

Applying $\mathcal{L}$, we get

$$\mathcal{L}(x^n) = \mathcal{L}\left(\sum_{i=0}^{n} m_{n,i}p_i(x)\right) = \sum_{i=0}^{n} m_{n,i}\mathcal{L}(p_i(x)) = \sum_{i=0}^{n} m_{n,i}\delta_{i,0} = m_{n,0} = [x^n]\mu(x)$$

as required. □

We shall be interested in calculating the so-called "Hankel transform" of moment sequences that we shall encounter in this chapter. The *Hankel transform* of a given sequence $A = \{a_0, a_1, a_2, ...\}$ is the sequence of Hankel determinants $\{h_0, h_1, h_2, \ldots\}$, where $h_n = |a_{i+j}|_{i,j=0}^n$, i.e.,

$$A = \{a_n\}_{n\in\mathbb{N}} \quad \to \quad h = \{h_n\}_{n\in\mathbb{N}}: \quad h_n = \begin{vmatrix} a_0 & a_1 & \cdots & a_n \\ a_1 & a_2 & & a_{n+1} \\ \vdots & & \ddots & \\ a_n & a_{n+1} & & a_{2n} \end{vmatrix}. \tag{9.1.3}$$

The Hankel transform of a sequence a_n and its binomial transform are equal.

In the case that a_n has a generating function $g(x)$ expressible in the form

$$g(x) = \cfrac{a_0}{1-\alpha_0 x - \cfrac{\beta_1 x^2}{1-\alpha_1 x - \cfrac{\beta_2 x^2}{1-\alpha_2 x - \cfrac{\beta_3 x^2}{1-\alpha_3 x - \cdots}}}}$$

then we have [25]

$$h_n = a_0^{n+1}\beta_1^n\beta_2^{n-1}\cdots\beta_{n-1}^2\beta_n = a_0^{n+1}\prod_{k=1}^{n}\beta_k^{n+1-k}. \tag{9.1.4}$$

Note that this is independent from α_n.

We note that α_n and β_n are in general not integers. Now let $H\begin{pmatrix} u_1 \ldots u_k \\ v_1 \ldots v_k \end{pmatrix}$ be the determinant of Hankel type with (i, j)th term $\mu_{u_i+v_j}$. Let

$$\Delta_n = H\begin{pmatrix} 0\ 1 \ldots n \\ 0\ 1 \ldots n \end{pmatrix}, \qquad \Delta_n' = H_n\begin{pmatrix} 0\ 1 \ldots n-1 & n \\ 0\ 1 \ldots n-1 & n+1 \end{pmatrix}.$$

Then we have

$$\alpha_n = \frac{\Delta_n'}{\Delta_n} - \frac{\Delta_{n-1}'}{\Delta_{n-1}}, \qquad \beta_n = \frac{\Delta_{n-2}\Delta_n}{\Delta_{n-1}^2}. \tag{9.1.5}$$

Classical and semi-classical orthogonal polynomials

The classical orthogonal polynomials of mathematical science are the Jacobi, Laguerre and Hermite polynomials, defined by the weights $w_J(x) = (1-x)^\alpha(1+x)^\beta$ on $[-1, 1]$, $w_L(x) = x^\alpha e^{-x}$ on $[0, \infty)$, and $w_H(x) = e^{-x^2}$ on $(-\infty, \infty)$, respectively. In particular, these orthogonal polynomials [10, 18, 37] are associated with measures that are absolutely continuous. We have

$$\frac{w_J'(x)}{w_J(x)} = \frac{x(\alpha+\beta)+\alpha-\beta}{x^2-1},$$

$$\frac{w_L'(x)}{w_L(x)} = \frac{\alpha-x}{x},$$

and

$$\frac{w_H'(x)}{w_H(x)} = -2x.$$

We approach the general case as follows. Suppose that

$$\frac{w'(x)}{w(x)} = \frac{U(x)}{V(x)}$$

where $U(x)$ and $V(x)$ are relatively prime polynomials. If $deg(V) \leq 2$ and $deg(U) \leq 1$ then the family of polynomials is said to be classical; otherwise, it is called semi-classical. In the classical case, the polynomials $y = P_n(x)$ satisfy the differential equation

$$\frac{w'(x)}{w(x)} = \frac{U(x)}{V(x)} = \frac{u_0+u_1x}{v_0+v_1x+v_2x^2},$$

where $U(x)/V(x)$ can be expressed as

$$\frac{U(x)}{V(x)} = \frac{u_0+u_1x}{v_0+v_1x+v_2x^2}.$$

Note that all orthogonal polynomials that we shall consider later will be *monic* (the coefficient of x^n in $P_n(x)$ is 1).

The classical orthogonal polynomials defined by ordinary Riordan arrays

We have seen that for an ordinary Riordan array to define a family of orthogonal polynomials, it must be of the form

$$R = \left(\frac{1+cx+dx^2}{1+ax+bx^2}, \frac{x}{1+ax+bx^2}\right).$$

This matrix will then be the coefficient array of the family of polynomials. The procedure to find the weight function associated with this family is as follows. First, we form the moment matrix $M = R^{-1}$ given by

$$\left(-\frac{(b-d)\sqrt{1-2ax+x^2(a^2-4b)}+x(a(b+d)-2bc)-b-d}{2(x^2(a^2d-ac(b+d)+b^2+b(c^2-2d)+d^2)+x(c(b+d)-2ad)+d)}, \frac{1-ax-\sqrt{1-2ax+x^2(a^2-4b)}}{2bx}\right).$$

The first element of this array is the generating function $\mu(x)$ of the moments of the family of orthogonal polynomials $P_n(x)$. These moments begin

$$1, a-c, a^2-2ac+b+c^2-d, a^3-3a^2c+a(3b+3c^2-3d)-c(2b+c^2-2d), \ldots$$

$$\mu(x) = \cfrac{1}{1-(a-c)x-\cfrac{(b-d)x^2}{1-ax-\cfrac{bx^2}{1-ax-\cfrac{bx^2}{1-\cdots}}}}.$$

From this we can see that the Hankel transform [25, 26] of this sequence of moments is given by

$$h_n = (b-d)^n b^{\binom{n}{2}}.$$

Our next step is to use the Stieltjes-Perron theorem [4, 20] to derive the associated measure. We find that the measure sought is given by $w(x)dx$ where

$$w(x) = \frac{1}{2\pi}\frac{(b-d)\sqrt{4b-(x-a)^2}}{dx^2+x(c(b+d)-2ad)+a^2d-ac(b+d)+b^2+b(c^2-2d)+d^2}.$$

Finally, we form the ratio $\frac{w'(x)}{w(x)}$ to obtain the expression

$$-\frac{dx^3-3adx^2+x(3a^2d-b^2-b(c^2+6d)-d^2)-a^3d+a(b^2+b(c^2+6d)+d^2)-4bc(b+d)}{((x-a)^2-4b)(dx^2+x(c(b+d)-2ad)+a^2d-ac(b+d)+b^2+b(c^2-2d)+d^2)}.$$

The form of this ratio now tells us that the orthogonal polynomials defined by ordinary Riordan arrays are at least *semi-classical*. Inspection of the above ratio allows us to announce the following results.

Proposition 9.5 *The ordinary Riordan array*

$$\left(\frac{1+cx+dx^2}{1+ax+bx^2}, \frac{x}{1+ax+bx^2}\right)$$

defines a family of classical orthogonal polynomials in the case that either $c = d = 0$ *or* $c = 0, d = -b$.

Corollary 9.2 *When* $c = d = 0$, *we have*

$$w(x) = \frac{1}{2\pi}\frac{\sqrt{4b-(x-a)^2}}{b}$$

on the interval

$$[a - 2\sqrt{b}, a + 2\sqrt{b}],$$

with

$$\frac{w'(x)}{w(x)} = \frac{x-a}{(x-a)^2-4b}.$$

The moments μ_n *have integral representation*

$$\mu_n = \frac{1}{2\pi}\int_{a-2\sqrt{b}}^{a+2\sqrt{b}} x^n \frac{\sqrt{4b-(x-a)^2}}{b}\,dx.$$

The moments have generating function

$$\mu(x) = \frac{1-ax-\sqrt{(1-ax)^2-4bx^2}}{2bx^2}$$

given by

$$\mu(x) = \cfrac{1}{1-ax-\cfrac{bx^2}{1-ax-\cfrac{bx^2}{1-ax-\cfrac{bx^2}{1-\cdots}}}}.$$

By an application of Lagrange inversion, we obtain

$$\begin{aligned}\mu_n &= \frac{1}{n+1}[x^n](1+ax+bx^2)^{n+1}\\ &= \frac{1}{n+1}\sum_{k=0}^{n}\binom{n+1}{j}\binom{j}{n-j}a^{2j-n}b^{n-j}\\ &= \frac{1}{n+1}\sum_{k=0}^{n}\binom{n+1}{n-k}\binom{n-k}{k}a^{n-2k}b^{k}.\end{aligned}$$

The moments have their Hankel transform given by

$$h_n = b^{\binom{n+1}{2}}.$$

The polynomials $P_n(x)$ satisfy the three-term recurrence

$$P_n(x) = (x-a)P_{n-1}(x) - bP_{n-2}(x), \quad n > 1,$$

with $P_0(x) = 1$, $P_1(x) = x - a$.

If $y = P_n(x)$ then y satisfies the differential equation

$$(x^2 - 2ax + a^2 - 4b)y'' + 3(x-a)y' - n(n+2)y = 0.$$

The form of the differential equation satisfied by these polynomials follows from the form of the weight function [36]. For a proof of the other statements, the reader is referred to [7].

Corollary 9.3 *When $c = 0$ and $d = -b$, we have*

$$w(x) = \frac{1}{\pi}\frac{1}{\sqrt{4b-(x-a)^2}}$$

on the interval

$$[a - 2\sqrt{b}, a + 2\sqrt{b}],$$

with

$$\frac{w'(x)}{w(x)} = \frac{a-x}{(x-a)^2 - 4b}.$$

The moments μ_n have integral representation

$$\mu_n = \frac{1}{\pi}\int_{a-2\sqrt{b}}^{a+2\sqrt{b}} x^n \frac{1}{\sqrt{4b-(x-a)^2}}\,dx.$$

The moments have generating function

$$\mu(x) = \frac{1}{\sqrt{(1-ax)^2 - 4bx^2}}$$

given by

$$\mu(x) = \cfrac{1}{1 - ax - \cfrac{2bx^2}{1 - ax - \cfrac{bx^2}{1 - ax - \cfrac{bx^2}{1 - \cdots}}}}.$$

We have the closed form expression for the moments

$$\mu_n = \sum_{i=0}^{n} \binom{n-i}{i}\binom{n-i-1/2}{n-i}(-1)^i (a^2-4b)^i (2a)^{n-2i}$$
$$= \frac{1}{4^n}\sum_{k=0}^{n}\binom{2n-2k}{n-k}\binom{2k}{k}(a+2\sqrt{b})^k(a-2\sqrt{b})^{n-k}.$$

The moments have Hankel transform

$$h_n = 2^n b^{\binom{n+1}{2}}.$$

The polynomials $P_n(x)$ satisfy the three-term recurrence

$$P_n(x) = (x-a)P_{n-1}(x) - bP_{n-2}(x), \quad n > 2,$$

with $P_0(x) = 1$, $P_1(x) = x - a$, and $P_2(x) = (x-a)^2 - b(b+1)$.
If $y = P_n(x)$ then y satisfies the differential equation

$$(x^2 - 2ax + a^2 - 4b)y'' + (x-a)y' - n^2 y = 0.$$

We note that in the case $c = 0$ and $d = -b$, the generating function

$$\frac{1}{\sqrt{(1-ax)^2 - 4bx^2}} = \frac{1}{\sqrt{1 - 2ax + x^2(a^2-4b)}}$$

can be compared with the generating function

$$\frac{1}{\sqrt{1-2xt+t^2}} = \sum_{n=0}^{\infty} \mathcal{P}_n(x)t^n$$

of the Legendre polynomials $\mathcal{P}_n(x)$. Then we get [33]

$$\mu_n = (a^2-4b)^{n/2}\mathcal{P}_n\left(\frac{a}{\sqrt{a^2-4b}}\right).$$

For the next result, we note that

$$C_n = \frac{1}{n+1}\binom{2n}{n}$$

is the nth Catalan number A000108. The generating function of the Catalan numbers is given by

$$c(x) = \frac{1-\sqrt{1-4x}}{2x}.$$

Proposition 9.6 *The ordinary Riordan array* $\left(\frac{1}{1+ax}, \frac{x}{(1+ax)^2}\right)$ *$(a \neq 0)$ is the coefficient array of a family of classical orthogonal polynomials. We have*

$$w(x) = \frac{1}{2\pi} \frac{\sqrt{x(4a-x)}}{2ax}$$

on the interval

$$[0, 4a],$$

with

$$\frac{w'(x)}{w(x)} = \frac{2a}{x(x-4a)}.$$

The moments have integral representation

$$\mu_n = \frac{1}{2\pi} \int_0^{4a} x^n \frac{\sqrt{x(4a-x)}}{2ax}\, dx = a^n C_n.$$

The moments have generating function

$$\mu(x) = \frac{1-\sqrt{1-4ax}}{2ax},$$

with

$$\mu(x) = \cfrac{1}{1-\cfrac{ax}{1-\cfrac{ax}{1-\cfrac{ax}{1-\cdots}}}},$$

or equivalently,

$$\mu(x) = \cfrac{1}{1-ax-\cfrac{a^2x^2}{1-2ax-\cfrac{a^2x^2}{1-2ax-\cdots}}}.$$

The moments μ_n have Hankel transform

$$h_n = a^{n(n+1)}.$$

The polynomials $P_n(x)$ satisfy the three-term recurrence

$$P_n(x) = (x - 2a)P_{n-1}(x) - a^2 P_{n-2} \quad n > 1,$$

with $P_0(x) = 1$, $P_1(x) = x - a$. If $y = P_n(x)$ then y satisfies the differential equation

$$x(x - 4a)y'' + 2(x - a)y' - n(n + 1)y = 0.$$

Example 9.1 The Riordan array $\left(\frac{1}{1+x^2}, \frac{x}{1+x^2}\right)$ is the coefficient array of the scaled Chebyshev polynomials of the second kind $P_n(x) = U_n(x/2)$, with their moments being the aerated Catalan numbers

$$1, 0, 1, 0, 2, 0, 5, 0, 14, 0, \ldots$$

given by

$$\mu_n = \frac{1}{2\pi} \int_{-2}^{2} x^n \sqrt{4 - x^2}\, dx.$$

This is closely related to Wigner's semicircle distribution [41, 42]. Note that if we define

$$P_{n-1}^{(1)}(x) = \frac{1}{2\pi} \int_{-2}^{2} \frac{P_n(x) - P_n(x)}{x - z} \sqrt{4 - z^2}\, dz$$

then we find that

$$P_n^{(1)}(x) = P_n(x) = U_n(x/2).$$

Here,

$$U_n(x) = \sum_{k=0}^{\lfloor \frac{n}{2} \rfloor} \binom{n-k}{k} (-1)^k (2x)^{n-2k}$$

are the Chebyshev polynomials of the second kind [29].

Example 9.2 The Riordan array $\left(\frac{1-x^2}{1+x^2}, \frac{x}{1+x^2}\right)$ is closely related to the Chebyshev polynomials of the first kind. This array is the coefficient array of a family of orthogonal polynomials $P_n(x)$ whose moments are the aerated central binomial numbers (A000984)

$$1, 0, 2, 0, 6, 0, 20, 0, 70, 0, 252, \ldots$$

given by

$$\frac{1}{\pi} \int_{-2}^{2} x^n \frac{1}{\sqrt{4 - x^2}}\, dx.$$

In this case we have

$$P_n(x) = U_n(x/2) - U_{n-2}(x/2).$$

We note that the Chebyshev polynomials of the first kind are defined not by a Riordan array, but by a related structure known as an almost Riordan array [8].

The semi-classical case

By the results of the last section, if an ordinary Riordan array is the coefficient array of a family of polynomials that is not of classical type, then the ratio $\frac{\tilde{w}'}{\tilde{w}}$ is of semi-classical type, for the absolutely continuous part of the measure.

We begin this section by looking at a family of orthogonal polynomials said to be of "restricted Chebyshev Boubaker type" [5]. These are ordinary Riordan arrays of the form

$$\left(\frac{1+rx^2}{1+x^2}, \frac{x}{1+x^2}\right).$$

We exclude the case $r=-1$, which is of classical type. The polynomials $P_n(x;r) = P_n(x)$ defined by these arrays are given by

$$P_n(x;r) = \sum_{k=0}^{\lfloor \frac{n}{2} \rfloor} \binom{n-k}{k} \frac{n-(r+1)k}{n-k} (-1)^k x^{n-2k}.$$

The case $r=3$ corresponds to the family of Boubaker polynomials. The moments $\mu_n(r)$ of this family of orthogonal polynomials have generating function

$$\mu(x;r) = \frac{\sqrt{1-4x^2}(r-1)+r+1}{2(r+x^2(r-1)^2)},$$

which can be expressed as the continued fraction [40]

$$\mu(x;r) = \cfrac{1}{1-\cfrac{(1-r)x^2}{1-\cfrac{x^2}{1-\cfrac{x^2}{1-\cdots}}}}.$$

We note that the Hankel transform of $\mu_n(r)$, which by the above is an aerated sequence, and that of its un-aerated version, is given by

$$h_n(r) = (1-r)^n.$$

Further, the un-aerated moments

$$1, 1-r, r^2-3r+2, -r^3+5r^2-9r+5, \ldots$$

are themselves moments for the family of orthogonal polynomials that have coefficient matrix given by

$$\left(\frac{(1+x)(1+rx)}{(1+x)^2}, \frac{x}{(1+x)^2}\right).$$

We have the following integral representation of the moment sequence $\mu_n(r)$.

$$\mu_n(r) = \frac{-1}{\pi}\int_2^2 x^n \frac{\sqrt{4-x^2}(r-1)}{2(rx^2+(r-1)^2)}\,dx + \frac{r+1}{2r}\left(-\frac{r-1}{\sqrt{-r}}\right)^n + \frac{r+1}{2r}\left(\frac{r-1}{\sqrt{-r}}\right)^n.$$

Thus in this case, the measure defining the orthogonal polynomial is no longer absolutely continuous, but it takes into account the zeros of the denominator term $rx^2+(r-1)^2$. Note that we have

$$\frac{\tilde{w}'(x)}{\tilde{w}} = \frac{x(rx^2-r^2-6r-1)}{(4-x^2)(rx^2+(r-1)^2)}$$

in this case.

We now move to a more general example.

Example 9.3 We consider the Riordan array

$$\left(\frac{1-x-x^2}{1-3x-4x^2}, \frac{x}{1-3x-4x^2}\right)$$

which begins

$$\left[\begin{array}{ccccccc} 1 & 0 & 0 & 0 & 0 & 0 & 0 \\ 2 & 1 & 0 & 0 & 0 & 0 & 0 \\ 9 & 5 & 1 & 0 & 0 & 0 & 0 \\ 35 & 28 & 8 & 1 & 0 & 0 & 0 \\ 141 & 139 & 56 & 11 & 1 & 0 & 0 \\ 563 & 670 & 339 & 93 & 14 & 1 & 0 \\ 2253 & 3129 & 1911 & 662 & 139 & 17 & 1 \end{array}\right],$$

with inverse

$$\left(\frac{5+7x-3\sqrt{1+6x+25x^2}}{2(1+x-11x^2)}, \frac{\sqrt{1+6x+25x^2}-3x-1}{8x}\right)$$

that begins

$$\begin{bmatrix} 1 & 0 & 0 & 0 & 0 & 0 & 0 \\ -2 & 1 & 0 & 0 & 0 & 0 & 0 \\ 1 & -5 & 1 & 0 & 0 & 0 & 0 \\ 13 & 12 & -8 & 1 & 0 & 0 & 0 \\ -62 & 9 & 32 & -11 & 1 & 0 & 0 \\ 97 & -217 & -43 & 61 & -14 & 1 & 0 \\ 457 & 920 & -332 & -170 & 99 & -17 & 1 \end{bmatrix}.$$

The moment sequence μ_n thus begins

$$1, -2, 1, 13, -62, 97, 457, \ldots.$$

The inverse matrix has a production matrix that begins

$$\begin{bmatrix} -2 & 1 & 0 & 0 & 0 & 0 & 0 \\ -3 & -3 & 1 & 0 & 0 & 0 & 0 \\ 0 & -4 & -3 & 1 & 0 & 0 & 0 \\ 0 & 0 & -4 & -3 & 1 & 0 & 0 \\ 0 & 0 & 0 & -4 & -3 & 1 & 0 \\ 0 & 0 & 0 & 0 & -4 & -3 & 1 \\ 0 & 0 & 0 & 0 & 0 & -4 & -3 \end{bmatrix},$$

and hence the generating function $\frac{5+7x-3\sqrt{1+6x+25x^2}}{2(1+x-11x^2}$ of the moment sequence has a continued fraction expression as

$$\cfrac{1}{1+2x+\cfrac{3x^2}{1+3x+\cfrac{4x^2}{1+3x+\cfrac{4x^2}{1+\cdots}}}}.$$

This implies that the moments μ_n have Hankel transform given by

$$h_n = (-3)^n(-4)^{\binom{n}{2}}.$$

We find that the moments have integral representation

$$\mu_n = \frac{1}{\pi}\int_{-3-4i}^{-3+4i} x^n \frac{3i\sqrt{x^2+6x+25}}{2(x^2+x-11)}\,dx + \left(\frac{3\sqrt{5}}{10}+\frac{5}{2}\right)\left(\frac{3\sqrt{5}}{2}-\frac{1}{2}\right)^n.$$

Thus the support for the measure for the corresponding family of orthogonal polynomials has an absolutely continuous part supported by the imaginary line segment $[-3-4i, -3+4i]$ and an atomic mass on the real axis at $x = \frac{3\sqrt{5}}{2} - \frac{1}{2}$. In this case

we have

$$\frac{\tilde{w}'(x)}{\tilde{w}(x)} = -\frac{x^3+9x^2+64x+58}{(x^2+x-11)(x^2+6x+25)}.$$

The corresponding family of orthogonal polynomials $P_n(x)$ satisfy the three-term recurrence

$$P_n(x) = (x+3)P_{n-1}(x) + 4P_{n-2},$$

with $P_0(x) = 1$, $P_1(x) = x+2$.

9.2 Exponential Riordan Arrays and Classical Orthogonal Polynomials

We now turn our attention to exponential Riordan arrays. We recall that an *exponential* Riordan array R is an invertible lower triangular matrix defined by two power series

$$g(x) = 1 + g_1\frac{x}{1!} + g_2\frac{x^2}{2!} + \cdots$$

and

$$f(x) = \frac{x}{1!} + f_2\frac{x^2}{2!} + f_3\frac{x^3}{3!} + \cdots,$$

where the (n,k)th element of the corresponding matrix is given by

$$r_{n,k} = \frac{n!}{k!}[x^n]g(x)f(x)^k.$$

Note that we have chosen $g_0 = 1$ and $f_1 = 1$ here, to simplify the exposition. We denote the exponential array defined by the pair g, f by $[g, f]$. All orthogonal polynomials in this section will be *monic* (the coefficient of x^n in $P_n(x)$ is 1).

In order for a Riordan array R to be the coefficient array of a family of orthogonal polynomials, we require that the production matrix $P_M = M^{-1}\overline{M}$ of the inverse matrix $M = R^{-1}$ be tri-diagonal. If $M = [u, v]$ then this production matrix is generated by two power series, the A series and the Z series. We have

$$A(x) = v'(\bar{v}(x)), \quad Z(x) = \frac{u'(\bar{v}(x))}{u(\bar{v}(x))}.$$

The matrix P_M then has its bivariate generating function given by

$$e^{xy}(Z(x) + yA(x)).$$

The most general bivariate generating function of the production matrix of an exponential Riordan array M, in order for that matrix to have a tri-diagonal production matrix, is given by

$$e^{xy}(\alpha+\beta x+y(1+\gamma x+\delta x^2)),$$

where we have

$$Z(x)=\alpha+\beta x, \quad A(x)=1+\gamma x+\delta x^2.$$

This leads to a production matrix that begins

$$\left[\begin{array}{ccccccc}
\alpha & 1 & 0 & 0 & 0 & 0 & 0 \\
\beta & \alpha+\gamma & 1 & 0 & 0 & 0 & 0 \\
0 & 2(\beta+\delta) & \alpha+2\gamma & 1 & 0 & 0 & 0 \\
0 & 0 & 3\beta+6\delta & \alpha+3\gamma & 1 & 0 & 0 \\
0 & 0 & 0 & 4\beta+12\delta & \alpha+4\gamma & 1 & 0 \\
0 & 0 & 0 & 0 & 5\beta+20\delta & \alpha+5\gamma & 1 \\
0 & 0 & 0 & 0 & 0 & 6\beta+30\delta & \alpha+6\gamma
\end{array}\right].$$

In this case where the production matrix of $M=R^{-1}$ is tri-diagonal, we call M the moment matrix of the family of orthogonal polynomials whose coefficient array is given by the Riordan array R.

We then have that $R=M^{-1}$, the coefficient array of the associated orthogonal polynomials, will be defined by [6]

$$\begin{aligned}
R &= \left[e^{-\int_0^x \frac{Z(t)}{A(t)}\,dt}, \int_0^x \frac{dt}{A(t)}\right] \\
&= \left[e^{-\int_0^x \frac{\alpha+\beta t}{1+\gamma t+\delta t^2}\,dt}, \int_0^x \frac{dt}{1+\gamma t+\delta t^2}\right] \\
&= \left[\frac{e^{\frac{\beta\gamma/\delta-2\alpha}{4\delta-\gamma^2}\left(2\tan^{-1}\left(\frac{\gamma+2\delta x}{\sqrt{4\delta-\gamma^2}}\right)-2\sin^{-1}\left(\frac{\gamma}{2\sqrt{\delta}}\right)\right)}}{(1+\gamma x+\delta x^2)^{\beta/(2\delta)}}, \frac{1}{4\delta-\gamma^2}\left(2\tan^{-1}\left(\frac{\gamma+2\delta x}{\sqrt{4\delta-\gamma^2}}\right)-2\sin^{-1}\left(\frac{\gamma}{2\sqrt{\delta}}\right)\right)\right].
\end{aligned}$$

This is the most general form that an exponential Riordan array can have for it to be the coefficient array of a family $P_n(x)$ of orthogonal polynomials. These polynomials satisfy the three-term recurrence

$$P_n(x)=(x-(\alpha+(n-1)\gamma))P_{n-1}(x)-(n-1)(\beta+(n-2)\delta)P_{n-2}(x),$$

with $P_0(x)=1$, $P_1(x)=x-\alpha$.

Example 9.4 We let $A(x)=1+x+x^2/2$, $Z(x)=1+x$. We find that R is given by

$$R=M^{-1}=\left[\frac{2}{2+2x+x^2}, 2\tan^{-1}(1+x)-\frac{\pi}{2}\right].$$

The moment matrix M is then given by

$$M = \left[(1+\sin(x))\sec(x)^2, \tan(x)+\sin(x)-1\right].$$

The moments in this case are the shifted Euler or up/down numbers

$$1, 1, 2, 5, 16, 61, 272, 1385, 7936, 50521, \ldots.$$

We have

$$P_n(x) = (x-n)P_{n-1}(x) - \frac{n(n-1)}{2}P_{n-2}(x).$$

The production matrix of the moment matrix begins

$$\begin{bmatrix} 1 & 1 & 0 & 0 & 0 & 0 \\ 1 & 2 & 1 & 0 & 0 & 0 \\ 0 & 3 & 3 & 1 & 0 & 0 \\ 0 & 0 & 6 & 4 & 1 & 0 \\ 0 & 0 & 0 & 10 & 5 & 1 \\ 0 & 0 & 0 & 0 & 15 & 6 \end{bmatrix}.$$

This indicates that the Hankel transform of the moment sequence is given by

$$h_n = \prod_{k=0}^{n} \binom{k+2}{2}^{n-k}.$$

Example 9.5 We let $A(x) = 1 + 2x + x^2$, $Z(x) = 1 + x$. We find that R is given by

$$R = M^{-1} = \left[\frac{1}{1+x}, \frac{x}{1+x}\right]$$

with moment matrix

$$M = \left[\frac{1}{1-x}, \frac{x}{1-x}\right].$$

The moments are thus $n!$ and the polynomials are the scaled Laguerre polynomials

$$n!\sum_{k=0}^{n} \binom{n}{k} \frac{(-1)^{n-k}}{k!} x^k.$$

Classically, we have

$$n! = \int_0^\infty x^n e^{-x}\, dx.$$

Then $w(x) = e^{-x}$ and $\frac{w'(x)}{w(x)} = -1$. The polynomials satisfy the recurrence

$$P_n(x) = (x - (2n-1))P_{n-1}(x) - (n-1)^2 P_{n-1}(x).$$

The production matrix in this case begins

$$\begin{bmatrix} 1 & 1 & 0 & 0 & 0 & 0 \\ 1 & 3 & 1 & 0 & 0 & 0 \\ 0 & 4 & 5 & 1 & 0 & 0 \\ 0 & 0 & 9 & 7 & 1 & 0 \\ 0 & 0 & 0 & 16 & 9 & 1 \\ 0 & 0 & 0 & 0 & 25 & 11 \end{bmatrix}.$$

This indicates that the Hankel transform of the moment sequence is given by

$$h_n = \prod_{k=1}^{n} k^{2(n-k+1)} = \prod_{k=0}^{n} k!^2.$$

Example 9.6 Let $Z(x) = 1 + 2x$ and $A(x) = 1 + x + x^2$. The exponential Riordan array whose production matrix is defined by $A(x)$ and $Z(x)$ is given by

$$M = \left[\frac{3}{2\left(\cos\left(\sqrt{3}x + \frac{\pi}{3}\right) + 1\right)}, \frac{\sqrt{3}}{2}\tan\left(\frac{\sqrt{3}x}{2} + \frac{\pi}{6}\right) - \frac{1}{2}\right].$$

The production matrix of this array begins

$$\begin{bmatrix} 1 & 1 & 0 & 0 & 0 & 0 \\ 2 & 2 & 1 & 0 & 0 & 0 \\ 0 & 6 & 3 & 1 & 0 & 0 \\ 0 & 0 & 12 & 4 & 1 & 0 \\ 0 & 0 & 0 & 20 & 5 & 1 \\ 0 & 0 & 0 & 0 & 30 & 6 \end{bmatrix}.$$

The inverse array $R = M^{-1}$ is given by

$$\left[\frac{1}{1+x+x^2}, \frac{2}{\sqrt{3}}\tan^{-1}\left(\frac{1+2x}{\sqrt{3}}\right) - \frac{\pi}{3\sqrt{3}}\right].$$

This is the coefficient array of the family of orthogonal polynomials

$$P_n(x) = (x-n)P_{n-1}(x) - n(n-1)P_{n-2}(x),$$

with $P_0(x) = 1$, $P_1(x) = x - 1$.

In the general case, we have

$$h_n = \prod_{k=1}^{n} (k(\beta + (k-1)\delta))^{n-k+1}.$$

Proposition 9.7 *The exponential Riordan array* $\left[\frac{1}{(1+x)^{r+1}}, \frac{x}{1+x}\right]$ *is the coefficient array of a family of classical orthogonal polynomials. We have*

$$w(x) = e^{-x}\frac{x^r}{r!}$$

on the interval

$$[0, \infty).$$

The moments have integral representation

$$\mu_n = \int_0^\infty x^n e^{-x}\frac{x^r}{r!}\,dx = n!\binom{n+r}{r} = \prod_{k=1}^{n} r+k.$$

The moments have generating function given by

$$\mu(x) = \cfrac{1}{1-(r+1)x-\cfrac{(r+1)x^2}{1-(r+3)x-\cfrac{2(r+2)x^2}{1-(r+5)x-\cfrac{3(r+3)x^2}{1-(r+7)x-\cdots}}}}.$$

The Hankel transform of the moments is given by

$$h_n = \prod_{k=0}^{n} (k(r+k))^{n-k+1}.$$

The polynomials $P_n(x)$ *satisfy the three-term recurrence*

$$P_n(x) = (x-(r+2n-1))P_{n-1}(x) - (n-1)(r+n-1)P_{n-2}(x), \quad n > 1,$$

with $P_0(x) = 1$, $P_1(x) = x-(r+1)$.

If $y = P_n(x)$, *then* y *satisfies the differential equation*

$$xy'' + (r+1-x)y' + ny = 0.$$

Proof The main conclusion follows from the fact that

$$\frac{w'(x)}{w(x)} = \frac{r-x}{x},$$

where

$$w(x) = e^{-x}\frac{x^r}{r!}.$$

For $M = \left[\frac{1}{(1+x)^{r+1}}, \frac{x}{1+x}\right]$ we have $M^{-1} = \left[\frac{1}{(1-x)^{r+1}}, \frac{x}{1-x}\right]$. We find that $P_{M^{-1}}$ is generated by

$$e^{xy}((r+1)(1+x) + y(1+x)^2).$$

This matrix is therefore tri-diagonal and begins

$$\left[\begin{array}{ccccccc} r+1 & 1 & 0 & 0 & 0 & 0 & 0 \\ r+1 & r+3 & 1 & 0 & 0 & 0 & 0 \\ 0 & 2r+4 & r+5 & 1 & 0 & 0 & 0 \\ 0 & 0 & 3(r+3) & r+7 & 1 & 0 & 0 \\ 0 & 0 & 0 & 4(r+4) & r+9 & 1 & 0 \\ 0 & 0 & 0 & 0 & 5(r+5) & r+11 & 1 \\ 0 & 0 & 0 & 0 & 0 & 6(r+6) & r+13 \end{array}\right].$$

The continued fraction, the Hankel transform, and three-term recurrence now follow. □

Proposition 9.8 *The exponential Riordan array* $\left[e^{-\frac{rx^2}{2}}, x\right]$ *is the coefficient array of a family of classical orthogonal polynomials. We have*

$$w(x) = e^{-\frac{x^2}{2r}}$$

on the interval

$$(-\infty, \infty).$$

The moments have integral representation

$$\mu_n = \int_{-\infty}^{\infty} x^n e^{-\frac{x^2}{2r}}\, dx.$$

These begin

$$1, 0, r, 0, 3r^2, 0, 15r^3, 0, 105r^4, 0, 945r^5, 0, \ldots.$$

The moments have generating function given by

$$\mu(x) = \cfrac{1}{1 - \cfrac{rx^2}{1 - \cfrac{2rx^2}{1 - \cfrac{3rx^2}{1 - \cdots}}}}.$$

The Hankel transform of the moments is given by

$$h_n = r^{\binom{n+1}{2}} \prod_{k=1}^{n} k^{n-k+1}.$$

The polynomials $P_n(x)$ satisfy the three-term recurrence

$$P_n(x) = xP_{n-1}(x) - r(n-1)P_{n-2}(x), \quad n > 1,$$

with $P_0(x) = 1$, $P_1(x) = x$.

If $y = P_n(x)$, then y satisfies the differential equation

$$ry'' - xy' + ny = 0.$$

Proof The main conclusion follows from the fact that

$$\frac{w'(x)}{w(x)} = \frac{-x}{r},$$

where

$$w(x) = e^{-\frac{x^2}{r}}.$$

The inverse coefficient matrix $[u, v] = \left[e^{\frac{rx^2}{2}}, x\right]$, and hence the production matrix is generated by $e^{xy}(rx + y)$. This matrix begins

$$\begin{bmatrix} 0 & 1 & 0 & 0 & 0 & 0 & 0 & 0 & 0 \\ r & 0 & 1 & 0 & 0 & 0 & 0 & 0 & 0 \\ 0 & 2r & 0 & 1 & 0 & 0 & 0 & 0 & 0 \\ 0 & 0 & 3r & 0 & 1 & 0 & 0 & 0 & 0 \\ 0 & 0 & 0 & 4r & 0 & 1 & 0 & 0 & 0 \\ 0 & 0 & 0 & 0 & 5r & 0 & 1 & 0 & 0 \\ 0 & 0 & 0 & 0 & 0 & 6r & 0 & 1 & 0 \\ 0 & 0 & 0 & 0 & 0 & 0 & 7r & 0 & 1 \\ 0 & 0 & 0 & 0 & 0 & 0 & 0 & 8r & 0 \end{bmatrix}.$$

The continued fraction, the Hankel transform, and three-term recurrence now follow.

□

The case $r = 1$ is related to the Hermite polynomials. The Riordan array $\left[e^{-\frac{t^2}{2}}, t\right]$ is the coefficient array of the (Probabilist) Hermite polynomials $He(n, x)$ given by

$$He_n(x) = \sum_{k=0}^{n} \frac{n!}{(-2)^{\frac{n-k}{2}} k! \left(\frac{n-k}{2}\right)!} \frac{1 + (-1)^{n-k}}{2} x^k.$$

The Physicists' Hermite polynomials are given by

$$H_n(x) = n! \sum_{k=0}^{\lfloor \frac{n}{2} \rfloor} \frac{(-1)^k (2x)^{n-2k}}{k!(n-2k)!}.$$

The coefficient array of the Physicists' Hermite polynomials is the exponential Riordan array

$$\left[e^{-t^2}, 2t\right].$$

We have

$$He_n(x) = 2^{-\frac{n}{2}} H_n(\sqrt{2}x).$$

We have

$$He_n(x) = x He_{n-1}(x) - (n-1) He_{n-2}(x),$$

and

$$H_n(x) = 2x H_{n-1}(x) - 2(n-1) H_{n-1}(x).$$

9.3 Orthogonal Polynomials as Moments

In [23], the authors give functionals whose moments are the Hermite, Laguerre, and various Meixner families of polynomials. In this section, we shall show how we can use suitable Riordan arrays to derive these results for Legendre and Hermite polynomials.

Legendre polynomials as moments

We recall that the Legendre polynomials $P_n(x)$ can be defined by

$$P_n(x) = \sum_{k=0}^{n} (-1)^k \binom{n}{k}^2 \left(\frac{1+x}{2}\right)^{n-k} \left(\frac{1-x}{2}\right)^k.$$

Their generating function is given by

$$\frac{1}{\sqrt{1-2xt+t^2}} = \sum_{n=0}^{\infty} P_n(x)t^n.$$

We note that the production matrix of the inverse of the coefficient array of these polynomials is given by

$$\left[\begin{array}{ccccccc} 0 & 1 & 0 & 0 & 0 & 0 & \dots \\ \frac{1}{3} & 0 & \frac{2}{3} & 0 & 0 & 0 & \dots \\ 0 & \frac{2}{5} & 0 & \frac{3}{5} & 0 & 0 & \dots \\ 0 & 0 & \frac{3}{7} & 0 & \frac{4}{7} & 0 & \dots \\ 0 & 0 & 0 & \frac{4}{9} & 0 & \frac{5}{9} & \dots \\ 0 & 0 & 0 & 0 & \frac{5}{11} & 0 & \dots \\ \vdots & \vdots & \vdots & \vdots & \vdots & \vdots & \ddots \end{array}\right],$$

which corresponds to the fact that the $P_n(x)$ satisfy the following three-term recurrence

$$(n+1)P_{n+1}(x) = (2n+1)xP_n(x) - nP_{n-1}(x).$$

The shifted Legendre polynomials $\tilde{P}_n(x)$ are defined by

$$\tilde{P}_n(x) = P_n(2x-1).$$

They satisfy

$$\tilde{P}_n(x) = (-1)^n \sum_{k=0}^{n} \binom{n}{k}\binom{n+k}{k}(-x)^k = \sum_{k=0}^{n}(-1)^{n-k}\binom{n+k}{2k}\binom{2k}{k}x^k.$$

Their coefficient array begins

$$\left[\begin{array}{ccccccc} 1 & 0 & 0 & 0 & 0 & 0 & \dots \\ -1 & 2 & 0 & 0 & 0 & 0 & \dots \\ 1 & -6 & 6 & 0 & 0 & 0 & \dots \\ -1 & 12 & -30 & 20 & 0 & 0 & \dots \\ 1 & -20 & 90 & -140 & 70 & 0 & \dots \\ -1 & 30 & -210 & 560 & -630 & 252 & \dots \\ \vdots & \vdots & \vdots & \vdots & \vdots & \vdots & \ddots \end{array}\right],$$

and so the initial terms are

$$1, 2x-1, 6x^2-6x+1, 20x^3-30x^2+12x-1, \dots$$

We clearly have

$$\frac{1}{\sqrt{1-2(2x-1)t+t^2}} = \sum_{n=0}^{\infty} \tilde{P}_n(x)t^n.$$

Our goal is to represent the Legendre polynomials as the first column of a Riordan array whose production matrix is tri-diagonal. We first of all consider the so-called shifted Legendre polynomials. We have the following proposition.

Proposition 9.9 *The inverse* $\mathbf{L}$ *of the Riordan array*

$$\left(\frac{1+r(1-r)x^2}{1+(2r-1)x+r(r-1)x^2}, \frac{x}{1+(2r-1)x+r(r-1)x^2}\right)$$

has as its first column the shifted Legendre polynomials $\tilde{P}_n(r)$. *The production matrix of* $\mathbf{L}$ *is tri-diagonal.*

Proof Indeed, standard Riordan array techniques show that we have

$$\begin{aligned}\mathbf{L} &= \left(\frac{1+r(1-r)x^2}{1+(2r-1)x+r(r-1)x^2}, \frac{x}{1+(2r-1)x+r(r-1)x^2}\right)^{-1} \\ &= \left(\frac{1}{\sqrt{1-2(2r-1)x+x^2}}, \frac{1-(2r-1)x-\sqrt{1-2(2r-1)x+x^2}}{2r(r-1)x}\right).\end{aligned}$$

This establishes the first part. Starting with the equations

$$f(x) = \frac{1-(2r-1)x-\sqrt{1-2(2r-1)x+x^2}}{2r(r-1)x}, \quad \bar{f}(x) = \frac{x}{1+(2r-1)x+r(r-1)x^2},$$

and

$$g(x) = \frac{1}{\sqrt{1-2(2r-1)x+x^2}},$$

we find that

$$Z(x) = (2r-1)+2rx(r-1), \quad A(x) = 1+x(2r-1)+x^2r(r-1).$$

Hence the production matrix $P = S_L$ of $\mathbf{L}$ is given by

$$\begin{bmatrix} 2r-1 & 1 & 0 & 0 & 0 & 0 & \cdots \\ 2r(r-1) & 2r-1 & 1 & 0 & 0 & 0 & \cdots \\ 0 & r(r-1) & 2r-1 & 1 & 0 & 0 & \cdots \\ 0 & 0 & r(r-1) & 2r-1 & 1 & 0 & \cdots \\ 0 & 0 & 0 & r(r-1) & 2r-1 & 1 & \cdots \\ 0 & 0 & 0 & 0 & r(r-1) & 2r-1 & \cdots \\ \vdots & \vdots & \vdots & \vdots & \vdots & \vdots & \ddots \end{bmatrix}$$

and this completes the proof. □

Corollary 9.4 *The shifted Legendre polynomials are moments of the family of orthogonal polynomials whose coefficient array is given by*

$$\mathbf{L}^{-1} = \left(\frac{1+r(1-r)x^2}{1+(2r-1)x+r(r-1)x^2}, \frac{x}{1+(2r-1)x+r(r-1)x^2} \right).$$

Proof This follows from the above result and Proposition 9.4. □

Proposition 9.10 *The Hankel transform of the sequence $\tilde{P}_n(r)$ is given by $2^n(r(r-1))^{\binom{n+1}{2}}$.*

Proof From the above, the generating function of $\tilde{P}_n(r)$ is given by

$$\cfrac{1}{1-(2r-1)x-\cfrac{2r(r-1)x^2}{1-(2r-1)x-\cfrac{r(r-1)x^2}{1-(2r-1)x-\cfrac{r(r-1)x^2}{1-\cdots}}}}.$$

The result now follows from Eq. (9.1.4). □

We note that

$$\tilde{P}_n(r) = \frac{1}{\pi} \int_{-2\sqrt{r(r-1)}+2r-1}^{2\sqrt{r(r-1)}+2r-1} \frac{x^n}{\sqrt{-x^2+2(2r-1)x-1}}\, dx$$

gives an explicit moment representation for $\tilde{P}_n(r)$.

Turning now to the Legendre polynomials $P_n(x)$, we have the following result.

Proposition 9.11 *The inverse* $\mathbf{L}$ *of the Riordan array*

$$\left(\frac{1+\frac{1-r^2}{4}x^2}{1+rx+\frac{r^2-1}{2}x^2}, \frac{x}{1+rx+\frac{r^2-1}{2}x^2} \right)$$

has as its first column the Legendre polynomials $P_n(r)$. The production matrix of $\mathbf{L}$ *is tri-diagonal.*

Proof We have

$$\mathbf{L} = \left(\frac{1+\frac{1-r^2}{4}x^2}{1+rx+\frac{r^2-1}{2}x^2}, \frac{x}{1+rx+\frac{r^2-1}{2}x^2} \right)^{-1}$$
$$= \left(\frac{1}{\sqrt{1-2rx+x^2}}, \frac{2(1-rx-\sqrt{1-2rx+x^2})}{x(r^2-1)} \right).$$

This proves the first assertion. In this case, we obtain the following tri-diagonal matrix as the production matrix of $\mathbf{L}$:

$$\begin{bmatrix} r & 1 & 0 & 0 & 0 & 0 \ldots \\ \frac{r^2-1}{2} & r & 1 & 0 & 0 & 0 \ldots \\ 0 & \frac{r^2-1}{4} & r & 1 & 0 & 0 \ldots \\ 0 & 0 & \frac{r^2-1}{4} & r & 1 & 0 \ldots \\ 0 & 0 & 0 & \frac{r^2-1}{4} & r & 1 \ldots \\ 0 & 0 & 0 & 0 & \frac{r^2-1}{4} & r \ldots \\ \vdots & \vdots & \vdots & \vdots & \vdots & \vdots \ddots \end{bmatrix}$$

and this completes the proof. □

Corollary 9.5

$$\mathbf{L}^{-1} = \left(\frac{1+\frac{1-r^2}{4}x^2}{1+rx+\frac{r^2-1}{2}x^2}, \frac{x}{1+rx+\frac{r^2-1}{2}x^2} \right)$$

is the coefficient array of a set of orthogonal polynomials for which the Legendre polynomials are moments.

Proposition 9.12 *The Hankel transform of $P_n(r)$ is given by*

$$\frac{(r^2-1)^{\binom{n+1}{2}}}{2^{n^2}}.$$

Proof From the above, we obtain that the generating function of $P_n(r)$ can be expressed as

$$\cfrac{1}{1-rx-\cfrac{\frac{r^2-1}{2}x^2}{1-rx-\cfrac{\frac{r^2-1}{4}x^2}{1-rx-\cfrac{\frac{r^2-1}{4}x^2}{1-\cdots}}}}.$$

The result now follows from Eq. (9.1.4). □

We end this section by noting that

$$P_n(r) = \frac{1}{\pi}\int_{r-\sqrt{r^2-1}}^{r+\sqrt{r^2-1}} \frac{x^n}{\sqrt{-x^2+2rx-1}}\,dx$$

gives an explicit moment representation for $P_n(r)$.

Hermite polynomials as moments

The Hermite polynomials may be defined as

$$H_n(x) = \sum_{k=0}^{\lfloor \frac{n}{2} \rfloor} \frac{(-1)^k (2x)^{n-2k}}{k!(n-2k)!}.$$

The generating function for $H_n(x)$ is given by

$$e^{2xt-t^2} = \sum_{n=0}^{\infty} H_n(x)\frac{t^n}{n!}.$$

The Probabilist Hermite polynomials (also called normalized Hermite polynomials) are given by

$$He_n(x) = 2^{-\frac{n}{2}} H_n(\sqrt{2}x) = \sum_{k=0}^{n} \frac{n!}{(-2)^{\frac{n-k}{2}} k! \left(\frac{n-k}{2}\right)!} \frac{1+(-1)^{n-k}}{2} x^k.$$

Their generating function is given by

$$e^{xt-\frac{t^2}{2}} = \sum_{n=0}^{\infty} He_n(x)\frac{t^n}{n!}.$$

We note that the coefficient array of He_n is given by the exponential Riordan array

$$\left[e^{-\frac{x^2}{2}}, x\right].$$

This array A066325 begins

$$\begin{bmatrix} 1 & 0 & 0 & 0 & 0 & 0 & \dots \\ 0 & 1 & 0 & 0 & 0 & 0 & \dots \\ -1 & 0 & 1 & 0 & 0 & 0 & \dots \\ 0 & -3 & 0 & 1 & 0 & 0 & \dots \\ 3 & 0 & -6 & 0 & 1 & 0 & \dots \\ 0 & 15 & 0 & -10 & 0 & 1 & \dots \\ \vdots & \vdots & \vdots & \vdots & \vdots & \vdots & \ddots \end{bmatrix}.$$

It is the aeration of the alternating sign version of the Bessel coefficient array A001497. The inverse of this matrix has a production matrix the begins as follows:

$$\begin{bmatrix} 0 & 1 & 0 & 0 & 0 & 0 & \dots \\ 1 & 0 & 1 & 0 & 0 & 0 & \dots \\ 0 & 2 & 0 & 1 & 0 & 0 & \dots \\ 0 & 0 & 3 & 0 & 1 & 0 & \dots \\ 0 & 0 & 0 & 4 & 0 & 1 & \dots \\ 0 & 0 & 0 & 0 & 5 & 0 & \dots \\ \vdots & \vdots & \vdots & \vdots & \vdots & \vdots & \ddots \end{bmatrix}.$$

This corresponds to the fact that we have the following three-term recurrence for He_n:

$$He_{n+1}(x) = xHe_n(x) - nHe_{n-1}(x).$$

Proposition 9.13 *The exponential Riordan array*

$$\mathbf{L} = \left[e^{rx-\frac{x^2}{2}}, x\right]$$

has as its first column the unitary Hermite polynomials $He_n(r)$. The array **L** *has a tri-diagonal production array.*

Proof The first column of **L** has generating function $e^{rx-\frac{x^2}{2}}$, from which the first assertion follows. We now calculate the production matrix $P = S_L$. We have $f(x) = x$, so that $\bar{f}(x) = x$ and $f'(\bar{f}(x)) = 1 = A(x)$. Also, $g(x) = e^{rx-\frac{x^2}{2}}$ implies that $g'(x) = g(x)(r - x)$, and hence

$$Z(x) = \frac{g'(\bar{f}(x))}{g(\bar{f}(x))} = r - x.$$

Thus the production array of **L** is indeed tri-diagonal, beginning

$$\begin{bmatrix} r & 1 & 0 & 0 & 0 & 0 & \dots \\ -1 & r & 1 & 0 & 0 & 0 & \dots \\ 0 & -2 & r & 1 & 0 & 0 & \dots \\ 0 & 0 & -3 & r & 1 & 0 & \dots \\ 0 & 0 & 0 & -4 & r & 1 & \dots \\ 0 & 0 & 0 & 0 & -5 & r & \dots \\ \vdots & \vdots & \vdots & \vdots & \vdots & \vdots & \ddots \end{bmatrix}$$

and we have the proof. □

We note that $\mathbf{L}$ starts

$$\left[\begin{array}{cccccc} 1 & 0 & 0 & 0 & 0 & 0 \ldots \\ r & 1 & 0 & 0 & 0 & 0 \ldots \\ r^2-1 & 2r & 1 & 0 & 0 & 0 \ldots \\ r(r^2-3) & 3(r^2-1) & 3r & 1 & 0 & 0 \ldots \\ r^4-6r^2+3 & 4r(r^2-3) & 6(r^2-1) & 4r & 1 & 0 \ldots \\ r(r^4-10r^2+15) & 5(r^4-6r^2+3) & 10r(r^2-3) & 10(r^2-1) & 5r & 1 \ldots \\ \vdots & \vdots & \vdots & \vdots & \vdots & \vdots \ddots \end{array}\right].$$

Thus

$$\mathbf{L}^{-1} = \left[e^{-rx+\frac{x^2}{2}}, x\right]$$

is the coefficient array of a set of orthogonal polynomials which have as moments the unitary Hermite polynomials. These new orthogonal polynomials satisfy the three-term recurrence

$$\mathfrak{H}_{n+1}(x) = (x-r)\mathfrak{H}_n(x) + n\mathfrak{H}_{n-1}(x),$$

with $\mathfrak{H}_0 = 1$, $\mathfrak{H}_1 = x - r$.

Proposition 9.14 *The Hankel transform of the sequence $He_n(r)$ is given by*

$$(-1)^{\binom{n+1}{2}} \prod_{k=0}^{n} k!.$$

Proof By the above, the generating function of $He_n(r)$ is given by

$$\cfrac{1}{1-rx+\cfrac{x^2}{1-rx+\cfrac{2x^2}{1-rx+\cfrac{3x^2}{1-\ldots}}}}.$$

The result now follows from Eq. (9.1.4). □

Turning now to the Hermite polynomials $H_n(x)$, we have the following result.

Proposition 9.15 *The proper exponential Riordan array*

$$\mathbf{L} = \left[e^{2rx-x^2}, x\right]$$

has as its first column the Hermite polynomials $H_n(r)$. This array has a tri-diagonal production array.

Proof The first column of $\mathbf{L}$ has generating function e^{2rx-x^2}, from which the first assertion follows. Standard Riordan array techniques show us that the production array P of $\mathbf{L}$ is indeed tri-diagonal, beginning with

$$\begin{bmatrix} 2r & 1 & 0 & 0 & 0 & 0 & \dots \\ -2 & 2r & 1 & 0 & 0 & 0 & \dots \\ 0 & -4 & 2r & 1 & 0 & 0 & \dots \\ 0 & 0 & -6 & 2r & 1 & 0 & \dots \\ 0 & 0 & 0 & -8 & 2r & 1 & \dots \\ 0 & 0 & 0 & 0 & -10 & 2r & \dots \\ \vdots & \vdots & \vdots & \vdots & \vdots & \vdots & \ddots \end{bmatrix}.$$

This is so since $f(x) = x$ gives us $\bar{f}(x) = x$ and $f'(x) = 1$, and thus

$$A(x) = f'(\bar{f}(x)) = f'(x) = 1.$$

Similarly,

$$Z(x) = \frac{g'(\bar{f}(x))}{g(\bar{f}(x))} = 2(r - x),$$

and thus the bivariate generating function $\phi_P(y, w)$ of P is given by

$$\phi_P(y, z) = e^{yz}(2(r - z) + y)$$

as required. □

We note that $\mathbf{L}$ starts

$$\begin{bmatrix} 1 & 0 & 0 & 0 & 0 & 0 & \dots \\ 2r & 1 & 0 & 0 & 0 & 0 & \dots \\ 2(2r^2-1) & 4r & 1 & 0 & 0 & 0 & \dots \\ 4r(2r^2-3) & 6(2r^2-1) & 6r & 1 & 0 & 0 & \dots \\ 4(4r^3-12r^2+3) & 16r(2r^2-3) & 12(2r^2-1) & 8r & 1 & 0 & \dots \\ 8r(4r^4-20r^2+15) & 20(4r^4-12r^2+3) & 40r(2r^2-3) & 20(2r^2-1) & 10r & 1 & \dots \\ \vdots & \vdots & \vdots & \vdots & \vdots & \vdots & \ddots \end{bmatrix}.$$

Thus

$$\mathbf{L}^{-1} = \left[e^{-2rx+x^2}, x\right]$$

is the coefficient array of a set of orthogonal polynomials which have as moments the Hermite polynomials. These new orthogonal polynomials satisfy the three-term recurrence

$$\mathfrak{H}_{n+1}(x) = (x - 2r)\mathfrak{H}_n(x) + 2n\mathfrak{H}_{n-1}(x),$$

with $\mathfrak{H}_0 = 1$, $\mathfrak{H}_1 = x - 2r$.

Proposition 9.16 *The Hankel transform of the sequence $H_n(r)$ is given by*

$$(-1)^{\binom{n+1}{2}} \prod_{k=0}^{n} 2^k k!.$$

Proof By the above, the generating function of $H_n(r)$ is given by

$$\cfrac{1}{1-2rx+\cfrac{2x^2}{1-2rx+\cfrac{4x^2}{1-2rx+\cfrac{6x^2}{1-\dots}}}}.$$

The result now follows from Eq. (9.1.4). □

An application to the Toda lattice equations

The restricted Toda chain equation [31, 39] is simply described by

$$\dot{u}_n = u_n(b_n - b_{n-1}), \quad n = 1, 2, \dots; \quad \dot{b}_n = u_{n+1} - u_n, \quad n = 0, 1, \dots \tag{9.3.1}$$

with $u_0 = 0$, where the dot indicates differentiation with respect to t. In this section, we shall show how solutions to this equation can be formulated in the context of exponential Riordan arrays. The Riordan arrays we shall consider may be thought of as parameterized (or "time-dependent") Riordan arrays. We have already considered parameterized Riordan arrays, exploring the links between these Riordan arrays and orthogonal polynomials.

The restricted Toda chain equation is closely related to orthogonal polynomials, since the functions u_n and b_n can be considered as the coefficients in the usual three-term recurrence [10, 18, 37] satisfied by orthogonal polynomials:

$$P_{n+1}(x) + b_n P_n(x) + u_n P_{n-1}(x) = x P_n(x), \quad n = 1, 2, \dots \tag{9.3.2}$$

with initial conditions $P_0(x) = 1$ and $P_1(x) = x - b_0$.

Nakamura and Zhedanov [31, 32] give solutions of the restricted Toda chain equation in terms of Sheffer polynomials. Given the close links between Sheffer class polynomials and Riordan arrays [12, 21], we find it instructive to re-interpret these results in terms of Riordan arrays. Thus in the cases discussed in this section, we find that the coefficient arrays of the orthogonal polynomials linked to solutions of the restricted Toda chain are exponential Riordan arrays. In particular, the parameters of the solution are obtained by calculating the terms of the corresponding production matrices, using the procedures of Deutsch, Ferrari, and Rinaldi [13, 14].

Hermite polynomials and the Toda chain

We recall that the Hermite polynomials may be defined as

$$H_n(x) = n! \sum_{k=0}^{\lfloor \frac{n}{2} \rfloor} \frac{(-1)^k (2x)^{n-2k}}{k!(n-2k)!}.$$

The generating function for $H_n(x)$ is given by

$$e^{2xt-t^2} = \sum_{n=0}^{\infty} H_n(x) \frac{t^n}{n!}.$$

We have seen previously that the exponential Riordan array

$$\mathbf{L} = \left[e^{2rx-x^2}, x \right]$$

has as first column the Hermite polynomials $H_n(r)$. The array $\mathbf{L}$ has a tri-diagonal production array, so that $\mathbf{L}^{-1}$ is the coefficient array of a family of orthogonal polynomials, whose moments are the Hermite polynomials.
We note that $\mathbf{L}$ starts with

$$\begin{bmatrix} 1 & 0 & 0 & 0 & 0 & 0 & \dots \\ 2r & 1 & 0 & 0 & 0 & 0 & \dots \\ 2(2r^2-1) & 4r & 1 & 0 & 0 & 0 & \dots \\ 4r(2r^2-3) & 6(2r^2-1) & 6r & 1 & 0 & 0 & \dots \\ 4(4r^3-12r^2+3) & 16r(2r^2-3) & 12(2r^2-1) & 8r & 1 & 0 & \dots \\ 8r(4r^4-20r^2+15) & 20(4r^4-12r^2+3) & 40r(2r^2-3) & 20(2r^2-1) & 10r & 1 & \dots \\ \vdots & \vdots & \vdots & \vdots & \vdots & \vdots & \ddots \end{bmatrix}.$$

We have

$$\mathbf{L}^{-1} = \left[e^{-2rx+x^2}, x \right].$$

The orthogonal polynomials whose coefficient array is given by $\mathbf{L}^{-1}$ satisfy the three-term recurrence

$$\mathfrak{H}_{n+1}(x) = (x-2r)\mathfrak{H}_n(x) + 2n\mathfrak{H}_{n-1}(x),$$

with $\mathfrak{H}_0 = 1$, $\mathfrak{H}_1 = x - 2r$.

We can now modify this result to give us our first Toda chain result.

Proposition 9.17 *The exponential Riordan array*

$$\left[e^{-2(z-t)x+x^2}, x \right]$$

is the coefficient array of a family of orthogonal polynomials $P_n(x)$ with

$$P_{n+1}(x) + b_n P_n(x) + u_n P_{n-1}(x) = x P_n(x),$$

where (u_n, b_n) *is a solution to the restricted Toda chain.*

Proof We easily determine that the inverse matrix

$$\left[e^{2(z-t)x-x^2}, x\right]$$

has the production matrix P given by

$$\begin{bmatrix} 2(z-t) & 1 & 0 & 0 & 0 & 0 & \dots \\ -2 & 2(z-t) & 1 & 0 & 0 & 0 & \dots \\ 0 & -4 & 2(z-t) & 1 & 0 & 0 & \dots \\ 0 & 0 & -6 & 2(z-t) & 1 & 0 & \dots \\ 0 & 0 & 0 & -8 & 2(z-t) & 1 & \dots \\ 0 & 0 & 0 & 0 & -10 & 2(z-t) & \dots \\ \vdots & \vdots & \vdots & \vdots & \vdots & \vdots & \ddots \end{bmatrix}.$$

This is so since $f(x) = x$ gives us $\bar{f}(x) = x$ and $f'(x) = 1$, and thus

$$A(x) = f'(\bar{f}(x)) = f'(x) = 1.$$

Similarly,

$$Z(x) = \frac{g'(\bar{f}(x))}{g(\bar{f}(x))} = -2(x - z + t),$$

and thus the bivariate generating function $\phi_P(y, w)$ of P is given by

$$\phi_P(y, w) = e^{yw}(-2(w - z + t) + y),$$

as required.

This verifies that $P_n(x)$ is indeed a family of orthogonal polynomials, for which

$$u_n(t) = -2n, \quad b_n(t) = 2(z - t).$$

It is immediate that these satisfy Eq. (9.3.1). □

We now note that the moments of this polynomial family (the first column of the inverse matrix) m_n satisfy the following relation:

$$\begin{aligned} m_n &= n![x^n]e^{2(z-t)x-x^2} \\ &= \frac{1}{e^{-t^2+2tz}}\frac{d^n}{dt^n}e^{-t^2+2tz} \\ &= n!\sum_{k=0}^{\lfloor \frac{n}{2} \rfloor}\binom{n-k}{k}(-1)^k\frac{(2(z-t))^{n-2k}}{(n-k)!} \\ &= \sum_{k=0}^{\lfloor \frac{n}{2} \rfloor}\binom{n}{k}\binom{n-k}{k}k!(2(z-t))^{n-2k} \\ &= H_n(z-t). \end{aligned}$$

From the form of the production matrix P, we infer that the generating function of m_n may be expressed as

$$\cfrac{1}{1-2(z-t)x+\cfrac{2x^2}{1-2(z-t)x+\cfrac{4x^2}{1-2(z-t)x+\cfrac{6x^2}{1-2(z-t)x+\cdots}}}}.$$

The Hankel transform of m_n, defined as the sequence h_n where

$$h_n = |m_{i+j}|_{0 \le i,j \le n}$$

is then given by

$$h_n = (-2)^{\binom{n+1}{2}}\prod_{k=0}^{n} k!.$$

Charlier polynomials and the Toda chain

Proposition 9.18 *The exponential Riordan array*

$$\left[e^{-xe^t}, \ln(1+x)\right]$$

is the coefficient array of a family of orthogonal polynomials $P_n(x)$ with

$$P_{n+1}(x) + b_n P_n(x) + u_n P_{n-1}(x) = x P_n(x),$$

where (u_n, b_n) is a solution to the restricted Toda chain.

Proof We determine that the inverse matrix

$$\left[e^{e^{t+x}-e^t}, e^x - 1\right]$$

has the production matrix

$$\begin{bmatrix} e^t & 1 & 0 & 0 & 0 & 0 & \dots \\ e^t & e^t+1 & 1 & 0 & 0 & 0 & \dots \\ 0 & 2e^t & e^t+2 & 1 & 0 & 0 & \dots \\ 0 & 0 & 3e^t & e^t+3 & 1 & 0 & \dots \\ 0 & 0 & 0 & 4e^t & e^t+4 & 1 & \dots \\ 0 & 0 & 0 & 0 & 5e^t & e^t+5 & \dots \\ \vdots & \vdots & \vdots & \vdots & \vdots & \vdots & \ddots \end{bmatrix}$$

as follows. We have $\bar{f}(x) = \ln(1+x)$ and $f'(x) = e^x$, and thus

$$A(x) = f'(\bar{f}(x)) = 1 + x.$$

Similarly,

$$Z(x) = \frac{g'(\bar{f}(x))}{g(\bar{f}(x))} = \frac{e^{xe^t+t}(1+x)}{e^{xe^t}} = e^t(1+x).$$

Thus the bivariate generating function for the production matrix P is given by

$$\phi_P(y,z) = e^{yz}(e^t(1+z) + y(1+z)),$$

as required. This verifies that $P_n(x)$ is indeed a family of orthogonal polynomials, for which

$$u_n(t) = ne^t, \quad b_n(t) = n + e^t.$$

It is easy now to verify that, with these values, (u_n, b_n) satisfies the Toda chain equations Eq. (9.3.1). □

The moments m_n of this family of orthogonal polynomials may be expressed as

$$\begin{aligned} m_n &= n![x^n]e^{e^{t+x}-e^t} \\ &= \frac{1}{e^{e^t-1}}\frac{d^n}{dt^n}e^{e^t-1} \\ &= \sum_{k=0}^{n} \left\{ {n \atop k} \right\} e^{kt}, \end{aligned}$$

where

$$\left\{ {n \atop k} \right\} = \frac{1}{k!}\sum_{j=0}^{k}(-1)^{k-j}\binom{k}{j} j^n$$

are the Stirling numbers of the second kind. From the form of the production matrix P, we infer that the generating function of m_n may be expressed as

$$\cfrac{1}{1-e^t x-\cfrac{e^t}{1-(e^t+1)x-\cfrac{2e^t}{1-(e^t+2)x-\cfrac{3e^t}{1-(e^t+3)x-\cdots}}}}.$$

The Hankel transform of m_n is then given by

$$h_n=(e^t)^{\binom{n+1}{2}}\prod_{k=0}^{n} k!.$$

In particular we have

$$m_n(0)=\sum_{k=0}^{n}\left\{ {n \atop k} \right\}=\text{Bell}(n),$$

where $\text{Bell}(n)$ denotes the nth Bell number, with Hankel transform

$$h_n=\prod_{k=0}^{n} k!.$$

Laguerre polynomials and the Toda chain

We have seen that the exponential Riordan array

$$\left[\frac{1}{1-x},\frac{x}{1-x}\right]$$

is closely related to the Laguerre polynomials. We introduce a "time" parameter t as follows:

$$\left[\frac{1}{1-\frac{x}{1+t}},\frac{x}{1-\frac{x}{1+t}}\right]=\left[\left(1-\frac{x}{1+t}\right)^{-1},\frac{x}{1-\frac{x}{1+t}}\right],$$

and further generalize this array by introducing a general power factor α, to get the array

$$\left[\left(1-\frac{x}{1+t}\right)^{\alpha},\frac{x}{1-\frac{x}{1+t}}\right].$$

We then have the following proposition.

Proposition 9.19 *The exponential Riordan array*

$$\left[\left(1-\frac{x}{1+t}\right)^{\alpha}, \frac{x}{1-\frac{x}{1+t}}\right]$$

is the coefficient array of a family of orthogonal polynomials $P_n(x)$ with

$$P_{n+1}(x) + b_n P_n(x) + u_n P_{n-1}(x) = x P_n(x),$$

where (u_n, b_n) is a solution to the restricted Toda chain.

Proof The inverse matrix

$$\left[\left(\frac{1+t+x}{1+t}\right)^{\alpha}, \frac{(1+t)x}{1+t+x}\right]$$

has the production matrix

$$\begin{bmatrix} \frac{\alpha}{1+t} & 1 & 0 & 0 & 0 & 0 & \cdots \\ \frac{-\alpha}{(1+t)^2} & \frac{\alpha-2}{1+t} & 1 & 0 & 0 & 0 & \cdots \\ 0 & \frac{2(1-\alpha)}{(1+t)^2} & \frac{\alpha-4}{1+t} & 1 & 0 & 0 & \cdots \\ 0 & 0 & \frac{3(2-\alpha)}{(1+t)^2} & \frac{\alpha-6}{1+t} & 1 & 0 & \cdots \\ 0 & 0 & 0 & \frac{4(3-\alpha)}{(1+t)^2} & \frac{\alpha-8}{1+t} & 1 & \cdots \\ 0 & 0 & 0 & 0 & \frac{5(4-\alpha)}{(1+t)^2} & \frac{\alpha-10}{1+t} & \cdots \\ \vdots & \vdots & \vdots & \vdots & \vdots & \vdots & \ddots \end{bmatrix}.$$

This verifies that $P_n(x)$ is indeed a family of orthogonal polynomials, for which

$$u_n(t) = \frac{n(n-\alpha-1)}{(1+t)^2}, \quad b_n(t) = \frac{\alpha-2n}{1+t}.$$

It is easy now to verify that, with these values, the pair (u_n, b_n) satisfies the Toda chain equations Eq. (9.3.1). □

For this family of orthogonal polynomials, the moments m_n may be expressed as

$$m_n = n![x^n]\left(1+\frac{x}{1+t}\right)^{\alpha} = \frac{1}{(1+t)^{\alpha}}\frac{d^n}{dt^n}(1+t)^{\alpha} = \frac{(\alpha)_n}{(1+t)^n}, \tag{9.3.3}$$

where

$$(\alpha)_n = \alpha(\alpha-1)\cdots(\alpha-n+1).$$

From the form of the production matrix, we infer that the generating function of m_n may be expressed as

$$\cfrac{1}{1-\frac{\alpha}{1+t}x-\cfrac{\frac{0-\alpha}{(1+t)^2}x^2}{1-\frac{\alpha-2}{1+t}x-\cfrac{\frac{2(1-\alpha)}{(1+t)^2}x^2}{1-\frac{\alpha-4}{1+t}x-\cfrac{\frac{3(2-\alpha)}{(1+t)^2}x^2}{1-\frac{\alpha-6}{1+t}x-\cdots}}}}.$$

The Hankel transform of m_n is then given by

$$h_n = \frac{(-1)^{\binom{n+1}{2}}}{(1+t)^{n(n+1)}} \prod_{k=0}^{n} k!(\alpha-k)^{n-k}.$$

Noticing that

$$\lim_{c\to 1} \frac{c(1+t)}{c-1} \ln\left(\frac{1-\frac{x}{c(1+t)}}{1-\frac{x}{1+t}}\right) = \frac{x}{1-\frac{x}{1+t}},$$

we can "embed" this previous result in the following:

Proposition 9.20 *The exponential Riordan array*

$$\left[\left(1-\frac{x}{1+t}\right)^{\alpha}, \frac{c(1+t)}{c-1} \ln\left(\frac{1-\frac{x}{c(1+t)}}{1-\frac{x}{1+t}}\right)\right]$$

is the coefficient array of a family of orthogonal polynomials $P_n(x)$ with

$$P_{n+1}(x) + b_n P_n(x) + u_n P_{n-1}(x) = x P_n(x),$$

where for $c = 1$, (u_n, b_n) is a solution to the restricted Toda chain.

Proof We can show that the inverse array has a tri-diagonal production matrix with

$$b(n) = \frac{\alpha c - n(c+1)}{c(1+t)},$$

and

$$u(n) = \frac{n(n-1-\alpha)}{c(1+t)^2}.$$

Letting $c \to 1$ now gives us the result. □

Another direction of generalization is given by looking at the related exponential Riordan array

$$\left[e^{\alpha x}, \frac{x}{1+\frac{x}{1+t}}\right],$$

which has inverse

$$\left[e^{\frac{\alpha x}{1-\frac{x}{1+t}}}, \frac{x}{1-\frac{x}{1+t}}\right].$$

For this, we have the following proposition.

Proposition 9.21 *The exponential Riordan array*

$$\left[e^{\frac{\alpha x}{1-\frac{x}{1+t}}}, \frac{x}{1-\frac{x}{1+t}}\right]$$

is the coefficient array of a family of orthogonal polynomials $P_n(x)$ *with*

$$P_{n+1}(x) + b_n P_n(x) + u_n P_{n-1}(x) = x P_n(x),$$

where (u_n, b_n) *is a solution to the restricted Toda chain.*

Proof We find that

$$u_n(t) = \frac{n(n-1)}{(1+t)^2}$$

and

$$b_n(t) = -\frac{\alpha(1+t) + 2n}{1+t}$$

and we have the proof. □

In this case, the moments are simply

$$m_n = (-\alpha)^n.$$

Meixner polynomials and the Toda chain

Proposition 9.22 *The exponential Riordan array*

$$\left[\frac{1}{\sqrt{1-2x\tanh(t)-x^2\operatorname{sech}(t)^2}}, \ln\left(\sqrt{\frac{1+xe^{-t}\operatorname{sech}(t)}{1-xe^{t}\operatorname{sech}(t)}}\right)\right]$$

is the coefficient array of a family of orthogonal polynomials $P_n(x)$ *with*

$$P_{n+1}(x) + b_n P_n(x) + u_n P_{n-1}(x) = x P_n(x),$$

where (u_n, b_n) *is a solution to the restricted Toda chain.*

Proof The inverse matrix

$$\left[\frac{\operatorname{sech}(x+t)}{\operatorname{sech}(t)}, \sinh(x)\frac{\operatorname{sech}(x+t)}{\operatorname{sech}(t)}\right]$$

has the production matrix

$$\begin{bmatrix} -\tanh(t) & 1 & 0 & 0 & 0 & 0 & \dots \\ -\operatorname{sech}^2(t) & -3\tanh(t) & 1 & 0 & 0 & 0 & \dots \\ 0 & -4\operatorname{sech}^2(t) & -5\tanh(t) & 1 & 0 & 0 & \dots \\ 0 & 0 & -9\operatorname{sech}^2(t) & -7\tanh(t) & 1 & 0 & \dots \\ 0 & 0 & 0 & -16\operatorname{sech}^2(t) & -9\tanh(t) & 1 & \dots \\ 0 & 0 & 0 & 0 & -25\operatorname{sech}^2(t) & -11\tanh(t) & \dots \\ \vdots & \vdots & \vdots & \vdots & \vdots & \vdots & \ddots \end{bmatrix}.$$

This verifies that $P_n(x)$ is indeed a family of orthogonal polynomials, for which

$$u_n(t) = -n^2\operatorname{sech}^2(t), \quad b_n(t) = -(2n+1)\tanh(t).$$

It is easy now to verify that, with these values, the pair (u_n, b_n) satisfies the Toda chain equations Eq. (9.3.1). □

We may describe the moments m_n of this family of polynomials by

$$m_n = n![x^n]\frac{\operatorname{sech}(x+t)}{\operatorname{sech}(t)} = \frac{1}{\operatorname{sech}(t)}\frac{d^n}{dt^n}\operatorname{sech}(t). \tag{9.3.4}$$

The Hankel transform of m_n is then given by

$$h_n = (-1)^{\binom{n+1}{2}}\operatorname{sech}(t)^{n(n+1)}\prod_{k=0}^{n}(k!)^2.$$

9.4 Combinatorial Polynomials as Moments of Riordan Arrays

Many polynomials of combinatorial significance can be represented as moment sequences corresponding to Riordan arrays that are the coefficient arrays of appropriate families of orthogonal polynomials. In this section we look at two such sets of polynomials, each with three variants, that correspond, respectively, to ordinary and exponential Riordan arrays.

Narayana polynomials

The first set of polynomials that we consider are the Narayana polynomials, which correspond to the three number triangles that begin as follows.

The triangle $\mathbf{N}_1$ with general term

$$N_1(n,k) = 0^{n+k} + \frac{1}{n+0^n}\binom{n}{k}\binom{n}{k+1} = \frac{1}{k+1}\binom{n-1}{k}\binom{n}{k} \quad (9.4.1)$$

which begins

$$\mathbf{N}_1 = \begin{bmatrix} 1 & 0 & 0 & 0 & 0 & 0 & \dots \\ 1 & 0 & 0 & 0 & 0 & 0 & \dots \\ 1 & 1 & 0 & 0 & 0 & 0 & \dots \\ 1 & 3 & 1 & 0 & 0 & 0 & \dots \\ 1 & 6 & 6 & 1 & 0 & 0 & \dots \\ 1 & 10 & 20 & 10 & 1 & 0 & \dots \\ \vdots & \vdots & \vdots & \vdots & \vdots & \vdots & \ddots \end{bmatrix},$$

has the generating function

$$\phi_1(x,y) = 1 + \phi_0(x,y) = \frac{1 - x(1-y) - \sqrt{1-2x(1+y)+x^2(1-y)^2}}{2xy}. \quad (9.4.2)$$

The triangle $\mathbf{N}_2$, which is the reversal of $\mathbf{N}_1$, has general term

$$N_2(n,k) = [k \le n]N_1(n,n-k) = 0^{n+k} + \frac{1}{n+0^{nk}}\binom{n}{k}\binom{n}{k-1} \quad (9.4.3)$$

$$= \frac{1}{n-k+1}\binom{n-1}{n-k}\binom{n}{k}, \quad (9.4.4)$$

and begins

$$\mathbf{N}_2 = \begin{bmatrix} 1 & 0 & 0 & 0 & 0 & 0 & \dots \\ 0 & 1 & 0 & 0 & 0 & 0 & \dots \\ 0 & 1 & 1 & 0 & 0 & 0 & \dots \\ 0 & 1 & 3 & 1 & 0 & 0 & \dots \\ 0 & 1 & 6 & 6 & 1 & 0 & \dots \\ 0 & 1 & 10 & 20 & 10 & 1 & \dots \\ \vdots & \vdots & \vdots & \vdots & \vdots & \vdots & \ddots \end{bmatrix}.$$

This triangle has generating function

$$\phi_2(x,y) = 1 + y\phi_0(x,y) = \frac{1 + x(1-y) - \sqrt{1-2x(1+y)+x^2(1-y)^2}}{2x}. \quad (9.4.5)$$

Finally the "Pascal-like" variant $\mathbf{N}_3$ with general term

$$N_3(n,k) = N_0(n+1,k) = \frac{1}{n+1}\binom{n+1}{k}\binom{n+1}{k+1} \tag{9.4.6}$$

which begins

$$\mathbf{N}_3 = \begin{bmatrix} 1 & 0 & 0 & 0 & 0 & 0 & \dots \\ 1 & 1 & 0 & 0 & 0 & 0 & \dots \\ 1 & 3 & 1 & 0 & 0 & 0 & \dots \\ 1 & 6 & 6 & 1 & 0 & 0 & \dots \\ 1 & 10 & 20 & 10 & 1 & 0 & \dots \\ 1 & 15 & 50 & 50 & 15 & 1 & \dots \\ \vdots & \vdots & \vdots & \vdots & \vdots & \vdots & \ddots \end{bmatrix},$$

has generating function

$$\phi_3(x,y) = \frac{\phi_0(x,y)}{x} = \frac{1 - x(1+y) - \sqrt{1-2x(1+y)+x^2(1-y)^2}}{2x^2y}. \tag{9.4.7}$$

Triangle	A-number	Generating function
$\mathbf{N}_1$	A131198	$\phi_1(x,y) = \frac{1-x(1-y)-\sqrt{1-2x(1+y)+x^2(1-y)^2}}{2xy}$
$\mathbf{N}_2$	A090181	$\phi_2(x,y) = \frac{1+x(1-y)-\sqrt{1-2x(1+y)+x^2(1-y)^2}}{2x}$
$\mathbf{N}_3$	A001263	$\phi_3(x,y) = \frac{1-x(1+y)-\sqrt{1-2x(1+y)+x^2(1-y)^2}}{2x^2y}$

To each of the above triangles, there is a family of "Narayana" polynomials, where the triangles take on the role of coefficient arrays. Thus we get the polynomials

$$\mathcal{N}_{1,n}(y) = \sum_{k=0}^n N_1(n,k)y^k$$

$$\mathcal{N}_{2,n}(y) = \sum_{k=0}^n N_2(n,k)y^k$$

$$\mathcal{N}_{3,n}(y) = \sum_{k=0}^n N_3(n,k)y^k.$$

Note that since $\mathbf{N}_2$ is the reversal of $\mathbf{N}_1$, we have

$$\mathcal{N}_{2,n}(y) = \sum_{k=0}^n N_2(n,k)y^k = \sum_{k=0}^n N_1(n,k)y^{n-k}.$$

Example 9.7 The first terms of the sequence $(\mathcal{N}_{1,n}(y))_{n\geq 0}$ are

$$1, 1, 1+y, 1+3y+y^2, 1+6y+6y^2+y^3, 1+10y+20y^2+10y^3+y^4, \ldots$$

We can then show that

$$\mathcal{N}_{1,n}(y) = [x^{n+1}]\mathrm{Rev}_x \frac{x(1-xy)}{1-(y-1)x}$$
$$\mathcal{N}_{2,n}(y) = [x^{n+1}]\mathrm{Rev}_x \frac{x(1-x)}{1-(1-y)x}$$
$$\mathcal{N}_{3,n}(y) = [x^{n+1}]\mathrm{Rev}_x \frac{x}{1+(1+y)x+yx^2}.$$

Values of these polynomials are often of significant combinatorial interest. Sample values for these polynomials are tabulated below.

y	$\mathcal{N}_{1,0}(y), \mathcal{N}_{1,1}(y), \mathcal{N}_{1,2}(y), \ldots$	A-number	Name
1	$1, 1, 2, 5, 14, 42, \ldots$	A000108	Catalan numbers
2	$1, 1, 3, 11, 45, 197 \ldots$	A001003	little Schröder numbers
3	$1, 1, 4, 19, 100, 562, \ldots$	A007564	
4	$1, 1, 5, 29, 185, 1257, \ldots$	A059231	

y	$\mathcal{N}_{2,0}(y), \mathcal{N}_{2,1}(y), \mathcal{N}_{2,2}(y), \ldots$	A-number	Name
1	$1, 1, 2, 5, 14, 42, \ldots$	A000108	Catalan numbers
2	$1, 2, 6, 22, 90, 394, \ldots$	A006318	Large Schröder numbers
3	$1, 3, 12, 57, 300, 1686, \ldots$	A047891	
4	$1, 4, 20, 116, 740, 5028, \ldots$	A082298	

y	$\mathcal{N}_{3,0}(y), \mathcal{N}_{3,1}(y), \mathcal{N}_{3,2}(y), \ldots$	A-number	Name
1	$1, 2, 5, 14, 42, 132, \ldots$	A000108(n+1)	shifted Catalan numbers
2	$1, 3, 11, 45, 197, 903, \ldots$	A001003(n+1)	shifted little Schröder numbers
3	$1, 4, 19, 100, 562, 3304, \ldots$	A007564(n+1)	
4	$1, 5, 29, 185, 1257, 8925, \ldots$	A059231(n+1)	

We can derive a moment representation for these polynomials using the generating functions above and the Stieltjes-Perron transform. We obtain the following:

Proposition 9.23 *The families of polynomials*

$$(\mathcal{N}_{1,n}(y))_{n\geq 0}, \quad (\mathcal{N}_{2,n}(y))_{n\geq 0}, \quad (\mathcal{N}_{3,n}(y))_{n\geq 0}$$

are each a family of moments corresponding to an associated family of orthogonal functions.

Proof Using the established generating functions $\phi_1(x, y)$, $\phi_2(x, y)$ and $\phi_3(x, y)$, and the Stieltjes-Perron transform, we can establish the following moment representations, for the densities shown.

$$\mathcal{N}_{1,n}(y) = \frac{y-1}{y}0^n + \frac{1}{2\pi}\int_{y-2\sqrt{y}+1}^{y+2\sqrt{y}+1} x^n \frac{\sqrt{-x^2+2x(1+y)-(1-y)^2}}{2y}dx,$$

$$\mathcal{N}_{2,n}(y) = \frac{1}{2\pi}\int_{y-2\sqrt{y}+1}^{y+2\sqrt{y}+1} x^n \frac{\sqrt{-x^2+2x(1+y)-(1-y)^2}}{x}dx,$$

$$\mathcal{N}_{3,n}(y) = \frac{1}{2\pi}\int_{y-2\sqrt{y}+1}^{y+2\sqrt{y}+1} x^n \frac{\sqrt{-x^2+2x(1+y)-(1-y)^2}}{y}dx.$$

The associated orthogonal polynomials are determined by the densities shown. □

Using the theory developed in [4, 7], we can exhibit these families of polynomials as the first columns of three related Riordan arrays. More precisely, we have

Proposition 9.24 *The elements of the three families of polynomials $(\mathcal{N}_{1,n}(y))_{n\ge0}$, $(\mathcal{N}_{2,n}(y))_{n\ge0}$, $(\mathcal{N}_{3,n}(y))_{n\ge0}$ are given by the first column of the inverse Riordan arrays given by $\left(\frac{1}{1+x}, \frac{x}{(1+x)(1+yx)}\right)$, $\left(\frac{1}{1+yx}, \frac{x}{(1+x)(1+yx)}\right)$, and $\left(\frac{1}{(1+x)(1+yx)}, \frac{x}{(1+x)(1+yx)}\right)$, respectively. These Riordan arrays are the coefficient arrays of the corresponding families of orthogonal polynomials. Thus*

$$\mathcal{N}_{1,n}(y) \quad \textit{is given by the first column of} \quad \left(\frac{1}{1+x}, \frac{x}{(1+x)(1+yx)}\right)^{-1},$$

$$\mathcal{N}_{2,n}(y) \quad \textit{is given by the first column of} \quad \left(\frac{1}{1+yx}, \frac{x}{(1+x)(1+yx)}\right)^{-1}$$

$$\mathcal{N}_{3,n}(y) \quad \textit{is given by the first column of} \quad \left(\frac{1}{(1+x)(1+yx)}, \frac{x}{(1+x)(1+yx)}\right)^{-1}.$$

Proof We look at the case of $\mathcal{N}_{1,n}$, as the other cases are proved in similar manner. Thus we let

$$\left(\frac{1}{1+x}, \frac{x}{(1+x)(1+yx)}\right) = (g, f).$$

We wish then to show that

$$\phi_1(x, y) = \frac{1}{g(\text{Rev}_x f(x, y))}.$$

For $f(x, y) = \frac{x}{(1+x)(1+yx)}$, we find that

$$\text{Rev}_x f(x, y) = \frac{1 - x(1+y) - \sqrt{1-2x(1+y)+x^2(1-y)^2}}{2xy}.$$

Then since $g(x) = \frac{1}{1+x}$, we find that

$$\frac{1}{g(\text{Rev}_x f(x,y))} = 1 + \text{Rev}_x f(x,y)$$
$$= 1 + \frac{1 - x(1+y) - \sqrt{1 - 2x(1+y) + x^2(1-y)^2}}{2xy} = \phi_1(x,y)$$

as required. Now

$$\left(\frac{1}{1+x}, \frac{x}{(1+x)(1+yx)}\right) = \left(\frac{1+yx}{(1+x)(1+yx)}, \frac{x}{(1+x)(1+yx)}\right)$$
$$= \left(\frac{1+yx}{1+(1+y)x+yx^2}, \frac{x}{1+(1+y)x+yx^2}\right),$$

and hence $\left(\frac{1}{1+x}, \frac{x}{(1+x)(1+yx)}\right)$ is the coefficient array of a family of orthogonal polynomials. □

The Eulerian numbers

As with the Narayana numbers and their associated number triangles, we can distinguish between three distinct but related triangles of Eulerian numbers. Thus we have the triangle A173018 [19, 22]

$$\begin{bmatrix} 1 & 0 & 0 & 0 & 0 & 0 & \dots \\ 1 & 0 & 0 & 0 & 0 & 0 & \dots \\ 1 & 1 & 0 & 0 & 0 & 0 & \dots \\ 1 & 4 & 1 & 0 & 0 & 0 & \dots \\ 1 & 11 & 11 & 1 & 0 & 0 & \dots \\ 1 & 26 & 66 & 26 & 1 & 0 & \dots \\ \vdots & \vdots & \vdots & \vdots & \vdots & \vdots & \ddots \end{bmatrix}$$

of Eulerian numbers $W_{n,k}$ that obey the recurrence

$$W_{n,k} = (k+1)W_{n-1,k} + (n-k)W_{n-1,k-1}$$

with appropriate boundary conditions, for which the closed form expression

$$W_{n,k} = \sum_{i=0}^{n-k} (-1)^i \binom{n+1}{i} (n-k-i)^n$$

holds. We have the reversal of this triangle, which is the triangle A123125 of the coefficients $A_{n,k}$ [1] where

$$A_{n,k}=\sum_{i=0}^{k}(-1)^i\binom{n+1}{i}(k-i)^n,$$

which begins

$$\begin{bmatrix} 1 & 0 & 0 & 0 & 0 & 0 & \dots \\ 0 & 1 & 0 & 0 & 0 & 0 & \dots \\ 0 & 1 & 1 & 0 & 0 & 0 & \dots \\ 0 & 1 & 4 & 1 & 0 & 0 & \dots \\ 0 & 1 & 11 & 11 & 1 & 0 & \dots \\ 0 & 1 & 26 & 66 & 26 & 1 & \dots \\ \vdots & \vdots & \vdots & \vdots & \vdots & \vdots & \ddots \end{bmatrix},$$

and finally we have the Pascal-like triangle of coefficients

$$\tilde{A}_{n,k}=A_{n+1,k+1}=\sum_{i=0}^{k+1}(-1)^i\binom{n+2}{i}(k-i)^{n+1},$$

which begins

$$\begin{bmatrix} 1 & 0 & 0 & 0 & 0 & 0 & \dots \\ 1 & 1 & 0 & 0 & 0 & 0 & \dots \\ 1 & 4 & 1 & 0 & 0 & 0 & \dots \\ 1 & 11 & 11 & 1 & 0 & 0 & \dots \\ 1 & 26 & 66 & 26 & 1 & 0 & \dots \\ 1 & 57 & 302 & 302 & 57 & 1 & \dots \\ \vdots & \vdots & \vdots & \vdots & \vdots & \vdots & \ddots \end{bmatrix}.$$

This is A008292.

We have

$$\tilde{A}_{n,k}=(n-k+1)\tilde{A}_{n-1,k-1}+(k+1)\tilde{A}_{n-1,k},$$

with appropriate boundary conditions. As with the Narayana numbers, each of these triangles has significant combinatorial applications and it is often important to distinguish one from the other.

Example 9.8 The sequence $a_n=\sum_{k=0}^{n}W_{n,k}2^k$ is the sequence A000670 of preferential arrangements, or rankings of competitors in a race, with ties [30]. The sequence

$$b_n=\sum_{k=0}^{n}A_{n,k}2^k=\sum_{k=0}^{n}W_{n,n-k}2^k$$

or A000629 is the sequence of rankings of competitors in a race, with ties and dropouts [28]. Note that from our results above, the sequence a_n has generating function given by

$$\cfrac{1}{1-x-\cfrac{2x^2}{1-4x-\cfrac{8x^2}{1-7x-\cfrac{18x^2}{1-\cdots}}}}.$$

The generating function of the sequence a_{n+1} is given by

$$\cfrac{1}{1-3x-\cfrac{4x^2}{1-6x-\cfrac{12x^2}{1-9x-\cfrac{24x^2}{1-\cdots}}}}.$$

In this case it happens that b_n is the binomial transform of a_n, and hence [3] its generating function has a continued fraction expression

$$\cfrac{1}{1-2x-\cfrac{2x^2}{1-5x-\cfrac{8x^2}{1-8x-\cfrac{18x^2}{1-\cdots}}}}.$$

We deduce from this that the Hankel transform of b_n is

$$h_n = 2^{\binom{n+1}{2}} \prod_{k=0}^{n} k!^2,$$

which is the sequence A091804 that begins

$$1, 2, 32, 9216, 84934656, 39137889484800, \ldots.$$

The Eulerian polynomials

The Eulerian polynomials [11, 17, 22, 27]

$$P_n(x) = \sum_{k=0}^{n} W_{n,k} x^k$$

form the sequence $P_n(x)$ which begins

$$P_0(x) = 1,\ P_1(x) = 1,\ P_2(x) = 1 + x,\ P_3(x) = 1 + 4x + x^2, \ldots,$$

with the well-known triangle of Eulerian numbers [19]

$$\begin{bmatrix} 1 & 0 & 0 & 0 & 0 & 0 & \dots \\ 1 & 0 & 0 & 0 & 0 & 0 & \dots \\ 1 & 1 & 0 & 0 & 0 & 0 & \dots \\ 1 & 4 & 1 & 0 & 0 & 0 & \dots \\ 1 & 11 & 11 & 1 & 0 & 0 & \dots \\ 1 & 26 & 66 & 26 & 1 & 0 & \dots \\ \vdots & \vdots & \vdots & \vdots & \vdots & \vdots & \ddots \end{bmatrix}$$

as coefficient array. The polynomials $P_n(x)$ were introduced by Euler [15] in the form

$$\sum_{k=0}^{\infty}(k+1)^n t^k = \frac{P_n(t)}{(1-t)^{n+1}}.$$

They have exponential generating function

$$\sum_{n=0}^{\infty} P_n(x)\frac{t^n}{n!} = \frac{(1-x)e^{(1-x)t}}{1-xe^{(1-x)t}}.$$

We consider the sequence with exponential generating function

$$\frac{(\alpha-\beta)e^{(\alpha-\beta)t}}{\alpha-\beta e^{(\alpha-\beta)t}}.$$

This is the sequence that begins

$$1, \alpha, \alpha(\alpha+\beta), \alpha(\alpha^2+4\alpha\beta+\beta^2), \alpha(\alpha^3+11\alpha^2\beta+11\alpha\beta^2+\beta^3), \dots.$$

Setting $\alpha = 1$ and $\beta = x$ gives us the Eulerian polynomials $P_n(x)$. We then have the following proposition.

Proposition 9.25 *The production matrix of the exponential Riordan array*

$$\left[\frac{(\alpha-\beta)e^{(\alpha-\beta)t}}{\alpha-\beta e^{(\alpha-\beta)t}}, \frac{e^{(\alpha-\beta)t}-1}{\alpha-\beta e^{(\alpha-\beta)t}}\right]$$

is tri-diagonal.

Proof Writing the above exponential Riordan array as $[g, f]$, we have

$$f(t) = \frac{e^{(\alpha-\beta)t}-1}{\alpha-\beta e^{(\alpha-\beta)t}}$$

and hence

$$f'(t) = \frac{e^{(\alpha+\beta)t}(\alpha-\beta)^2}{\beta e^{\alpha t} - \alpha e^{\beta t}},$$

and

$$\bar{f}(t) = \frac{1}{\alpha-\beta}\ln\left(\frac{\alpha t+1}{\beta t+1}\right).$$

Then

$$A(t) = f'(\bar{f}(t)) = (\alpha t+1)(\beta t+1) = 1 + (\alpha+\beta)t + \alpha\beta t^2.$$

We have

$$g(t) = \frac{(\alpha-\beta)e^{(\alpha-\beta)t}}{\alpha - \beta e^{(\alpha-\beta)t}}$$

and hence

$$g'(t) = \frac{\alpha e^{(\alpha+\beta)t}(\alpha-\beta)^2}{(\beta e^{\alpha t} - \alpha e^{\beta t})^2},$$

and so

$$Z(t) = \frac{g'(\bar{f}(t))}{g(\bar{f}(t))} = \alpha(\beta t+1) = \alpha + \alpha\beta t.$$

Thus the production matrix, which has bivariate generating function given by

$$e^{ty}(\alpha + \alpha\beta t + (1 + (\alpha+\beta)t + \alpha\beta t^2)y),$$

is tri-diagonal. □

The production matrix takes the form

$$\left[\begin{array}{ccccccc}
\alpha & 1 & 0 & 0 & 0 & 0 & \cdots \\
\alpha\beta & 2\alpha+\beta & 1 & 0 & 0 & 0 & \cdots \\
0 & 4\alpha\beta & 3\alpha+2\beta & 1 & 0 & 0 & \cdots \\
0 & 0 & 9\alpha\beta & 4\alpha+3\beta & 1 & 0 & \cdots \\
0 & 0 & 0 & 16\alpha\beta & 5\alpha+4\beta & 1 & \cdots \\
0 & 0 & 0 & 0 & 25\alpha\beta & 6\alpha+5\beta & \cdots \\
\vdots & \vdots & \vdots & \vdots & \vdots & \vdots & \ddots
\end{array}\right].$$

For completeness, we observe that while in the special case $\alpha = \beta$ the Riordan array is not obviously well-defined, the production matrix is, and it leads in this special case to the exponential Riordan array

$$\left[\frac{1}{1-\alpha t}, \frac{t}{1-\alpha t}\right]$$

which has general element $\binom{n}{k}\frac{n!}{k!}\alpha^{n-k}$. In the case $\alpha = \beta = 1$, we get the exponential Riordan array

$$\left[\frac{1}{1-t}, \frac{t}{1-t}\right]$$

whose inverse is the coefficient array of the Laguerre polynomials [2].

Returning now to the Eulerian polynomials, we set $\alpha = 1$ and $\beta = x$, to get

Theorem 9.3 *The Eulerian polynomials $P_n(x)$ are the moments of the family of orthogonal polynomials $Q_n(t)$ defined by $Q_0(t) = 1$, $Q_1(t) = t - 1$, and*

$$Q_n(t) = (t - ((n-1)x + n))Q_{n-1}(t) - (n-1)^2 x Q_{n-2}(t).$$

Proof The initial polynomial terms of the sequence $Q_n(t)$ can be read from the elements of

$$\left[\frac{(1-x)e^{(1-x)t}}{1-xe^{(1-x)t}}, \frac{e^{(1-x)t}-1}{1-xe^{(1-x)t}}\right]^{-1} = \left[\frac{1}{1+t}, \frac{1}{1-x}\ln\left(\frac{1+t}{1+xt}\right)\right],$$

which begins

$$\begin{bmatrix} 1 & 0 & 0 & 0 \dots \\ -1 & 1 & 0 & 0 \dots \\ 2 & -x-3 & 1 & 0 \dots \\ -6 & 2x^2+5x+11 & -3(x+2) & 1 \dots \\ \vdots & \vdots & \vdots & \vdots \ddots \end{bmatrix}.$$

Hence in particular $Q_0(t) = 1$ and $Q_1(t) = t - 1$. The three-term recurrence is derived from the production matrix, which in this case is

$$\begin{bmatrix} 1 & 1 & 0 & 0 & 0 & 0 & \dots \\ x & 2+x & 1 & 0 & 0 & 0 & \dots \\ 0 & 4x & 3+2x & 1 & 0 & 0 & \dots \\ 0 & 0 & 9x & 4+3x & 1 & 0 & \dots \\ 0 & 0 & 0 & 16x & 5+4x & 1 & \dots \\ 0 & 0 & 0 & 0 & 25x & 6+5x & \dots \\ \vdots & \vdots & \vdots & \vdots & \vdots & \vdots & \ddots \end{bmatrix}$$

and we have the proof. □

Corollary 9.6 *The sequence of Eulerian polynomials $P_n(x)$ has ordinary generating function given by the continued fraction*

$$\cfrac{1}{1-t-\cfrac{xt^2}{1-(2+x)t-\cfrac{4xt^2}{1-(3+2x)t-\cfrac{9xt^2}{1-\cdots}}}}.$$

Corollary 9.7 *The Hankel transform of the sequence of Eulerian polynomials $P_n(x)$ is given by*

$$h_n = x^{\binom{n+1}{2}} \prod_{k=1}^{n} k!^2.$$

The shifted Eulerian polynomials $P_{n+1}(x)$

For the shifted Eulerian polynomials $P_{n+1}(x)$, we consider the exponential Riordan array

$$[g'(t), f(t)],$$

where

$$g'(t) = \frac{(\alpha-\beta)^2 e^{(\alpha+\beta)t}}{\beta e^{\alpha t} - \alpha e^{\beta t}},$$

where we retain the use of $g(t) = \frac{(\alpha-\beta)e^{(\alpha-\beta)t}}{\alpha-\beta e^{(\alpha-\beta)t}}$ from the previous section.

When $\alpha = 1$ and $\beta = x$, $g'(t)$ generates the shifted sequence $P_{n+1}(x)$. We then have

Proposition 9.26 *The production matrix of the exponential Riordan array*

$$\left[\frac{(\alpha-\beta)^2 e^{(\alpha+\beta)t}}{\beta e^{\alpha t} - \alpha e^{\beta t}}, \frac{e^{(\alpha-\beta)t}-1}{\alpha-\beta e^{(\alpha-\beta)t}}\right]$$

is tri-diagonal.

Proof As in the previous proposition, we obtain

$$A(t) = f'(\bar{f}(t)) = (\alpha t+1)(\beta t+1) = 1 + (\alpha+\beta)t + \alpha\beta t^2,$$

where

$$\bar{f}(t) = \frac{1}{\alpha-\beta} \ln\left(\frac{\alpha t+1}{\beta t+1}\right).$$

Then

$$Z(t) = \frac{g''(\bar{f}(t))}{g'(\bar{f}(t))} = (\alpha+\beta) + 2\alpha\beta t.$$

The bivariate generating function of the production matrix is then

$$e^{ty}((\alpha+\beta)+2\alpha\beta t+(1+(\alpha+\beta)t+\alpha\beta t^2)y),$$

and hence the production matrix is tri-diagonal. □

The production matrix in this case begins

$$\begin{bmatrix} \alpha+\beta & 1 & 0 & 0 & \dots \\ 2\alpha\beta & 2(\alpha+\beta) & 1 & 0 & \dots \\ 0 & 6\alpha\beta & 3(\alpha+\beta) & 1 & \dots \\ 0 & 0 & 12\alpha\beta & 4(\alpha+\beta) & \dots \\ \vdots & \vdots & \vdots & \vdots & \ddots \end{bmatrix}.$$

In the case $\alpha=\beta$, we obtain the exponential Riordan array

$$\left[\frac{1}{(1-\alpha t)^2}, \frac{t}{1-\alpha t}\right],$$

with (n,k)th element $\binom{n+1}{k+1}\frac{n!}{k!}\alpha^{n-k}$. For $\alpha=\beta=1$ this gives us

$$\left[\frac{1}{(1-t)^2}, \frac{t}{1-t}\right],$$

which is A105278.

Specializing to the values $\alpha=1$, $\beta=x$, we get the following result.

Theorem 9.4 *The shifted Eulerian polynomials $P_{n+1}(x)$ are the moments of the family of orthogonal polynomials $R_n(t)$ given by $R_0(t)=1$, $R_1(t)=t-x-1$, and for $n>1$,*

$$R_n(t)=(t-n(1+x))R_{n-1}(t)-n(n-1)xR_{n-2}(t).$$

Proof The initial terms of the polynomial sequence $R_n(t)$ can be read from the elements of the inverse matrix

$$\left[\frac{(\alpha-\beta)^2e^{(\alpha+\beta)t}}{\beta e^{\alpha t}-\alpha e^{\beta t}}, \frac{e^{(1-x)t}-1}{1-xe^{(1-x)t}}\right]^{-1}=\left[\frac{1}{(1+t)(1+tx)}, \frac{1}{1-x}\ln\left(\frac{1+t}{1+xt}\right)\right],$$

which begins

$$\begin{bmatrix} 1 & 0 & 0 & 0 & \dots \\ -x-1 & 1 & 0 & 0 & \dots \\ 2x^2+2x+2 & -3(x+1) & 1 & 0 & \dots \\ -6(x^3+x^2+x+1) & 11x^2+14x+11 & -6(x+1) & 1 & \dots \\ \vdots & \vdots & \vdots & \vdots & \ddots \end{bmatrix}.$$

The three-term recurrence is derived from the production matrix, which in this case is

$$\begin{bmatrix} 1+x & 1 & 0 & 0 & 0 & 0 & \cdots \\ 2x & 2(1+x) & 1 & 0 & 0 & 0 & \cdots \\ 0 & 6x & 3(1+x) & 1 & 0 & 0 & \cdots \\ 0 & 0 & 12x & 4(1+x) & 1 & 0 & \cdots \\ 0 & 0 & 0 & 20x & 5(1+x) & 1 & \cdots \\ 0 & 0 & 0 & 0 & 30x & 6(1+x) & \cdots \\ \vdots & \vdots & \vdots & \vdots & \vdots & \vdots & \ddots \end{bmatrix}$$

and we have the proof. □

Corollary 9.8 *The sequence of shifted Eulerian polynomials $P_{n+1}(x)$ has ordinary generating function given by the continued fraction*

$$\cfrac{1}{1-(1+x)t-\cfrac{2xt^2}{1-2(1+x)t-\cfrac{6xt^2}{1-3(1+x)t-\cfrac{12xt^2}{1-\cdots}}}}.$$

Corollary 9.9 *The Hankel transform of the shifted Eulerian polynomials $P_{n+1}(x)$ is given by*

$$h_n = (2x)^{\binom{n+1}{2}} \prod_{k=1}^{n} \binom{k+2}{2}^{n-k}.$$

In this chapter we have shown that both ordinary and exponential Riordan arrays define families of orthogonal polynomials when their production matrices are tridiagonal. In the case of ordinary Riordan arrays, these polynomials are generalized Chebyshev polynomials of the second kind. In the exponential case, we encounter such orthogonal polynomials as the Laguerre and Hermite polynomials. By parameterizing the Riordan arrays, we can obtain interesting polynomial sequences as the corresponding moment sequences, or indeed, other families of orthogonal polynomials.

9.5 Continued Fractions and Riordan Arrays

In the foregoing sections of this chapter, we have had occasion to use a variety of continued fractions, many related to moment sequences corresponding to families of orthogonal polynomials. We now look at some further links between the theory of continued fractions and Riordan arrays. The types of continued fractions that we

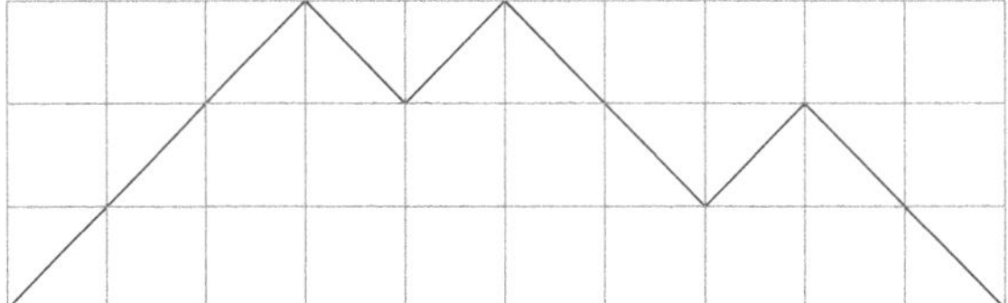

Fig. 9.1 A Dyck path

shall consider are related to lattice paths in the plane. We will be concerned with three types of lattice paths: Dyck paths, Motzkin paths, and Schröder paths.

A *Dyck path* (see Fig. 9.1) is a path in the first quadrant that begins at the origin $(0, 0)$, ends at $(2n, 0)$, and consist of steps $(1, 1)$ (North-East), called *rises* or *up steps*, and $(1, -1)$ (South-East), called *falls* or *down steps*. We refer to n as the *semi-length* of the path. Dyck paths of semi-length n are sometimes referred to as Dyck n-paths. A *peak* of a Dyck path is the joint node formed by a rise step immediately followed by a fall step. The *height* of a peak is the y-coordinate of this node.

Example 9.9 The Riordan array $(1, xc(x))$, where $c(x) = \frac{1-\sqrt{1-4x}}{2x}$ is the generating function of the Catalan numbers. The (n, k)th element of this array gives the number of Dyck paths of semi-length n that have k contact points on the x-axis. (see Fig. 9.2). The array $(1, xc(x))$ begins

$$\begin{bmatrix} 1 & 0 & 0 & 0 & 0 & 0 \\ 0 & 1 & 0 & 0 & 0 & 0 \\ 0 & 1 & 1 & 0 & 0 & 0 \\ 0 & 2 & 2 & 1 & 0 & 0 \\ 0 & 5 & 5 & 3 & 1 & 0 \\ 0 & 14 & 14 & 9 & 4 & 1 \end{bmatrix}.$$

It has a production matrix that begins

$$\begin{bmatrix} 0 & 1 & 0 & 0 & 0 & 0 \\ 0 & 1 & 1 & 0 & 0 & 0 \\ 0 & 1 & 1 & 1 & 0 & 0 \\ 0 & 1 & 1 & 1 & 1 & 0 \\ 0 & 1 & 1 & 1 & 1 & 1 \\ 0 & 1 & 1 & 1 & 1 & 1 \end{bmatrix}.$$

A *Motzkin path* (see Fig. 9.3) is a path in the first quadrant which begins at the origin $(0, 0)$, ends at $(n, 0)$, and consists of steps $(1, 1)$ (North-East), called *rises*, and (1,-1) (South-East), called *falls*, and steps $(1, 0)$ (East) called *horizontals*. A partial Motzkin path that starts from $(0, 0)$ and ends at the point (n, k) (not necessarily on the x-axis) is called a *left factor* of a Motzkin path.

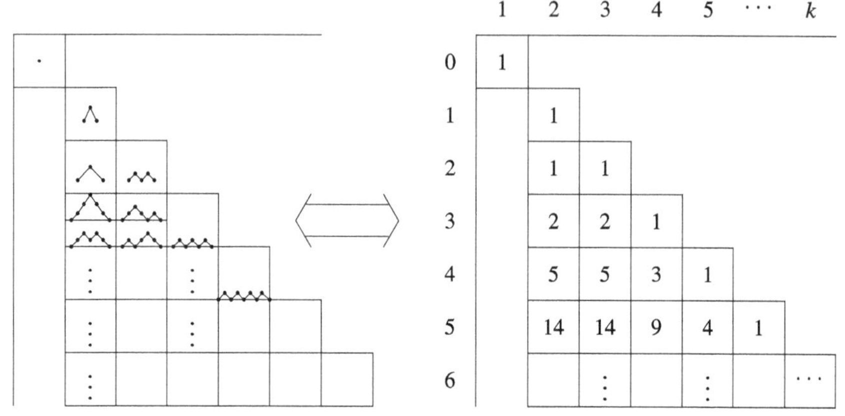

Fig. 9.2 Dyck paths counted per length and per x-axis points

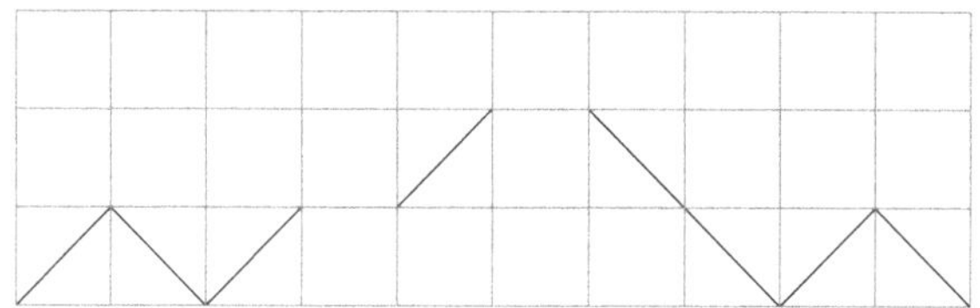

Fig. 9.3 A Motzkin path

The Riordan array

$$\left(\frac{1}{1+x+x^2}, \frac{x}{1+x+x^2}\right)^{-1} =$$

$$\left(\frac{1-x-\sqrt{1-2x-3x^2}}{2x^2}, \frac{1-x-\sqrt{1-2x-3x^2}}{2x}\right)$$

counts the number of Motzkin left factors from $(0, 0)$ to (n, k). See Fig. 9.4. This array begins

$$\begin{bmatrix} 1 & 0 & 0 & 0 & 0 & 0 \\ 1 & 1 & 0 & 0 & 0 & 0 \\ 2 & 2 & 1 & 0 & 0 & 0 \\ 4 & 5 & 3 & 1 & 0 & 0 \\ 9 & 12 & 9 & 4 & 1 & 0 \\ 21 & 30 & 25 & 14 & 5 & 1 \end{bmatrix}.$$

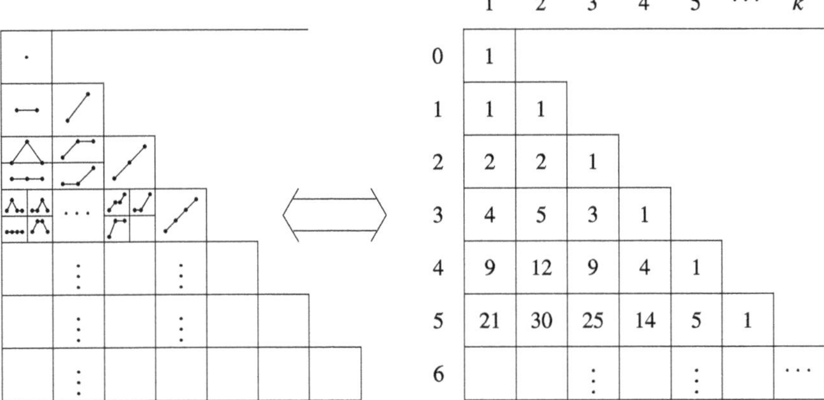

Fig. 9.4 Motzkin left factors to (n, k)

This is a member of the Bell subgroup of the Riordan group, and has a production matrix that begins

$$\begin{bmatrix} 1 & 1 & 0 & 0 & 0 & 0 \\ 1 & 1 & 1 & 0 & 0 & 0 \\ 0 & 1 & 1 & 1 & 0 & 0 \\ 0 & 0 & 1 & 1 & 1 & 0 \\ 0 & 0 & 0 & 1 & 1 & 1 \\ 0 & 0 & 0 & 0 & 1 & 1 \end{bmatrix}.$$

The first column of the above Riordan array consists of the Motzkin numbers. The Motzkin numbers count Motzkin paths of length n. These are then the moments for the family of orthogonal polynomials whose coefficient array is given by the Riordan array $\left(\frac{1}{1+x+x^2}, \frac{x}{1+x+x^2}\right)$.

The (n, k)th term of the exponential Riordan array $\left[\frac{I_1(2x)}{x}, x\right]$ counts the number of Motzkin paths of length n which have k horizontal steps.

Finally a *Schröder path* (see Fig. 9.5) is a path in the first quadrant which begins at the origin $(0, 0)$, ends at $(2n, 0)$, and consists of steps $(1, 1)$ (North-East), called *rises*, and $(1, -1)$ (South-East) called *falls*, and steps $(2, 0)$ (East) called *horizontals* or *level steps*.

Example 9.10 The Riordan array

$$\left(\frac{1-2x}{1-x}, \frac{x(1-2x)}{1-x}\right)^{-1} =$$

$$\left(\frac{1+x-\sqrt{1-6x+x^2}}{4x}, \frac{1+x-\sqrt{1-6x+x^2}}{4}\right)$$

Fig. 9.5 A Schröder path

counts the number of Schröder paths of length $2n$ that have k peaks at height 1. This member of the Bell subgroup of the Riordan group begins

$$\left[\begin{array}{cccccc} 1 & 0 & 0 & 0 & 0 & 0 \\ 1 & 1 & 0 & 0 & 0 & 0 \\ 3 & 2 & 1 & 0 & 0 & 0 \\ 11 & 7 & 3 & 1 & 0 & 0 \\ 45 & 28 & 12 & 4 & 1 & 0 \\ 197 & 121 & 52 & 18 & 5 & 1 \end{array}\right],$$

with a production matrix that begins

$$\left[\begin{array}{cccccc} 1 & 1 & 0 & 0 & 0 & 0 \\ 2 & 1 & 1 & 0 & 0 & 0 \\ 4 & 2 & 1 & 1 & 0 & 0 \\ 8 & 4 & 2 & 1 & 1 & 0 \\ 16 & 8 & 4 & 2 & 1 & 1 \\ 32 & 16 & 8 & 4 & 2 & 1 \end{array}\right].$$

Dyck paths of semi-length n are counted by the Catalan numbers $C_n = \frac{1}{n+1}\binom{2n}{n}$, since they both obey the same recurrence relation. The generating function of the Catalan numbers, $c(x) = \frac{1-\sqrt{1-4x}}{2x}$, can be expressed as a continued fraction as follows.

$$c(x) = \cfrac{1}{1 - \cfrac{x}{1 - \cfrac{x}{1 - \cdots}}}.$$

We can see this by solving the equation

$$u = \frac{1}{1 - xu},$$

equivalent to the continued fraction expression. We obtain

$$u = \frac{1-\sqrt{1-4x}}{2x} \quad \text{or} \quad u = \frac{1+\sqrt{1-4x}}{2x}.$$

The first form satisfies $u(0) = 0$, so this is the solution we seek.

Continued fractions of the form

$$\cfrac{1}{1-\cfrac{ax}{1-\cfrac{bx}{1-\cfrac{cx}{1-\cdots}}}}$$

are known as *Stieltjes continued fractions*. Flajolet [16] showed that

$$\cfrac{1}{1-\cfrac{\alpha_1 x}{1-\cfrac{\alpha_2 x}{1-\cdots}}} = \sum_{n=0}^{\infty} S_n(\alpha_1, \ldots, \alpha_n)x^n,$$

where $S_n(\alpha_1, \ldots, \alpha_n)$ is the generating function for Dyck paths of semi-length n in which each fall starting at height i has weight α_i (rises have weight 1).

Motzkin paths of length n are counted by the Motzkin numbers M_n, where

$$M_n = \sum_{k=0}^{\lfloor \frac{n}{2} \rfloor} \binom{n}{2k} C_k.$$

The generating function $m(x)$ of the Motzkin numbers is given by

$$m(x) = \frac{1-x-\sqrt{1-2x-3x^2}}{2x^2}.$$

We have

$$m(x) = \cfrac{1}{1-x-\cfrac{x^2}{1-x-\cfrac{x^2}{1-x-\cdots}}}.$$

Continued fractions of the form

$$\cfrac{1}{1-\alpha_0 x-\cfrac{\beta_1 x^2}{1-\alpha_1 x-\cfrac{\beta_2 x^2}{1-\alpha_2 x-\cdots}}}$$

are called *Jacobi continued fractions*. Flajolet [16] has shown that

$$\cfrac{1}{1-\alpha_0 x-\cfrac{\beta_1 x^2}{1-\alpha_1 x-\cfrac{\beta_2 x^2}{1-\alpha_2 x-\cdots}}}=\sum_{n=0}^{\infty} J_n(\alpha,\beta)x^n,$$

where $J_n(\alpha, \beta)$ is the generating function for Motzkin paths of length n in which each fall at height i has weight α_i, each horizontal step at height i has weight β_i, and rises have weight 1.

Schröder paths of length $2n$ are counted by the large Schröder numbers

$$S_n=\sum_{k=0}^{n}\binom{n+k}{2k}C_k.$$

The sequence S_n has its generating function $s(x)$ given by

$$s(x)=\frac{1-x-\sqrt{1-6x+x^2}}{2x}.$$

We can express this as the following continued fraction.

$$s(x)=\cfrac{1}{1-x-\cfrac{x}{1-x-\cfrac{x}{1-x-\cdots}}}.$$

This can be seen by solving the equation

$$u=\frac{1}{1-x-xu}$$

and taking the solution with $u(0)=0$.

Continued fractions of the form

$$\cfrac{1}{1-ax-\cfrac{bx}{1-cx-\cfrac{dx}{1-ex-\cdots}}}$$

are called Thron or T-continued fractions. We then have the following result [24, 34, 35]

$$\cfrac{1}{1-\alpha_0 x-\cfrac{\beta_1 x}{1-\alpha_1 x-\cfrac{\beta_2 x}{1-\alpha_2 x-\cdots}}}=\sum_{n=0}^{\infty} T_n(\alpha,\beta)x^n,$$

where $T_n(\alpha, \beta)$ is the generating function for Schröder paths of length $2n$ in which each rise has weight 1, each horizontal step at height i has weight α_i, and each fall from height i has weight β_i.

We are now in a position to describe certain families of Riordan arrays which are defined by Stieltjes, Jacobi, and Thron continued fractions.

Proposition 9.27 *The number triangle with bivariate generating function given by the Stieltjes continued fraction*

$$\cfrac{1}{1-\cfrac{(y+a)x}{1-\cfrac{bx}{1-\cfrac{ax}{1-\cfrac{bx}{1-\cdots}}}}}$$

corresponds to the Riordan array

$$\left(\frac{1}{1+ax}, \frac{x}{1+(a+b)x+abx^2}\right)^{-1}.$$

Proof The proof consists in finding closed form expressions for the continued fraction and for the bivariate generating function of $\left(\frac{1}{1+ax}, \frac{x}{1+(a+b)x+abx^2}\right)^{-1}$, and showing that they are equal. To find a closed form expression for the continued fraction, we first solve the equation

$$u=\cfrac{1}{1-\cfrac{bx}{1-axu}},$$

and then the expression we seek is equivalent to

$$\frac{1}{1-(y+a)xu}.$$

We find that

$$u=\frac{1-(b-a)x-\sqrt{1-2(a+b)x+(a-b)^2x^2}}{2ax},$$

from which it follows that the closed form for the continued fraction is given by

$$\frac{2a}{(y+a)\sqrt{1-2(a+b)x+(a-b)^2x^2}+(y+a)(b-a)x-y+a}. \tag{9.5.1}$$

Now the Riordan array $\left(\frac{1}{1+ax}, \frac{x}{1+(a+b)x+abx^2}\right)^{-1}$ is given by

$$\left(\frac{1-(a-b)x-\sqrt{1-2(a+b)x+(a-b)^2x^2}}{2bx}, \frac{1-(a+b)x-\sqrt{1-2(a+b)x+(a-b)^2x^2}}{2bx}\right).$$

From this we can calculate the bivariate generating function of this array, which turns out to be the same as in expression (9.5.1). □

Example 9.11 We consider the Riordan array $\left(\frac{1}{1+x}, \frac{1}{1+3x+2x^2}\right)^{-1}$ corresponding to $a=1, b=2$. This matrix begins

$$\begin{bmatrix} 1 & 0 & 0 & 0 & 0 & 0 \\ 1 & 1 & 0 & 0 & 0 & 0 \\ 3 & 4 & 1 & 0 & 0 & 0 \\ 11 & 17 & 7 & 1 & 0 & 0 \\ 45 & 76 & 40 & 10 & 1 & 0 \\ 197 & 353 & 216 & 72 & 13 & 1 \end{bmatrix}.$$

Corresponding to $y=1$, we have

$$\begin{bmatrix} 1 & 0 & 0 & 0 & 0 & 0 \\ 1 & 1 & 0 & 0 & 0 & 0 \\ 3 & 4 & 1 & 0 & 0 & 0 \\ 11 & 17 & 7 & 1 & 0 & 0 \\ 45 & 76 & 40 & 10 & 1 & 0 \\ 197 & 353 & 216 & 72 & 13 & 1 \end{bmatrix} \begin{bmatrix} 1 \\ 1 \\ 1 \\ 1 \\ 1 \\ 1 \end{bmatrix} = \begin{bmatrix} 1 \\ 2 \\ 8 \\ 36 \\ 172 \\ 852 \end{bmatrix}.$$

Thus the sequence

$$1, 2, 8, 36, 172, 852, \ldots$$

counts Dyck paths where the fall steps at level 0 come in $1+1=2$ colors, and then at subsequent levels the fall steps alternate between having 2 colors and 1 color.

The generating function of this sequence is given by $\frac{1}{x+\sqrt{1-6x+x^2}}$. Corresponding to $y=2$, we have

$$\begin{bmatrix} 1 & 0 & 0 & 0 & 0 & 0 \\ 1 & 1 & 0 & 0 & 0 & 0 \\ 3 & 4 & 1 & 0 & 0 & 0 \\ 11 & 17 & 7 & 1 & 0 & 0 \\ 45 & 76 & 40 & 10 & 1 & 0 \\ 197 & 353 & 216 & 72 & 13 & 1 \end{bmatrix} \begin{bmatrix} 1 \\ 2 \\ 4 \\ 8 \\ 16 \\ 32 \end{bmatrix} = \begin{bmatrix} 1 \\ 3 \\ 15 \\ 81 \\ 453 \\ 2583 \end{bmatrix}.$$

Thus the sequence that begins

$$1, 3, 15, 81, 453, 2583, \ldots$$

counts Dyck paths where the fall steps at level 0 come in $2+1=3$ colors, and then at subsequent levels the fall steps alternate between having 2 colors and 1 color.

We now consider the case of Jacobi continued fractions. We have the following result.

Proposition 9.28 *The generating function $G(x, y)$, expressed as a Jacobi continued fraction, given by*

$$G(x,y) = \cfrac{1}{1-(y+a)x - \cfrac{bx^2}{1-cx-\cfrac{dx^2}{1-ax-\cfrac{bx^2}{1-cx-\cdots}}}}$$

is the bivariate generating function of the Riordan array $(g(x), xg(x))$ where we have

$$g(x) = \left(\frac{1-cx}{1-(a+c)x+(ac-b+d)x^2}, \frac{dx^2(1-(a+c)x+acx^2)}{(1-(a+c)x+(ac-b+d)x^2)^2}\right) \cdot c(x).$$

Proof We express the generating function, given by the continued fraction, in closed form by solving the equation

$$u = \cfrac{1}{1-(y+a)x-\cfrac{bx^2}{1-cx-dx^2u}}.$$

Comparing this with the bivariate generating function $\frac{g(x)}{1-yxg(x)}$ of $(g(x), xg(x))$ shows them to be the same. This bivariate generating function can be expressed as follows.

$$\frac{1-(y+a+c)x+(c(y+a)-b+d)x^2-\sqrt{-4dx^2(1-cx)(1-(y+a)x)+(1-(y+a+c)x+(c(y+a)-b+d)x^2)^2}}{2dx^2(1-(y+a)x)}.$$

□

Example 9.12 We consider the case $a=b=1, c=d=2$. We find that

$$g(x)=\frac{1-3x+3x^2-\sqrt{1-6x+7x^2+6x^2-7x^4}}{4x^2(1-x)}.$$

The Riordan array $(g(x), xg(x))$ then begins

$$\begin{bmatrix} 1 & 0 & 0 & 0 & 0 & 0 \\ 1 & 1 & 0 & 0 & 0 & 0 \\ 2 & 2 & 1 & 0 & 0 & 0 \\ 5 & 5 & 3 & 1 & 0 & 0 \\ 15 & 14 & 9 & 4 & 1 & 0 \\ 48 & 44 & 28 & 14 & 5 & 1 \end{bmatrix}.$$

We have

$$\begin{bmatrix} 1 & 0 & 0 & 0 & 0 & 0 \\ 1 & 1 & 0 & 0 & 0 & 0 \\ 2 & 2 & 1 & 0 & 0 & 0 \\ 5 & 5 & 3 & 1 & 0 & 0 \\ 15 & 14 & 9 & 4 & 1 & 0 \\ 48 & 44 & 28 & 14 & 5 & 1 \end{bmatrix}\begin{bmatrix} 1 \\ 2 \\ 4 \\ 8 \\ 16 \\ 32 \end{bmatrix}=\begin{bmatrix} 1 \\ 3 \\ 10 \\ 35 \\ 127 \\ 472 \end{bmatrix}.$$

Thus the sequence

$$1, 3, 10, 35, 127, 472, \ldots$$

counts Motzkin paths where the horizontal steps at level 0 have $1+2=3$ colors, and thereafter the horizontal steps at alternate levels have 2 or 1 color, respectively, and the falls have alternately 1 or 2 colors, beginning at level 0. The generating function of this sequence is given by

$$\frac{g(x)}{1-2xg(x)}=\frac{\sqrt{1-6x+7x^2+6x^3-7x^4}-11x^2+7x-1}{4x(1-6x+8x^2)}.$$

Specializing to the case of $c=a, d=b$ we have the following result.

Proposition 9.29 *The generating function $G(x, y)$ for $c=a, d=b$ is the generating function of the moment sequence of the orthogonal polynomials whose coefficient array is given by the Riordan array*

$$\left(\frac{1-xy}{1+ax+bx^2}, \frac{x}{1+ax+bx^2}\right).$$

In effect, the production matrix of the Riordan array $\left(\frac{1-xy}{1+ax+bx^2}, \frac{x}{1+ax+bx^2}\right)^{-1}$ begins

$$\begin{bmatrix} y+a & 1 & 0 & 0 & 0 \\ b & a & 1 & 0 & 0 \\ 0 & b & a & 1 & 0 \\ 0 & 0 & b & a & 1 \\ 0 & 0 & 0 & b & a \end{bmatrix}.$$

Example 9.13 The Riordan array $\left(\frac{1}{1+x+x^2}, \frac{x}{1+x+x^2}\right)^{-1}$ begins

$$\begin{bmatrix} 1 & 0 & 0 & 0 & 0 & 0 \\ 1 & 1 & 0 & 0 & 0 & 0 \\ 2 & 2 & 1 & 0 & 0 & 0 \\ 4 & 5 & 3 & 1 & 0 & 0 \\ 9 & 12 & 9 & 4 & 1 & 0 \\ 21 & 30 & 25 & 14 & 5 & 1 \end{bmatrix}.$$

We have

$$\begin{bmatrix} 1 & 0 & 0 & 0 & 0 & 0 \\ 1 & 1 & 0 & 0 & 0 & 0 \\ 2 & 2 & 1 & 0 & 0 & 0 \\ 4 & 5 & 3 & 1 & 0 & 0 \\ 9 & 12 & 9 & 4 & 1 & 0 \\ 21 & 30 & 25 & 14 & 5 & 1 \end{bmatrix} \begin{bmatrix} 1 \\ 0 \\ 0 \\ 0 \\ 0 \\ 0 \end{bmatrix} = \begin{bmatrix} 1 \\ 1 \\ 2 \\ 4 \\ 9 \\ 21 \end{bmatrix}.$$

Thus the sequence $1, 1, 2, 4, 9, 21, \ldots$ counts Motzkin paths where the falls have weight 1, the horizontals at level 0 have weight $0 + 1 = 1$ and all other horizontals have weight 1. As expected, these are just the Motzkin numbers. We also have

$$\begin{bmatrix} 1 & 0 & 0 & 0 & 0 & 0 \\ 1 & 1 & 0 & 0 & 0 & 0 \\ 2 & 2 & 1 & 0 & 0 & 0 \\ 4 & 5 & 3 & 1 & 0 & 0 \\ 9 & 12 & 9 & 4 & 1 & 0 \\ 21 & 30 & 25 & 14 & 5 & 1 \end{bmatrix} \begin{bmatrix} 1 \\ -1 \\ 1 \\ -1 \\ 1 \\ -1 \end{bmatrix} = \begin{bmatrix} 1 \\ 0 \\ 1 \\ 3 \\ 6 \\ 15 \end{bmatrix}.$$

Thus the sequence

$$1, 0, 1, 3, 6, 15, \ldots$$

counts Motzkin paths where the falls have weight 1, the horizontals at level 0 have weight $-1 + 1 = 0$ (that is, there are no horizontals at level 0), and all other horizontals have weight 1. The numbers $1, 0, 1, 3, 6, 15, \ldots$ are the Riordan numbers, whose generating function is given by $\frac{1+x-\sqrt{1-2x-3x^2}}{2x(1+x)}$. Alternatively, taking $y = -1$, we have that the Riordan array $\left(\frac{1+x}{1+x+x^2}, \frac{x}{1+x+x^2}\right)^{-1}$ begins

$$\left[\begin{array}{cccccc} 1 & 0 & 0 & 0 & 0 & 0 \\ 0 & 1 & 0 & 0 & 0 & 0 \\ 1 & 1 & 1 & 0 & 0 & 0 \\ 1 & 3 & 2 & 1 & 0 & 0 \\ 3 & 6 & 6 & 3 & 1 & 0 \\ 6 & 15 & 15 & 10 & 4 & 1 \end{array}\right],$$

and then we have

$$\left[\begin{array}{cccccc} 1 & 0 & 0 & 0 & 0 & 0 \\ 0 & 1 & 0 & 0 & 0 & 0 \\ 1 & 1 & 1 & 0 & 0 & 0 \\ 1 & 3 & 2 & 1 & 0 & 0 \\ 3 & 6 & 6 & 3 & 1 & 0 \\ 6 & 15 & 15 & 10 & 4 & 1 \end{array}\right] \left[\begin{array}{c} 1 \\ 0 \\ 0 \\ 0 \\ 0 \\ 0 \end{array}\right] = \left[\begin{array}{c} 1 \\ 0 \\ 1 \\ 3 \\ 6 \\ 15 \end{array}\right].$$

Again, the interpretation for $y = -1$ is that there are no horizontals at level 0.

Example 9.14 The sequence that counts directed animals of size n begins

$$1, 1, 2, 5, 13, 35, 96, 267, 750, 2123, \ldots.$$

In the current context, we take $y = 1$. Thus we consider the matrix

$$\left(\frac{1-x}{1+x+x^2}, \frac{x}{1+x+x^2}\right)^{-1}.$$

This matrix begins

$$\left[\begin{array}{cccccc} 1 & 0 & 0 & 0 & 0 & 0 \\ 2 & 1 & 0 & 0 & 0 & 0 \\ 5 & 3 & 1 & 0 & 0 & 0 \\ 13 & 9 & 4 & 1 & 0 & 0 \\ 35 & 26 & 14 & 5 & 1 & 0 \\ 96 & 75 & 45 & 20 & 6 & 1 \end{array}\right].$$

This means that the shifted sequence of directed animals $1, 2, 5, 13, \ldots$ counts Motzkin paths whose falls have weight 1, whose horizontals at level 0 have weight 2, and whose horizontals above level 0 have weight 1.

Example 9.15 For the case $a = b = 1, c = 0, d = 2$ we find that

$$g(x) = \frac{1 - x + x^2 - \sqrt{1 - 2x - 5x^2 + 6x^3 + x^4}}{4x^2(1-x)}.$$

The Riordan array $(g(x), xg(x))$ then begins

$$\begin{bmatrix} 1 & 0 & 0 & 0 & 0 & 0 \\ 1 & 1 & 0 & 0 & 0 & 0 \\ 2 & 2 & 1 & 0 & 0 & 0 \\ 3 & 5 & 3 & 1 & 0 & 0 \\ 7 & 10 & 9 & 4 & 1 & 0 \\ 14 & 24 & 22 & 14 & 5 & 1 \end{bmatrix}.$$

From this we infer that the sequence $1, 1, 2, 3, 7, 14, \ldots$ counts Motzkin paths whose horizontal steps have weight 1 at even levels, and 0 at odd levels, while the falls have weight 1 at even levels, and 2 at odd levels. The generating function of this sequence is given by

$$\frac{1 - x + x^2 - \sqrt{1 - 2x - 5x^2 + 6x^2 + x^4}}{4x^2(1-x)} = \cfrac{1}{1 - x - \cfrac{x^2}{1 - \cfrac{2x^2}{1 - x - \cdots}}}.$$

For $y = 3$, we have

$$\begin{bmatrix} 1 & 0 & 0 & 0 & 0 & 0 \\ 1 & 1 & 0 & 0 & 0 & 0 \\ 2 & 2 & 1 & 0 & 0 & 0 \\ 3 & 5 & 3 & 1 & 0 & 0 \\ 7 & 10 & 9 & 4 & 1 & 0 \\ 14 & 24 & 22 & 14 & 5 & 1 \end{bmatrix} \begin{bmatrix} 1 \\ 3 \\ 9 \\ 27 \\ 81 \\ 243 \end{bmatrix} = \begin{bmatrix} 1 \\ 4 \\ 17 \\ 72 \\ 307 \\ 1310 \end{bmatrix}.$$

Thus the sequence $1, 4, 17, 72, 307, 1310, \ldots$ counts Motzkin paths whose falls have weight 1 at even levels, and 2 at odd levels, whose horizontal steps at level 0 have weight $3 + 1 = 4$, and subsequently have weight 1 at even levels, and 0 at odd levels.

We now examine the case of Thron continued fractions. We have the following result concerning Riordan arrays.

Proposition 9.30 *The generating function given by the Thron continued fraction*

$$\cfrac{1}{1 - (y+a)x - \cfrac{bx}{1 - ax - \cfrac{bx}{1 - ax - \cdots}}}$$

is the bivariate generating function of the Riordan array

$$\left(\frac{1-bx}{1+ax}, \frac{x(1-bx)}{1+ax}\right)^{-1}.$$

Proof We let $G(x, y)$ denote the continued fraction above. We then have

$$G(x, y) = \frac{1}{1 - (y + a)x - bxg(x)},$$

where $u = g(x)$ satisfies

$$u = \frac{1}{1 - ax - bxu}.$$

We find that

$$g(x) = \frac{1 - ax - \sqrt{1 - 2(a + 2b)x + a^2x^2}}{2bx}.$$

This allow us to solve for $G(x, y)$. We find that

$$G(x, y) = \frac{2}{1 - (2y + a)x + \sqrt{1 - 2(a + 2b)x + a^2x^2}}.$$

But we have

$$\left(\frac{1 - bx}{1 + ax}, \frac{x(1 - bx)}{1 + ax}\right)^{-1} = (g(x), xg(x))$$

and so the bivariate generating function of this Riordan array is given by

$$\frac{g(x)}{1 - xyg(x)}.$$

Simplification now shows that

$$G(x, y) = \frac{g(x)}{1 - xyg(x)}.$$

□

We can interpret this result as follows. The row sum polynomials $\sum_{k=0}^{n} a_{n,k} y^n$, where $a_{n,k}$ denotes the (n, k)th element of the Riordan array $\left(\frac{1-bx}{1+ax}, \frac{x(1-bx)}{1+ax}\right)^{-1}$, count Schröder paths where the fall steps come in b colors, and the horizontal steps come in a colors, except for those at level zero which come in $y + a$ colors.

Example 9.16 The Bell matrix $\left(\frac{1-x}{1+x}, \frac{x(1-x)}{1+x}\right)^{-1}$ begins

$$\left[\begin{array}{cccccc} 1 & 0 & 0 & 0 & 0 & 0 \\ 2 & 1 & 0 & 0 & 0 & 0 \\ 6 & 4 & 1 & 0 & 0 & 0 \\ 22 & 16 & 6 & 1 & 0 & 0 \\ 90 & 68 & 30 & 8 & 1 & 0 \\ 394 & 304 & 146 & 48 & 10 & 1 \end{array}\right].$$

We then have, for example,

$$\begin{bmatrix} 1 & 0 & 0 & 0 & 0 & 0 \\ 2 & 1 & 0 & 0 & 0 & 0 \\ 6 & 4 & 1 & 0 & 0 & 0 \\ 22 & 16 & 6 & 1 & 0 & 0 \\ 90 & 68 & 30 & 8 & 1 & 0 \\ 394 & 304 & 146 & 48 & 10 & 1 \end{bmatrix} \begin{bmatrix} 1 \\ 0 \\ 0 \\ 0 \\ 0 \\ 0 \end{bmatrix} = \begin{bmatrix} 1 \\ 2 \\ 6 \\ 22 \\ 90 \\ 394 \end{bmatrix}.$$

This indicates that the sequence of Schröder numbers

$$1, 2, 6, 22, 90, 394, \ldots$$

is given by $\sum_{k=0}^{n} a_{n,k} 0^k$. This corresponds to Schröder paths having falls with 1 color, horizontal steps at level 0 having $1+0=1$ color, and all other horizontal steps having 1 color. Similarly we have

$$(a_{n,k}) = \begin{bmatrix} 1 & 0 & 0 & 0 & 0 & 0 \\ 2 & 1 & 0 & 0 & 0 & 0 \\ 6 & 4 & 1 & 0 & 0 & 0 \\ 22 & 16 & 6 & 1 & 0 & 0 \\ 90 & 68 & 30 & 8 & 1 & 0 \\ 394 & 304 & 146 & 48 & 10 & 1 \end{bmatrix} \begin{bmatrix} 1 \\ 1 \\ 1 \\ 1 \\ 1 \\ 1 \end{bmatrix} = \begin{bmatrix} 1 \\ 3 \\ 11 \\ 45 \\ 197 \\ 903 \end{bmatrix}.$$

Thus the sequence

$$1, 3, 11, 45, 197, 903, \ldots$$

is given by $\sum_{k=0} a_{n,k} 1^k = \sum_{k=0} a_{n,k}$. This sequence of shifted little Schröder numbers counts Schröder paths where the rises have 1 color, the level 0 horizontal steps have $1+1=2$ colors, and all other horizontal steps have 1 color.

Example 9.17 We consider the Riordan array $\left(\frac{1-2x}{1+x}, \frac{x(1-2x)}{1+x}\right)^{-1}$. For this matrix, we have

$$\begin{bmatrix} 1 & 0 & 0 & 0 & 0 & 0 \\ 3 & 1 & 0 & 0 & 0 & 0 \\ 15 & 6 & 1 & 0 & 0 & 0 \\ 93 & 39 & 9 & 1 & 0 & 0 \\ 645 & 276 & 72 & 12 & 1 & 0 \\ 4791 & 2073 & 576 & 114 & 15 & 1 \end{bmatrix} \begin{bmatrix} 1 \\ 2 \\ 4 \\ 8 \\ 16 \\ 32 \end{bmatrix} = \begin{bmatrix} 1 \\ 5 \\ 31 \\ 215 \\ 1597 \\ 12425 \end{bmatrix}.$$

Thus the sequence that begins

$$1, 5, 31, 215, 1597, 12425, \ldots$$

counts Schröder paths whose falls come in 2 colors, whose horizontal steps at level 0 come in $1+2=3$ colors, and whose other horizontal steps come in 1 color.

The foregoing examples show that the bivariate generating functions $G(x, y)$ of certain families of Riordan arrays can correspond to the continued fraction generating functions of families of lattice paths, where the y dependence in each case comes from the level 0 steps. This applies in particular to Dyck paths, Motzkin paths, and Schröder paths.

Exercises

9.1 Show that the moments corresponding to the Riordan array $\left(\frac{1}{1+x^2}, \frac{x}{1+x^2}\right)$ are the aerated Catalan numbers.

9.2 Show that the moments corresponding to the Riordan array $\left(\frac{1}{1+x}, \frac{x}{(1+x)^2}\right)$ are the Catalan numbers C_n.

9.3 Show that the moments corresponding to the Riordan array $\left(\frac{1}{(1+x)^2}, \frac{x}{(1+x)^2}\right)$ are the shifted Catalan numbers C_{n+1}.

9.4 Show that the moments corresponding to the Riordan array $\left(\frac{1}{1+x+x^2}, \frac{x}{1+x+x^2}\right)$ are the Motzkin numbers $M_n = \sum_{k=0}^{\lfloor \frac{n}{2} \rfloor} \binom{n}{2k} C_k$.

9.5 Show that the moments corresponding to the Riordan array $\left(\frac{1+x}{1+x+x^2}, \frac{x}{1+x+x^2}\right)$ are the Riordan numbers, A005043.

9.6 Show that the moments corresponding to the Riordan array $\left(\frac{1}{1+x}, \frac{x}{1+3x+2x^2}\right)$ are the little Schröder numbers.

9.7 Show that the moments corresponding to the Riordan array $\left(\frac{1}{1+2x}, \frac{x}{1+3x+2x^2}\right)$ are the large Schröder numbers $S_n = \sum_{k=0}^{n} \binom{n+k}{2k} C_k$.

9.8 Show that the moments corresponding to the Riordan array $\left(\frac{1-x}{1+x}, \frac{x}{(1+x)^2}\right)$ are the central binomial numbers $\binom{2n}{n}$.

9.9 Show that the moments corresponding to the Riordan array $\left(\frac{1-x}{1+x^2}, \frac{x}{1+x^2}\right)$ are the central binomial coefficients $\binom{n}{\lfloor \frac{n}{2} \rfloor}$.

9.10 Show that the moments corresponding to the Riordan array $\left(\frac{1-x}{(1+x)^2}, \frac{x}{(1+x)^2}\right)$ are the numbers $\binom{2n+1}{n+1}$.

9.11 An almost Riordan array of order 1 is a Riordan array (g, f) to which a column is added on the left, with generating function $a(x)$, to form a new lower triangular matrix. This is denoted by $(a(x), g(x), f(x))$. Show that the Chebyshev polynomials of the first kind, defined by $T_0(x) = 1$, $T_n(x) = \frac{n}{2}\sum_{k=0}^{\lfloor \frac{n}{2} \rfloor} \frac{(-1)^k}{n-k}\binom{n-k}{k}(2x)^{n-2k}$ for $n \geq 1$, have a coefficient array given by the almost Riordan array $\left(\frac{1}{1+x^2}, \frac{1-x^2}{(1+x^2)^2}, \frac{2x}{1+x^2}\right)$.

9.12 Let $\mu(x)$ denote the generating function of the moments of the orthogonal polynomials defined by the Riordan array $\left(\frac{1-x-x^2}{1+x^2}, \frac{x}{1+x^2}\right)$. Show that the generating function $\frac{1}{x}\text{Rev}\left(\frac{x}{\mu(x)}\right)$ expands to give the sequence with general term $F_{n+1}C_n$, that is the product of the non-negative Fibonacci numbers with the Catalan numbers.

9.13 Use Riordan arrays to count the number of Motzkin paths whose horizontals and falls alternate between having the weights 1 and 2, and 2 and 1, respectively, except that the horizontals at level 0 have weight 3.

9.14 Use Riordan arrays to count the number of Schröder paths whose horizontals and falls have weights 2 and 3, respectively, except that the horizontals at level 0 have weight 5.

References

1. M. Aigner, *A Course in Enumeration* (Springer, 2007)
2. P. Barry, Some observations on the Lah and Laguerre transforms of integer sequences. J. Integer Seq. **10**, Article 07.4.6 (2007)
3. P. Barry, Continued fractions and transformations of integer sequences. J. Integer Seq. **12**, Article 9.7.6 (2009)
4. P. Barry, Riordan arrays, orthogonal polynomials as moments and Hankel transforms. J. Integer Seq. **14**, Article 11.2.2 (2011)
5. P. Barry, On the restricted Chebyshev-Boubaker polynomials. Integr. Transform. Spec. Funct. **28**, 1–16 (2017)
6. P. Barry, *Riordan Arrays: A Primer* (Logic Press, 2017)
7. P. Barry, A. Hennessy, Meixner-type results for Riordan arrays and associated integer sequences. J. Integer Seq. **13**, Article 10.9.4 (2010)
8. P. Barry, N. Pantelidis, On pseudo-involutions, involutions and quasi-involutions in the group of almost Riordan array. J. Algebr. Comb. (2020)
9. G.-S. Cheon, J.-H. Jung, P. Barry, Horizontal and vertical formulas for exponential Riordan matrices and their applications. Linear Algebr. Appl. **541**, 266–284 (2018)
10. T.S. Chihara, *An Introduction to Orthogonal Polynomials* (Gordon and Breach, New York, 1978)
11. L. Comtet, *Advanced Combinatorics* (Reidel, Dordrecht, 1974)
12. G. Della Riccia, Riordan arrays, Sheffer sequences and "orthogonal" polynomials. J. Integer Seq. **11**, Article 08.5.3 (2008)
13. E. Deutsch, L. Ferrari, S. Rinaldi, Production matrices. Adv. Appl. Math. **34**, 101–122 (2005)
14. E. Deutsch, L. Ferrari, S. Rinaldi, Production matrices and Riordan arrays. Ann. Comb. **13**, 65–85 (2009)

15. L. Euler, Remarques sur un beau rapport entre les séries des puissances tant directes que réciproques. (E352 Eneström Index) (1768)
16. P. Flajolet, Combinatorial aspects of continued fractions. Discret. Math. **32**, 125–161 (1980)
17. D. Foata, Eulerian polynomials: from Euler's time to the present, in *The Legacy of Alladi Ramakrishnan in the Mathematical Sciences*, ed. by K. Alladi, J.R. Klauder, C.R. Rao (Springer, 2010)
18. W. Gautschi, *Orthogonal Polynomials: Computation and Approximation* (Clarendon Press, Oxford, 2003)
19. R.L. Graham, D.E. Knuth, O. Patashnik, *Concrete Mathematics* (Addison-Wesley, New York, 1989)
20. P. Henrici, *Applied and Computational Complex Analysis*, vol. 3 (Wiley, 1993)
21. T.-X. He, L.C. Hu, P.J.-S. Shiue, The Sheffer group and the Riordan group. Discret. Appl. Math. **155**, 1895–1909 (2007)
22. F. Hirzebruch, Eulerian polynomials. Munster J. Math. **1**, 9–14 (2008)
23. M.E.H. Ismail, D. Stanton, Classical orthogonal polynomials as moments. Can. J. Math. **49**, 520–542 (1997)
24. M. Josuat-Vergès, A q-analog of Schläfli and Gould identities on Stirling numbers. Ramanujan J. **46**, 483–507 (2018)
25. C. Krattenthaler, Advanced determinant calculus. Available electronically at http://www.mat.univie.ac.at/~slc/wpapers/s42kratt.pdf (2010)
26. J.W. Layman, The Hankel transform and some of its properties. J. Integer Seq. **4**, Article 01.1.5 (2001)
27. P. Luschny, Eulerian polynomials. Available electronically at http://www.luschny.de/math/euler/EulerianPolynomials.html (2018)
28. D. MacHale, Personal communication
29. J.C. Mason, D.C. Handscomb, *Chebyshev Polynomials* (Chapman and Hall/CRC, 2002)
30. E. Mendelson, Races with ties. Math. Mag. **55**, 170–175 (1982)
31. Y. Nakamura, A. Zhedanov, Toda chain, Sheffer class of orthogonal polynomials and combinatorial numbers. Pr. Inst. Mat. Nats. Akad. Nauk Ukr. Mat. Zastos **50**, 450–457 (2004)
32. Y. Nakamura, A. Zhedanov, Special solutions of the Toda chain and combinatorial numbers. J. Phys. A **37**, 5849–5862 (2004)
33. T.D. Noe, On the divisibility of generalized central trinomial coefficients. J. Integer Seq. **9**, Article 06.2.7 (2006)
34. R. Oste, J. Van der Jeugt, Motzkin paths, Motzkin polynomials and recurrence relations. Electron. J. Comb. **22**, #P2.8 (2015)
35. M. Pétréolle, A.D. Sokal, B.-X. Zhu, Lattice paths and branched continued fractions: an infinite sequence of generalizations of the Stieltjes–Rogers and Thron–Rogers polynomials, with coefficientwise Hankel-total positivity, Preprint, https://arxiv.org/pdf/1807.03271.pdf (2018)
36. P.K. Suetin, Classical Orthogonal Polynomials. *Encyclopedia of Mathematics* (2018)
37. G. Szegö, *Orthogonal Polynomials*, 4th edn. (American Mathematical Society, 1975)
38. G. Viennot, Une théorie combinatoire des polynômes orthogonaux généraux. UQAM, Montreal, Quebec (1983)
39. L. Vinet, A. Zhedanov, Elliptic solutions of the restricted Toda chain, Lamé polynomials and generalization of the elliptic Stieltjes polynomials. J. Phys. A: Math. Theor. **42**(45), 454024 (2009)
40. H.S. Wall, *Analytic Theory of Continued Fractions* (AMS Chelsea Publishing, 2000)
41. E.P. Wigner, Characteristic vectors of bordered matrices with infinite dimensions. Ann. Math. **62**, 548–564 (1955)
42. E.P. Wigner, On the distribution of the roots of certain symmetric matrices. Ann. Math. **67**, 325–327 (1958)

Solutions

Exercises of Chap. 1

1.1 Use $\mathfrak{B}_2^k = C^k$ and $\mathfrak{B}_3^k = g^k$ in identity (1.3.2). For (1) we have

$$v_{n,k} = [z^n]C(zC^2)^k = [z^{n-k}]C^{2k+1} = \frac{2k+1}{2n+1}\binom{2n+1}{n-k};$$

$$\ell_{n,k} = [z^n](zC^2)^k = [z^{n-k}]C^{2k} = \frac{k}{n}\binom{2n}{n-k}.$$

For (2) we have

$$v_{n,k} = [z^n]g^2(2(g-1))^k = 2^k\sum_{i=0}^{k}\binom{k}{i}(-1)^{k-i}[z^n]g^{i+2}$$

$$= 2^k\sum_{i=0}^{k}\binom{k}{i}(-1)^{k-i}\frac{i+2}{3n+i+2}\binom{3n+i+2}{n};$$

$$\ell_{n,k} = [z^n](2(g-1))^k = 2^k\sum_{i=0}^{k}\binom{k}{i}(-1)^{k-i}[z^n]g^i$$

$$= 2^k\sum_{i=0}^{k}\binom{k}{i}(-1)^{k-i}\frac{i}{3n+i}\binom{3n+i}{n}.$$

© Springer Nature Switzerland AG 2022

L. Shapiro et al., *The Riordan Group and Applications*, Springer Monographs in Mathematics, https://doi.org/10.1007/978-3-030-94151-2

1.2 Choose one of the k edges to be the leaf and attach trees to the other $k-1$ edges at the root. Thus

$$L_1 = \sum_{k\geq 1} a_k \binom{k}{1} z(zT)^{k-1} = zA'(zT).$$

1.3 The A-sequence is $(1, 1, 2, 3, 4, 5, \ldots)$. Thus $A(z) = 1 + \frac{z}{(1-z)^2}$.

1.4 See [1, p. 363].

Exercises of Chap. 2

2.1 Let $g_k = f_{k+1}$; by $(G2)$, $\mathcal{G}(g_k) = (\mathcal{G}(f_k) - f_0)/t$. Since $g_0 = f_1$, we have

$$\mathcal{G}(f_{k+2}) = \mathcal{G}(g_{k+1}) = \frac{\mathcal{G}(g_k) - g_0}{t} = \frac{\mathcal{G}(f_k) - f_0 - f_1 t}{t^2}.$$

By mathematical induction this result can be generalized to

$$\mathcal{G}\left(f_{k+j}\right) = \frac{\mathcal{G}(f_k) - f_0 - f_1 t - \cdots - f_{j-1}t^{j-1}}{t^j}.$$

2.2 If we set $g_k = kf_k$, we obtain

$$\mathcal{G}((k+1)f_{k+1}) = \mathcal{G}(g_{k+1}) = t^{-1}(\mathcal{G}(kf_k) - 0f_0) \overset{(G2)}{=} t^{-1}tD\mathcal{G}(f_k) = D\mathcal{G}(f_k).$$

2.3 By (G3) we have

$$\mathcal{G}\left(k^2 f_k\right) = tD\mathcal{G}(kf_k) = tD(tD\mathcal{G}(f_k)) = tD\mathcal{G}(f_k) + t^2D^2\mathcal{G}(f_k).$$

2.4 For the falling factorial $k^{\underline{r}} = k(k-1)\cdots(k-r+1)$ by applying (G3) it is immediate to find

$$\mathcal{G}\left(k^{\underline{r}} f_k\right) = t^r D^r \mathcal{G}(f_k).$$

2.5 By setting $g(t) = pt$ in $(G5)$ we have $f(pt) = \sum_{n=0}^{\infty} f_n(pt)^n = \sum_{n=0}^{\infty} p^n f_n t^n = \mathcal{G}\left(p^k f_k\right)$.

2.6 The generating function of harmonic numbers is given by

$$\mathcal{G}(H_n) = \frac{1}{1-t}\ln\frac{1}{1-t},$$

therefore, by (G3) and (2.4.4) we have

$$\mathcal{G}(nH_n) = tD\frac{1}{1-t}\ln\frac{1}{1-t} = \frac{t}{(1-t)^2}\left(\ln\frac{1}{1-t}+1\right)$$

$$\mathcal{G}\left(\frac{1}{n+1}H_n\right) = \frac{1}{t}\int_0^t \frac{1}{1-z}\ln\frac{1}{1-z}\,dz = \frac{1}{2t}\left(\ln\frac{1}{1-t}\right)^2.$$

2.7 By (G3), (2.4.3) and (2.4.4) we have

$$\mathcal{G}\left(k\binom{p}{k}\right) = tD\mathcal{G}\left(\binom{p}{k}\right) = tD(1+t)^p = pt(1+t)^{p-1}$$

$$\mathcal{G}\left(k^2\binom{p}{k}\right) = (tD+t^2D^2)\mathcal{G}\left(\binom{p}{k}\right) = pt(1+pt)(1+t)^{p-2}$$

$$\mathcal{G}\left(\frac{1}{k+1}\binom{p}{k}\right) = \frac{1}{t}\int_0^t \mathcal{G}\left(\binom{p}{k}\right)dz = \frac{1}{t}\left[\frac{(1+t)^{p+1}}{p+1}\right]_0^t = \frac{(1+t)^{p+1}-1}{(p+1)t}$$

$$\mathcal{G}\left(k\binom{k}{m}\right) = tD\frac{t^m}{(1-t)^{m+1}} = \frac{mt^m+t^{m+1}}{(1-t)^{m+2}}$$

$$\mathcal{G}\left(k^2\binom{k}{m}\right) = (tD+t^2D^2)\mathcal{G}\left(\binom{k}{m}\right) = \frac{m^2t^m+(3m+1)t^{m+1}+t^{m+2}}{(1-t)^{m+3}}$$

$$\mathcal{G}\left(\frac{1}{k}\binom{k}{m}\right) = \int_0^t\left(\mathcal{G}\left(\binom{k}{m}\right)-\binom{0}{m}\right)\frac{dz}{z} = \int_0^t \frac{z^{m-1}\,dz}{(1-z)^{m+1}} = \frac{t^m}{m(1-t)^m}.$$

The last integral can be solved by setting $y=(1-z)^{-1}$ and is valid for $m>0$; for $m=0$ it reduces to $\mathcal{G}(1/k) = -\ln(1-t)$.

2.8 We have

$$\mathcal{G}\left(\frac{n}{(n+1)!}\right) = \frac{te^t-e^t+1}{t}$$

therefore the following result follows:

$$\sum_{k=0}^{n}\frac{k}{(k+1)!} = [t^n]\frac{1}{1-t}\left(\frac{te^t-e^t+1}{t}\right) = [t^n]\left(\frac{1}{1-t}-\frac{e^t-1}{t}\right) = 1-\frac{1}{(n+1)!}.$$

2.9 The generating function of the central binomial coefficients is given by

$$\mathcal{G}\left(\binom{2n}{n}\right) = \frac{1}{\sqrt{1-4t}}$$

therefore by using (2.4.3), (G3) and (2.4.7) we have

$$\mathcal{G}\left(\frac{1}{n}\binom{2n}{n}\right) = \int_0^t \left(\frac{1}{\sqrt{1-4z}} - 1\right)\frac{dz}{z} = 2\ln\frac{1-\sqrt{1-4t}}{2t}$$
$$\mathcal{G}\left(n\binom{2n}{n}\right) = \frac{2t}{(1-4t)\sqrt{1-4t}}$$
$$\mathcal{G}\left(\frac{1}{2n+1}\binom{2n}{n}\right) = \frac{1}{\sqrt{t}}\int_0^t \frac{dz}{\sqrt{4z(1-4z)}} = \frac{1}{\sqrt{4t}}\arctan\sqrt{\frac{4t}{1-4t}}.$$

2.10 By setting $f_n = 4^n\binom{2n}{n}^{-1}$, since

$$4^{n+1}\binom{2n+2}{n+1}^{-1} = \frac{2n+2}{2n+1}4^n\binom{2n}{n}^{-1},$$

we have the recurrence: $(2n+1)f_{n+1} = 2(n+1)f_n$. By using the operator $\mathcal{G}$ and the rules of Sect. 2.4, the differential equation $2t(1-t)f'(t) - (1+2t)f(t) + 1 = 0$ is derived. The solution is

$$f(t) = \sqrt{\frac{t}{(1-t)^3}}\left(-\int_0^t \sqrt{\frac{(1-z)^3}{z}}\frac{dz}{2z(1-z)}\right).$$

By simplifying and using the change of variable $y = \sqrt{z/(1-z)}$, the integral can be computed without difficulty, and the final result is

$$\mathcal{G}\left(4^n\binom{2n}{n}^{-1}\right) = \sqrt{\frac{t}{(1-t)^3}}\arctan\sqrt{\frac{t}{1-t}} + \frac{1}{1-t}.$$

Some immediate consequences by (2.4.7) and (2.4.3) are

$$\mathcal{G}\left(\frac{4^n}{2n+1}\binom{2n}{n}^{-1}\right) = \frac{1}{\sqrt{t(1-t)}}\arctan\sqrt{\frac{t}{1-t}}$$
$$\mathcal{G}\left(\frac{4^n}{2n^2}\binom{2n}{n}^{-1}\right) = \left(\arctan\sqrt{\frac{t}{1-t}}\right)^2.$$

2.11 By using (2.4.5) and (2.4.6) we have, after rationalizing:

$$\mathcal{G}(F_{2n}) = \frac{F(\sqrt{t}) + F(-\sqrt{t})}{2} = \frac{1}{2}\left(\frac{\sqrt{t}}{1-\sqrt{t}-t} + \frac{-\sqrt{t}}{1+\sqrt{t}-t}\right) = \frac{t}{1-3t-t^2}$$
$$\mathcal{G}(F_{2n+1}) = \frac{F(\sqrt{t}) - F(-\sqrt{t})}{2\sqrt{t}} = \frac{1}{2\sqrt{t}}\left(\frac{\sqrt{t}}{1-\sqrt{t}-t} - \frac{-\sqrt{t}}{1+\sqrt{t}-t}\right) = \frac{1-t}{1-3t-t^2}.$$

2.12 By using the shifting and convolution properties of the $\mathcal{G}$ operator we have

$$\mathcal{G}(C_{n+1}) = \mathcal{G}\left(\sum_{k=0}^{n} C_k C_{n-k}\right)$$
$$\frac{\mathcal{G}(C_n) - C_0}{t} = \mathcal{G}(C_n)^2 .$$

Since $C_0 = 1$, solving the second degree equation we find

$$\mathcal{G}(C_n) = \frac{1 - \sqrt{1-4t}}{2t}$$

2.13 We have

$$n f_n = n\delta_{n,1} + 2\sum_{k=0}^{n-1} f_k, \quad f_0 = 0$$

hence

$$\mathcal{G}(nf_n) = \mathcal{G}\left(n\delta_{n,1}\right) + 2\mathcal{G}\left(\sum_{k=0}^{n-1} f_k\right)$$
$$tD\mathcal{G}(f_n) = t + 2t\mathcal{G}\left(\sum_{k=0}^{n} f_k\right) = t + \frac{2t}{1-t}\mathcal{G}(f_n)$$
$$D\mathcal{G}(f_n) = 1 + \frac{2}{1-t}\mathcal{G}(f_n) .$$

By solving the differential equation with initial condition $f_0 = 0$, we find

$$\mathcal{G}(f_n) = \frac{1}{3(1-t)^2} + \frac{1}{3}(t-1).$$

Exercises of Chap. 3

3.1 Consider the map $\phi : \mathcal{L} \to \mathcal{B}$ defined as $\phi((1, f)) = (f/t, f)$. Clearly, ϕ is one-to-one and onto. In addition,

$$\begin{aligned}\phi((1, f)(1, h)) &= \phi((1, h(f))) = (h(f)/t, h(f)) = (f/t, f)(h/t, h) \\ &= \phi((1, f)\phi((1, h)).\end{aligned}$$

3.2 Since $g(-t) = g(t)$ and $f(-t) = -f(t)$, we have

$$(g(t), f(t))(1, -t) = (g(t), -f(t)) = (g(-t), f(-t)) = (1, -t)(g(t), f(t)).$$

3.3 The elements of the checkerboard subgroup form a subset of the elements of the multi-checkerboard subgroup.

3.4 Let $A = (C(t)/C(f_a(t)), f_a(t))$ and $B = (C(t)/C(f_b(t)), f_b(t))$. Then $AB = (C(t)/C(f_b(f_a(t))), f_b(f_a(t)))$ and $A^{-1} = (C(t)/C(\bar{f}_a(t), \bar{f}_a(t))$.

3.5 Consider the map $\phi : \mathcal{D} \to \mathcal{B}$ defined as $\phi(f'(t), f(t)) = \left(\frac{1}{t}\int_0^t f'(x)dx, f(t)\right)$.

3.6 Letting $g(t) = g_0\hat{g}(t) \in \mathcal{F}_0$ and $f(t) = f_1\hat{f}(t) \in \mathcal{F}_1$, we obtain $(g, f) = (\hat{g}, \hat{f})$ $(g_0, f_1 t) \in \widehat{\mathcal{R}}\mathcal{D}$. Since $\widehat{\mathcal{R}} \lhd \mathcal{R}$ and $\widehat{\mathcal{R}} \cap \mathcal{D} = \{I\}$, we have $\mathcal{R} = \widehat{\mathcal{R}} \rtimes \mathcal{D}$.

3.7 $A^{-1} = MAM = (g(-t), -f(-t))$.

3.8 Since $PMP^{-1} = M$ and $PM = MP$, we have

$$(BM)^2 = P^{-1}(AMA)PM = P^{-1}(AMA)MP = P^{-1}(AM)^2P.$$

Thus $(AM)^2 = I$ if and only if $(BM)^2 = I$.

3.9 Let $M = [m_{nk}] \in \mathcal{H}$. since

$$\sum_{n\geq 0} m_{nk}t^n = tf'(t)f(t)^{k-1} = t\left(\frac{f^k(t)}{k}\right)',$$

we have $km_{nk} = [t^n]\left(t(f^k(t))'\right) = [t^n]t\left(\sum_{n\geq 0} d_{nk}t^n\right)' = nd_{nk}$.

Exercises of Chap. 4

4.1 The A- and Z- sequences are defined by the relation 4.1.2 and Theorem 4.5

$$h(t) = tA(h(t)), \quad d(t) = \frac{1}{1 - tZ(h(t))}.$$

By using these formulas we find $Z(t) = 1$ and

$$A(t) = \frac{1 + t + \sqrt{1 + 2t + 4rt + t^2}}{2}.$$

4.2 Proceeding as in Exercise 4.1, we find $A(t) = Z(t) = \frac{1+\sqrt{1+4t}}{2}$.

4.3 Proceeding as in Exercise 4.1, we find to $A(t) = 1 + t + t^2$ and $Z(t) = 1 + t$.

4.4 By formula (4.2.7), we have

$$P^{[0]}(t) = 1 + t^2 \quad Q^{[1]}(t) = 1$$

$$A(t) = P^{[0]}(t) + tA(t)Q^{[1]}(t)$$

$$A(t) = \frac{1+t^2}{1-t}, \quad h(t) = \frac{1-\sqrt{1-4t-4t^2}}{2(1+t)} \quad d(t) = h(t)/t.$$

4.5 From the solution of Exercise 4.2 we have $A(t) = \frac{1+t^2}{1-t}$, hence from Theorem 4.3 we obtain

$$\overline{h}(t) = \frac{t(1-t)}{1+t^2};$$

moreover, from Theorem 3.8 we have $\overline{d}(t) = \overline{h}(t)/t$.

4.6 The solution can be found as in Exercise 4.2 with

$$P^{[0]}(t) = 1 + 2t^2 \quad Q^{[1]}(t) = 1.$$

4.7 By formula (4.2.7) we have

$$A(t) = \frac{1-t+t^2-\sqrt{1+2t-5t^2+6t^2-3t^4}}{2(1-t)}.$$

A combinatorial interpretation can be found in [2, 3, Chap. 4].

4.8 By formula (4.2.6) we can find a third degree equation defining $A(t)$

$$\begin{aligned} A(t) &= \sum_{i\geq 0} t^i A(t)^{-i} P^{[i]}(t) + tA(t)Q(t) \\ &= 1 - t + t^2 + tA(t)^{-1}(1-t) + t^2 A(t)^{-2} + tA(t). \end{aligned}$$

A combinatorial interpretation can be found in [2, 3, Chap. 4].

4.9 The proof is based on the computation of the production matrix P corresponding to the Stirling number of first kind. As in Example 4.12, it is sufficient to compute the 8×8 truncation of the product $R^{-1} \cdot \overline{R}$:

$$P = R^{-1} \cdot \overline{R} = \begin{bmatrix} 1 & 1 & 0 & 0 & 0 & 0 & 0 & 0 \\ 1 & 2 & 1 & 0 & 0 & 0 & 0 & 0 \\ 1 & 3 & 3 & 1 & 0 & 0 & 0 & 0 \\ 1 & 4 & 6 & 4 & 1 & 0 & 0 & 0 \\ 1 & 5 & 10 & 10 & 5 & 1 & 0 & 0 \\ 1 & 6 & 15 & 20 & 15 & 6 & 1 & 0 \\ 1 & 7 & 21 & 35 & 35 & 21 & 7 & 1 \\ 1 & 8 & 28 & 56 & 70 & 56 & 28 & 8 \end{bmatrix}.$$

We observe that P corresponds to the Pascal triangle and has not the form (4.3.2).

Exercises of Chap. 5

5.1 We apply formula (5.1.1) to the Pascal triangle and $f(t) = \ln(1/(1+t))$ and then extract the coefficient:

$$\sum_{k=1}^{n} \binom{n}{k} \frac{(-1)^{k-1}}{k} = [t^n] \frac{1}{1-t} \left[\ln\left(\frac{1}{1+u}\right) \Big| u = \frac{t}{1-t} \right] =$$

$$= [t^n] \frac{1}{1-t} \ln\left(\frac{1}{1-t}\right) = H_n,$$

where H_n is the nth harmonic number.

5.2 We apply formula (5.1.1) to the Pascal triangle and $f(t) = \frac{1}{(1-t)^2}$ and then extract the following coefficient

$$[t^n] \frac{1-t}{(1-2t)^2} = (n+2)2^{n-1}.$$

5.3 We apply formula (5.1.1) by observing that

$$\left(\frac{1}{1+t}, \frac{t}{(1+t)^2}\right) * \frac{1}{1-4t} = \frac{1}{1+t} \cdot \frac{(1+t)^2}{1+2t+t^2-4t} = \frac{1+t}{(1-t)^2}.$$

5.4 The binomial coefficient $\binom{n+k}{m+2k}$ corresponds to the Riordan array

$$\left(\frac{t^m}{(1-t)^{m+1}}, \frac{1}{(1-t)^2}\right)$$

and from the generating function of Catalan numbers we have $f(t) = \frac{\sqrt{1+4t}-1}{2t}$; hence, by formula (5.1.1) and after some simplification we find the desired result by extracting the coefficient $[t^{n-m}]\frac{1}{(1-t)^m}$.

5.5 It is sufficient to observe that Stirling numbers of the first and second kind multiplied by the factor $k!/n!$ correspond to the Riordan arrays:

$$\left(1, \ln\frac{1}{1-t}\right), \quad (1, e^t - 1)$$

and apply formula (5.1.1) with $f_k = 1$.

5.6 By theorem 5.4, we find

$$[t^{n-k}]\frac{1-t}{(1+t)^3}\cdot(1+t)^{2n+3} = [t^{n-k}](1+t)^{2n} - [t^{n-k-1}](1+t)^{2n} =$$

$$= \binom{2n}{n-k} - \binom{2n}{n-k-1} = \frac{2k+1}{n+k+1}\binom{2n}{n-k}.$$

Hence, the inverse relation is

$$\sum_{k=0}^{n}\binom{2n}{n-k}\frac{(2k+1)^2}{n+k+1} = 4^n.$$

5.7 It is sufficient to prove that $\frac{k+1}{n+1}\binom{2n-k}{n-k}$ and $(-1)^{n-k}\binom{k+1}{n-k}$ correspond to the Riordan arrays

$$\left(\frac{1-\sqrt{1-4t}}{2t}, \frac{1-\sqrt{1-4t}}{2}\right), \quad (1-t, t-t^2)$$

which are one the inverse of the other.

5.8 This is the identity corresponding to $x = 0$ and $p = q = 1$ in Example 5.11.

5.9 By Theorem 5.8 with $d(t) = (1-t)^{-1}$ and $v(t) = (1-t)^{-s}$ we have the inverse identity:

$$b_n = \sum_{k=0}^{sn}(-1)^{sn-k}\binom{sn}{k}a_k.$$

5.10 By Theorem 5.8, the inverse identity is

$$b_n = \sum_{k=0}^{sn}(-1)^{sn-k}\binom{sn+p}{k+p}a_k.$$

Exercises of Chap. 6

6.1 (i) First, we determine a delta series $f(t)$ from the series $r(t)$ such that $r(t) = f'(\bar{f}(t))$. It is known that $f(\bar{f}(t)) = t$, which implies $f'(\bar{f}(t))\bar{f}'(t){=}r(t)\bar{f}'(t){=}1$. Hence, $\bar{f}'(t) = 1/r(t)$, so that

$$\bar{f}(t) = \int_0^t \bar{f}'(\tau)\mathrm{d}\tau = \int_0^t \frac{1}{r(\tau)}\mathrm{d}\tau\,.$$

Thus, we obtain $f(t)$. Next, let us determine an inverse series $g(t)$ such that $c(t) = g'(\bar{f}(t))/g(\bar{f}(t))$, or equivalently, $c(f(t)) = g'(t)/g(t)$. Because

$$\int_0^t c(f(\tau))\mathrm{d}\tau = \int_0^t \frac{g'(\tau)}{g(\tau)}\mathrm{d}\tau = \ln|g(t)| - \ln|g_0|\,,$$

we have

$$g(t) = g_0 \exp\left(\int_0^t c(f(\tau))\mathrm{d}\tau\right).$$

Alternatively, because

$$\frac{c(t)}{r(t)} = \frac{g'(\bar{f}(t))}{g(\bar{f}(t))f'(\bar{f}(t))} = \frac{g'(\bar{f}(t))\bar{f}'(t)}{g(\bar{f}(t))}\,,$$

we have

$$\int_0^t \frac{c(\tau)}{r(\tau)}\mathrm{d}\tau = \int_0^t \frac{\mathrm{d}g(\bar{f}(\tau))}{g(\bar{f}(\tau))} = \ln|g(\bar{f}(t))| - \ln|g_0|\,,$$

which gives

$$g(\bar{f}(t)) = g_0 \exp\left(\int_0^t \frac{c(\tau)}{r(\tau)}\mathrm{d}\tau\right) \quad \text{and} \quad g(t) = g_0 \exp\left(\int_0^{f(t)} \frac{c(\tau)}{r(\tau)}\mathrm{d}\tau\right).$$

Now, from the inverse series $g(t)$ and delta series $f(t)$, an exponential Riordan array $[g(t), f(t)]$ is obtained, which has $r(t)$ and $c(t)$ as its r- and c-sequence characterization.

(ii) In fact, from the above discussion, we have

$$\left[\frac{1}{g(\bar{f}(t))}, \bar{f}(t)\right] = \left[\frac{1}{g_0 \exp\left(\int_0^t \frac{c(\tau)}{r(\tau)}\mathrm{d}\tau\right)}, \int_0^t \frac{1}{r(\tau)}\mathrm{d}\tau\right].$$

(iii) By the results given in (i), it can be found that

$$\bar{f}(t) = \int_0^t \frac{1}{1+\tau^2} \mathrm{d}\tau = \arctan t\,,$$

$$g(\bar{f}(t)) = \exp\left(\int_0^t \frac{\tau}{1+\tau^2} \mathrm{d}\tau\right) = \sqrt{1+t^2}\,.$$

Therefore, $f(t) = \tan t$ and $g(t) = \sec t$, and the desired exponential Riordan array is $[\sec t, \tan t]$.

6.2 (i) Since the exponential Riordan array $[\{{n \atop k}\}]_{n,k\geq 0}$ is $[1, \mathrm{e}^t - 1]$, the generating functions of the corresponding c-sequence and r-sequence can be found by using formulae in (6.1.4) as $c(t) = 0$ and $r(t) = 1 + t$. Then the recurrence relation of Stirling numbers of the second kind can be obtained from (6.1.8):

$$\left\{{0 \atop 0}\right\} = 1\,, \quad \left\{{n \atop 0}\right\} = 0 \;\; (n \geq 1)\,, \quad \left\{{n \atop n}\right\} = 1 \;\; (n \geq 1),$$

$$\left\{{n+1 \atop k}\right\} = \left\{{n \atop k-1}\right\} + k\left\{{n \atop k}\right\}.$$

(ii) Noting $[S(n, k; \alpha, \beta, \gamma)]_{n,k\in\mathbb{N}} = [d(t), h(t)]$, where

$$d(t) = (1 + \alpha t)^{\gamma/\alpha} \quad \text{and} \quad h(t) = \frac{(1+\alpha t)^{\beta/\alpha} - 1}{\beta}\,,$$

from (6.1.4) we get

$$c(t) = \gamma(1 + \beta t)^{-\alpha/\beta} \quad \text{and} \quad r(t) = (1 + \beta t)^{1-\alpha/\beta}.$$

If $\alpha = \beta$, then $c(t) = \gamma/(1 + \alpha t)$ and $r(t) = 1$. Hence, by using (6.1.8) we obtain the following recurrence relation of Stirling numbers $S(n, k; \alpha, \alpha, \gamma)$:

$$S(n+1, 0; \alpha, \alpha, \gamma) = \gamma \sum_{i=0}^{n} i!(-\alpha)^i S(n, i; \alpha, \alpha, \gamma)\,,$$

$$S(n, n; \alpha, \alpha, \gamma) = 1 \quad (n \geq 0)\,,$$

$$S(n+1, k; \alpha, \alpha, \gamma) = S(n, k-1; \alpha, \alpha, \gamma) + \frac{\gamma}{k!} \sum_{i=k}^{n} i!(-\alpha)^{i-k} S(n, i; \alpha, \alpha, \gamma)$$

$$(n \geq k \geq 0)\,.$$

(iii) Denote the LHS of (6.6.1) by $\phi_k(t)$. Evidently, $\phi_k(0) = 0$ for $k \geq 1$ and $\phi_0(t) = (1 + \alpha t)^{\gamma/\alpha}$. Moreover, using elementary differentiation and algebraic computations, one can verify that

$$(1 + \alpha t)\frac{\mathrm{d}}{\mathrm{d}t}\phi_k(t) - (\gamma + k\beta)\phi_k(t) = \phi_{k-1}(t) \quad (k \geq 1). \tag{A.1}$$

Since the above differential equation has a unique solution satisfying the initial condition $\phi_k(0) = 0$, the RHS of (6.6.1), $\sum_{n\geq 0} S(n, k; \alpha, \beta, \gamma)t^n/n!$, also satisfies the above equation. Consequently,

$$(1+\alpha t)\sum_{n\geq 0} S(n,k;\alpha,\beta,\gamma)\frac{nt^{n-1}}{n!} - (\gamma + k\beta)\sum_{n\geq 0} S(n,k;\alpha,\beta,\gamma)$$
$$= \sum_{n\geq 0} S(n, k-1; \alpha, \beta, \gamma)\,.$$

The LHS of the last equation can be written as

$$\text{LHS} = \sum_{n\geq 1} S(n+1,k;\alpha,\beta,\gamma)\frac{t^n}{n!} - \sum_{n\geq 0}(\gamma + k\beta - n\alpha)S(n,k;\alpha,\beta,\gamma)\frac{t^n}{n!}\,.$$

Comparing the coefficients of the nth term on the both sides, we obtain (6.6.2). For $\alpha = 1$ and $\beta = 0$, we have the recursive relation for the unsigned Stirling numbers of the first kind,

$$\begin{bmatrix} n+1 \\ k \end{bmatrix} = \begin{bmatrix} n \\ k-1 \end{bmatrix} + n\begin{bmatrix} n \\ k \end{bmatrix} \quad (n \geq 1)\,,$$

while the recursive relation of the Stirling numbers of the second kind $\left\{{n \atop k}\right\} = S(n, k; 0, 1, 0)$ shown in (i) can also be derived from (6.6.2).

6.3 The given conditions ensure that there hold a pair of formal series expansions

$$f\left(a + \sum_{n\geq 1} \alpha_n t^n\right) = f(a) + \sum_{n\geq 1} \beta_n t^n, \tag{A.2}$$

$$\bar{f}\left(f(a) + \sum_{n\geq 1} \beta_n t^n\right) = a + \sum_{n\geq 1} \alpha_n t^n. \tag{A.3}$$

Thus, an application of Faá di Bruno's formula to $(f \circ \phi)(t)$,

$$[t^n](f\circ\phi) = \sum_{\sigma(n)} f^{(k)}(\phi(0)) \prod_{j=1}^{n} \frac{1}{k_j!}\left([t^i]\phi\right)^{k_j}\,,$$

on the LHS of (A.2) yields the expression (6.6.3) with $[t^i]\phi = \alpha_i$, $[t^n](f \circ \phi) = \beta_n$, and $\phi(0) = a$. Note that the LHS of (A.3) may be expressed as $\phi(t) = ((\bar{f} \circ f) \circ \phi)(t) = (\bar{f} \circ (f \circ \phi))(t)$. In a like manner, an application of Faá di Bruno's formula to the LHS of (A.3) gives precisely the equality (6.6.4). From the definitions of $\sigma(n)$, $\sigma(n, k)$, and $B_{n,k}(\ldots)$ we may write (6.6.3) and (6.6.4) into the forms of (6.6.5) and (6.6.6), respectively.

6.4 (i) They are the special cases of (6.6.3) and (6.6.4) for $f(x) = e^x$ and $\bar{f}(x) = \ln x$ with $f^{(k)}(0) = 1$ and $\bar{f}^{(k)}(1) = (-1)^k(k-1)!$, respectively.

(ii) They are the special cases of (6.6.3) and (6.6.4) for $f(x) = x^\alpha$ ($\alpha \neq 0$) and $\bar{f}(x) = x^{1/\alpha}$ with $f^{(k)}(1) = (\alpha)_k$ and $\bar{f}^{(k)}(1) = (1/\alpha)_k$, respectively.

(iii) Let $f(x) = e^x$ and

$$\begin{aligned}\phi(t) &= \ln\frac{1}{\sqrt{1-4t}} = -\frac{1}{2}\ln(1-4t)\\ &= \frac{1}{2}\left(4t + \frac{1}{2}(4t)^2 + \frac{1}{3}(4t)^3 + \cdots\right).\end{aligned}$$

Then $[t^n]\phi(t) = 4^n/(2n)$. Since

$$(f \circ \phi)(t) = \frac{1}{\sqrt{1-4t}} = \sum_{n\geq 0}\binom{2n}{n}t^n, \tag{A.4}$$

from (i) the coefficient of the nth term of $f \circ \phi = b_n$ can be written as

$$\begin{aligned}b_n = \binom{2n}{n} &= \sum_{\sigma(n)}\frac{\left(\frac{4}{2\cdot 1}\right)^{k_1}\left(\frac{4^2}{2\cdot 2}\right)^{k_2}\cdots\left(\frac{4^n}{2\cdot n}\right)^{k_n}}{k_1!k_2!\cdots k_n!}\\ &= \sum_{\sigma(n)}\frac{4^{k_1+2k_2+\cdots nk_n}}{2^{k_1+k_2+\cdots k_n}1^{k_1}k_1!2^{k_2}k_2!\cdots n^{k_n}k_n!}\\ &= \sum_{\sigma(n)}\frac{4^n}{2^k 1^{k_1}k_1!2^{k_2}k_2!\cdots n^{k_n}k_n!},\end{aligned}$$

which implies (6.6.7). (6.6.8) follows the second Faá di Bruno's relation shown in (i).

Let $f(x) = x^{-1/2}$ and $\phi(t) = 1-4t$. Then $(f \circ \phi)(t)$ can be expanded as (A.4). Thus, substituting $\beta_n = \binom{2n}{n}$, $\alpha_1 = 1$, and $\alpha_2 = -4$ into the first Faá di Bruno's relation in (ii) and noting $k_1 = 2k - n$ and $k_2 = n - k$, we obtain (6.6.9).

6.5 (i) The (n, k) entry of $(d(xt), h(xt)/x)$ is

$$\begin{aligned}[t^n]d(xt)\frac{h(xt)^k}{x^k} &= \left[\frac{(xt)^n}{x^n}\right]x^{-k}d(xt)(h(xt))^k\\ &= [(xt)^n]x^{n-k}d(xt)(h(xt))^k = d_{n,k}x^{n-k}.\end{aligned}$$

(ii) Since

$$P[x] = \left(\binom{n}{k}x^{n-k}\right)_{n\geq k\geq 0} = \left(\frac{1}{1-xt}, \frac{t}{1-xt}\right),$$

we have $P[x+y] = P[x]P[y]$ because

$$\left(\frac{1}{1-xt}, \frac{t}{1-xt}\right)\left(\frac{1}{1-yt}, \frac{t}{1-yt}\right) = \left(\frac{1}{1-(x+y)t}, \frac{t}{1-(x+y)t}\right).$$

(iii) Since $P[x]P[-x] = P[0] = I$, from $B(x) = P[x]B$ we obtain $P[-x]B(x) = P[-x]P[x]B = B$, which implies (6.6.10).

6.6 From the definition of Riordan arrays and noting $d(t) = h(t)/t$, we have

$$\begin{aligned}
&\sum_{n=0}^{s} \frac{c_{s+1}}{c_n c_{s-n}} \frac{c_k c_m}{c_{k+m+1}} d_{n,k} d_{s-n,m} \\
&= \sum_{n=0}^{s} \frac{c_{s+1}}{c_n c_{s-n}} \frac{c_k c_m}{c_{k+m+1}} \left[\frac{t^n}{c_n}\right] d(t) \frac{h(t)^k}{c_k} \left[\frac{t^{s-n}}{c_{s-n}}\right] d(t) \frac{h(t)^m}{c_m} \\
&= \frac{c_{s+1}}{c_{k+m+1}} \left[t^s\right] d(t)^2 h(t)^{k+m} = \left[\frac{t^{s+1}}{c_{s+1}}\right] d(t) \frac{h(t)^{k+m+1}}{c_{k+m+1}} = d_{s+1,k+m+1}.
\end{aligned}$$

Noting

$$d_{n,k} = [t^n] d(t) h(t)^k = \binom{n}{k}$$

when $c_k = 1$, $d(t) = 1/(1-t)$, and $h(t) = t/(1-t)$ (i.e., $(d(t), h(t))$ is the Pascal triangle), we immediately obtain (6.6.12) from (6.6.11).

6.7 Let $(d(t), h(t))$ be a Riordan array with A-sequence and Z-sequence, $A = (a_0, a_1, \ldots)$ and $Z = (z_0, z_1, \ldots)$. Thus

$$d_{n,k} = \sum_{j\geq 0} a_j d_{n-1,k+j-1}, \quad d_{n,0} = \sum_{j\geq 0} z_j d_{n-1,j} \tag{A.5}$$

for all $n, k \geq 0$. Substituting the above expressions into $p_n(x) = \sum_{n\geq k\geq 0} d_{n,k} x^k / k!$ and interchanging the order of summation, we have

$$\begin{aligned}
p_n(x) &= \sum_{k=0}^{n} \sum_{j=0}^{n-k} a_j d_{n-1,k+j-1} \frac{x^k}{k!} \\
&= \sum_{j=0}^{n} a_j \sum_{k=0}^{n-j} d_{n-1,k+j-1} \frac{x^k}{k!} \\
&= \sum_{j=0}^{n} a_j \sum_{k=j}^{n} d_{n-1,k-1} \frac{x^{k-j}}{(k-j)!}.
\end{aligned}$$

Conversely, if (6.6.13) holds, by comparing the coefficients of x^k term on the two sides of

$$p_n(x) = \sum_{k=0}^{n} d_{n,k} \frac{x^k}{k!} = \sum_{k=0}^{n} \sum_{j=0}^{n-k} a_j d_{n-1,k+j-1} \frac{x^k}{k!}$$

we obtain the first expression of (A.5).

Equation (6.6.14) comes from the second expression of (A.5) and the observation that

$$d_{n,j} = \frac{\mathrm{d}^j}{\mathrm{d}x^j} \sum_{k=0}^{n} d_{n,k} \frac{x^k}{k!}\bigg|_{x=0} = \frac{\mathrm{d}^j}{\mathrm{d}x^j} p_n(x)\bigg|_{x=0} .$$

Hence, we obtain the equivalence between the two expressions of (A.5) and (6.6.13) and (6.6.14), respectively. The remaining statement can be proved by using the isomorphism between the Riordan group and the Sheffer group.

6.8 For a Riordan array $(d(t), h(t))$ and a formal power series $g(t) = \sum_{k\geq 0} g_k t^k / c_k$, where the sequence $(c_k)_{k\geq 0}$ is defined before, we have

$$\sum_{k=0}^{n} d_{n,k} \frac{g_k}{c_k} = [t^n]d(t)g(h(t)) \tag{A.6}$$

because

$$[t^n]d(t)g(h(t)) = [t^n]d(t) \sum_{k\geq 0} g_k \frac{(h(t))^k}{c_k} = \sum_{k\geq 0} \frac{g_k}{c_k} [t^n]d(t)(h(t))^k .$$

Denote $f(t) = g(h(t))$. Then

$$g_k = \left[\frac{t^k}{c_k}\right] g(t) = c_k [t^k] f(y)\big|_{y=\bar{h}(t)} ,$$

where $\bar{h}(t)$ is the compositional inverse of $h(t)$. Using the Lagrange inversion formula and noting $y = ty/h(y)$ when $y = \bar{h}(t)$, we may calculate g_k as

$$g_k = \frac{c_k}{k} [y^{k-1}] \left[f'(y) \left(\frac{y}{h(y)} \right)^k \right]. \tag{A.7}$$

Let $(d(t), h(t)) = ((1 + at)^\alpha, t(1 + at)^\beta)$, and let $f(t) = t^b(1 + at)^\gamma$, where $\alpha, \beta, \gamma, n \in \mathbb{N}$, and let $b \in \mathbb{N}$ satisfying $b \leq n$. Then (A.7) yields

$$\begin{aligned} g_k &= \frac{c_k}{k}[y^{k-1}][by^{b-1}(1+ay)^{\gamma-\beta k} + a\gamma y^b(1+ay)^{\gamma-\beta k-1}] \\ &= \frac{c_k}{k}\left[b\binom{\gamma-\beta k}{k-b}a^{k-b} + \gamma\binom{\gamma-\beta k-1}{k-b-1}a^{k-b}\right] \\ &= c_k\frac{\gamma-\beta b}{\gamma-\beta k}\binom{\gamma-\beta k}{k-b}a^{k-b}. \end{aligned}$$

Since

$$d(t)(h(t))^k = [t^n](1+at)^\alpha\left(t(1+at)^\beta\right)^k = \binom{\alpha+\beta k}{n-k}a^{n-k}$$

and

$$[t^n]d(t)f(t) = [t^n](1+at)^\alpha t^b(1+at)^\gamma = \binom{\alpha+\gamma}{n-b}a^{n-b},$$

from (A.6) and using the concept of "natural range of summation" in [R. Sprugnoli, Negation of binomial coefficients, *Discrete Math.* 308 (2008), 5070–5077], we obtain the bth order Gould's identity, which includes Gould's identity as the case of $b = 0$.

6.9 Let $(d; h_1, h_2)$ be a double Riordan array, where even function $d \in \mathcal{F}_0$ and odd functions $h_1, h_2 \in \mathcal{F}_1$. Since for $k \geq 1$

$$\begin{aligned} (d; h_1, h_2)\frac{1}{1-kt} &= (d; h_1, h_2)\frac{1}{1-k^2t^2} + (d; h_1, h_2)\frac{kt}{1-k^2t^2} \\ &= \frac{d}{1-k^2h_1h_2} + \frac{kdh_1}{1-k^2h_1h_2} = \frac{d(1+kh_1)}{1-k^2h_1h_2}. \end{aligned}$$

From the request $(d; h_1, h_2)(1/(1-kt)) = 1/(1-kt)$ we have

$$d(1+kh_1)(1-kt) = 1-k^2h_1h_2,$$

where the function on the right-hand side is even. Hence the left-hand side function $d(1+kh_1)(1-kt) = d(1+kh_1-kt-k^2th_1)$ and its fact $(1+kh_1)(1-kt) = 1+kh_1-kt-k^2th_1$ are also even because d is even. Consequently, $h_1 = t$ and d can be presented as (6.6.15). Let $d(t) = d_0 + d_2t^2 + \text{higher powers of } t$, and $h_2'(t) = h_2'(0)t + \text{higher powers of } t$. Then from (6.6.15) we have

$$(1-k^2t^2)(d_0 + d_2t^2 + \text{higher powers of } t) = 1 - t(k^2h_2'(0)t + \text{higher powers of } t).$$

Comparing both sides of the above equation, we obtain $d(0) = 1$ and $d_2 - k^2d_0 = -k^2h_2'(0)$. The requirement of $h_2'(0) \neq 0$ implies that $d_2 \neq k^2$, i.e., the constant term and the coefficient of the quadratic term of d satisfy $d_0 = 1$ and $d_2 \neq k^2$. The converse is obviously true.

6.10 (i) If $d = C^2(t^2)$, where C is the generating function of the Catalan numbers, then $d_0 = 1$ and $d_2 = 2$. To find $(C^2(t^2); h_1, h_2) \in \mathcal{DS}_{1/(1-t)}$, we use (6.6.15) for

$k = 1$, namely,

$$C^2(t^2) = \frac{1 - th_2(t)}{1 - t^2},$$

to get

$$h_2(t) = \frac{1}{t}(1 - C^2(t^2)(1 - t^2)) = \frac{C(t^2) - C^2(t^2)}{t}.$$

Then from $C = 1 + tC^2$ we have

$$h_2 = \frac{C(t^2)(1 - C(t^2))}{t} = -tC^3(t^2).$$

Thus $(C^2(t^2); t, -tC^3(t^2)) \in \mathcal{DS}_{1/(1-t)}$, i.e.,

$$(C^2(t^2); t, -tC^3(t^2))\frac{1}{1-t} = \frac{1}{1-t},$$

or equivalently

$$\frac{C^2(t^2)}{1 + tC^3(t^2)} = \frac{1}{1-t}.$$

(ii) From the result of Problem 7.9, we know that a double Riordan array $(d; t, h_2)$ with $d = 1/(1 - t^2)$ or $C(t^2)$, where both $d_0 = d_2 = 1$, is not in the subgroup $\mathcal{DS}_{1/(1-t)}$, because equations

$$\frac{1 + t}{(1 - t^2)(1 - th_2)} = \frac{1}{1-t} \quad \text{and} \quad \frac{C(t^2)(1 + t)}{1 - th_2} = \frac{1}{1-t}$$

have no odd solutions h_2 in $\mathcal{F}_1$.

(iii) We consider the case of $k = 2$ for (6.6.15). If $d = 1/(1 - t^2)$, then $d_0 = d_2 = 1$. From (6.6.15), we have $h_2 = 3t/(4(1 - t^2))$ and the corresponding $(1/(1 - t^2); t, h_2) \in \mathcal{DS}_{1/(1-2t)}$. If $d = C(t^2)$, then $d_0 = d_2 = 1$. From (6.6.15), we have

$$h_2 = \frac{1}{2t}(1 - C(t^2) + 4t^2C(t^2)) = \frac{tC(t^2)}{4}(4 - C(t^2))$$

and $(C(t^2); t, h_2) \in \mathcal{DS}_{1/(1-2t)}$.

Exercises of Chap. 7

7.1 (a)–(h) Use the binary operation (7.1.7) on the 3-D Riordan group $\mathcal{R}^{\langle 3\rangle}$ to show that each subset in the question is a subgroup of $\mathcal{R}^{\langle 3\rangle}$. The proofs are very similar to what we studied in Sect. 3.2, Some special subgroups.

7.2 $(g_1, f_1, h_1)(g, z, 1)(g_1, f_1, h_1)^{-1} = (g(f_1), z, 1) \in \mathcal{A}_1$. A similar argument shows that $\mathcal{A}_2$ and $\mathcal{M}$ are normal subgroups of $\mathcal{R}^{\langle 3\rangle}$.

7.3 Let $(g, f, zh)(G, F, zH) = (\hat{c}_{n,k,\ell})$. By (7.2.20) we obtain

$$\sum_{n\ge 0} \hat{c}_{n,k,\ell} z^n = \sum_{x\ge 0}\left(\sum_{n\ge 0} \hat{a}_{n,x,\ell} z^n\right) \hat{b}_{x+\ell,k,\ell} = g(zh)^\ell \sum_{x\ge 0} b_{x,k,\ell} f^x$$
$$= gG(f)F(f)^k(zhH(f))^\ell.$$

7.4 The identity is $(1, z, z)$ and $(g, f, zh)^{-1} = \left(\frac{1}{g(\bar{f})}, \bar{f}, \frac{z}{h(\bar{f})}\right)$.

7.5 A map $\varphi : \mathcal{R}^{(3)} \to \hat{\mathcal{R}}^{(3)}$ defined by $\varphi\Big((g, f, h)\Big) = (g, f, zh)$ is a group isomorphism.

7.6 By the Fundamental Theorem of 3-D Riordan Array, $(C, zC, C) * \left(\frac{1}{1-z}, \frac{z}{1-z}\right) = (C^2, zC^3) := [b_{n,\ell}]$. Since

$$c_{n,k,\ell} = [z^{n-k}]C^{k+\ell+1} = [z^{(n+\ell)-(k+\ell)}]C^{k+\ell+1} = \sum_{i=0}^{\lfloor \frac{k+\ell}{2}\rfloor} (-1)^i \binom{k-i+\ell}{i} C_{n-i+\ell},$$

$$b_{n,\ell} = \sum_{k\ge 0}\binom{k}{\ell} c_{n,k,\ell} = \sum_{k=0}^{n}\sum_{i=0}^{\lfloor \frac{k+\ell}{2}\rfloor} (-1)^i \frac{1}{n-i+\ell}\binom{k}{\ell}\binom{k-i+\ell}{i}\binom{2(n-i+\ell)}{n-i+\ell}.$$

Thus $b_{n,\ell} = [z^n]C^2\cdot(zC^3)^\ell = [z^{(n+2\ell+1)-(3\ell+1)}]C^{2+3\ell} = \frac{3\ell+2}{n+2\ell+1}\binom{2n+\ell+1}{n+2\ell+1}$.

7.7 The identity (7.2.26) for $i = m-1$ can be written as $\sum_{n\ge 0}^{m}\binom{2m-n-1}{m-n} = \binom{2m}{m}$, $m \ge 1$. Let $g = 1 + \sum_{m\ge 1}\binom{2m-1}{m-1}z^m$. Then $(g, zC)\frac{1}{1-z} = \frac{1}{\sqrt{1-4z}}$. Using the FTRA gives the required result.

7.8 (a) Consider $\left(\frac{1}{1-z}, \frac{z}{1-z}, \frac{z}{1-z}\right)$, $A(z) = (1+z)^n$ and $B(z) = (1+z)$.

(b) Consider $\left(\frac{1}{1-z}, \frac{z}{1-z}, \frac{z}{1-z}\right)$, $A(z) = B(z) = \frac{1}{1-z}$.

(c) Consider $\Big(C, zC, zC\Big)$, $A(z) = B(z) = \frac{1}{1-z}$.

7.9 Find the step sets P_{S_1} and P_{S_2} such that $P_{S_1} = (C, zC)$ and $P_{S_2} = \left(\frac{1}{1-z}, \frac{z}{1-z}\right)$.

7.10 By a similar argument used in the proof of Theorem 7.10, it is straightforward.

7.11 By a similar argument used in the proof of Theorem 7.10, it is straightforward.

7.12 Since $(G, zH)^2 = (GG(zH), zHH(zH)) = (1, z)$, we have

$$\begin{aligned} G(\varphi)\, G(\varphi H(\varphi)) &= G(\varphi)\, G\left(\sum_{k=1}^{d} a_k z_k H(\varphi)\right) = 1, \\ H(\varphi)\, H(\varphi H(\varphi)) &= H(\varphi)\, H\left(\sum_{k=1}^{d} a_k z_k H(\varphi)\right) = 1. \end{aligned}$$

It implies that $\left(G(\varphi), \mathbf{z}H(\varphi)\right)^2 = (1, \mathbf{z})$.

7.13 (a) $\begin{bmatrix} 1\,0\,0\,0\,0\,0 \\ 1\,1\,0\,0\,0\,0 \\ 1\,0\,1\,0\,0\,0 \\ 1\,2\,0\,1\,0\,0 \\ 2\,2\,2\,0\,1\,0 \\ 1\,0\,2\,0\,0\,1 \end{bmatrix}$.

(b) $\mathbb{P}^{-1} = \mathcal{M}\left(\frac{1}{1+z_1+\cdots+z_d}, \frac{\mathbf{z}}{1+z_1+\cdots+z_d}\right)$.

(c) By using the Taylor series, one can show that all diagonal blocks M_{nn} of $\mathbb{P}$ are lower triangular matrices.

Exercises of Chap. 8

8.1 $[n]_q! = \sum_{\sigma \in S_n} q^{\mathrm{inv}(\sigma)}$.

8.2 We give the proof for the q-product rule. The q-sum rule and q-quotient rule can be similarly proved.

$$\begin{aligned} D_q(fg) &= f(qx)g(qx) - fg = f(qx)g(qx) - f(qx)g(x) + f(qx)g(x) - fg \\ &= f(qx)D_q g(x) + g(x)D_q f(x). \end{aligned}$$

8.3 Since $[\alpha+1]_q = [k]_q + q^k[\alpha+1-k]_q$ we have

$$\begin{aligned} \binom{\alpha+1}{k}_q &= \frac{[\alpha+1]_q!}{[k]_q![\alpha+1-k]_q!} = \frac{[\alpha]_q!([k]_q + q^k[\alpha+1-k]_q)}{[k]_q![\alpha+1-k]_q!} \\ &= \binom{\alpha}{k-1}_q + q^k\binom{\alpha}{k}_q. \end{aligned}$$

The second equality follows from $\binom{\alpha+1}{k}_q = \binom{\alpha+1}{\alpha+1-k}_q$.

8.4 The Gaussian coefficient $\binom{n}{k}_q$ counts the subspaces of dimension k of an n-dimensional vector space over the finite field with q elements. Thus $G_n(q)$ counts all subspaces of V_n.

8.5 Clearly, $\mathcal{R}_0$ is the ordinary Riordan group and $\mathcal{R}_1$ is the exponential Riordan group.

Now let $A = (a(z), b(z))_q$, $B = (c(z), d(z))_q$ and $C = (e(z), f(z))_q$. Assume that $\mathcal{R}_q$ is a group. By associativity, $f[d[b(z)]] = (f[d(z)])[b(z)]$. Since

$$f[d[b(z)]] = \sum_{n\geq 1}\left(\sum_{k=1}^{n} f_k x_{n,k}\right)\frac{z^n}{[n]_q!}$$

$$(f[d(z)])[b(z)] = \sum_{n\geq 1}\left(\sum_{k=1}^{n} b_{n,k}\sum_{j=1}^{k} f_j d_{k,j}\right)\frac{z^n}{[n]_q!}$$

where

$$x(z) = \sum_{n\geq 1} x_n \frac{z^n}{[n]_q!} = \sum_{n\geq 1}\left(\sum_{k=1}^{n} d_k b_{n,k}\right)\frac{z^n}{[n]_q!},$$

we have for all $n \geq 1$

$$\sum_{k=1}^{n}\left(f_k x_{n,k} - b_{n,k}\sum_{j=1}^{k} f_j d_{k,j}\right) = 0.$$

By a simple computation, if $n = 4$ then

$$0 = \sum_{k=1}^{4}\left(f_k x_{4,k} - b_{n,k}\sum_{j=1}^{k} f_j d_{k,j}\right) = b_1^2 b_2 d_1 d_2 f_2 q(q-1)$$

where $b_1 b_2 d_1 d_2 f_2 \neq 0$. Hence $\mathcal{R}_q$ is not a group if $q \notin \{0, 1\}$.

The subloop $\{(g(z), z)_q : g(z) \in \mathcal{E}_q(0)\}$ forms a group with $(g(z), z)_q^{-1} = (1/g(z), z)_q$.

8.6

$$(1, e_q(z) - 1)_q = \begin{bmatrix} 1 & 0 & 0 & 0 & 0 \\ 0 & 1 & 0 & 0 & 0 \\ 0 & 1 & 1 & 0 & 0 \\ 0 & 1 & 2+q & 1 & 0 \\ 0 & 1 & 3+2q+2q^2 & 3+2q+q^2 & 1 \end{bmatrix}.$$

8.7 (1) It follows from Corollary 8.1 and the q-counting function that $f(m) = (m-1)!_q$ since $\sum_{\sigma \in \mathcal{S}_{m,1}} q^{\mathrm{cinv}(\sigma)} = (m-1)!_q$.

(2)

$$(1, -\mathrm{loq}_q(1+z))_q = \begin{bmatrix} 1 & 0 & 0 & 0 & 0 \\ 0 & 1 & 0 & 0 & 0 \\ 0 & 1 & 1 & 0 & 0 \\ 0 & 1+q & 2+q & 1 & 0 \\ 0 & 1+2q+2q^2+q^3 & 3+4q+3q^2+q^3 & 3+2q+q^2 & 1 \end{bmatrix}.$$

For instance, the 2-cyclic partitions of $\{1, 2, 3, 4\}$ are

$$\begin{aligned} &\sigma_1 = (1)(2,3,4), \quad \sigma_2 = (1)(2,4,3), \quad \sigma_3 = (1,2)(3,4), \quad \sigma_4 = (1,3)(2,4), \\ &\sigma_5 = (1,4)(2,3), \quad \sigma_6 = (1,2,3)(4), \quad \sigma_7 = (1,3,2)(4), \quad \sigma_8 = (1,2,4)(3), \\ &\sigma_9 = (1,4,2)(3), \quad \sigma_{10} = (1,3,4)(2), \quad \sigma_{11} = (1,4,3)(2). \end{aligned}$$

Thus we have

$$\sum_{\sigma \in \mathcal{S}_{4,2}} q^{\mathrm{cinv}(\sigma)} = \sum_{i=1}^{11} q^{\mathrm{cinv}(\sigma_i)} = 3 + 4q + 3q^2 + q^3 = c_q(4,2).$$

8.8 For $f(z) = \sum_{k \geq 1} f_k z^k / [k]_q!$, let $\overline{f}_R(z) = \sum_{k \geq 1} \overline{f}_k z^k / [k]_q!$ satisfy $f[\overline{f}_R(z)] = z$. Then use the similar method used in the proof of Theorem 8.2.

Exercises of Chap. 9

9.1 Let $(g, f) = \left(\frac{1}{1+x^2}, \frac{x}{1+x^2}\right)$. We require the generating function of the first column of the inverse $(g, f)^{-1}$. For this, we solve the equation $\frac{u}{1+u^2} = x$ to get the solution $u(x) = \bar{f}(x)$, where we choose the solution with $u(0) = 0$. We obtain $\bar{f} = \frac{1-\sqrt{1-4x^2}}{2x}$. The array is a Bell matrix and so the generating function of the first column is $\bar{f}/x = \frac{1-\sqrt{1-4x^2}}{2x^2}$. This is the generating function of the aerated Catalan numbers.

9.2 Let $(g, f) = \left(\frac{1}{1+x}, \frac{x}{(1+x)^2}\right)$. We find that $\bar{f}(x) = \frac{1-2x-\sqrt{1-4x}}{2x}$, and so the first column of the inverse matrix has generating function $\frac{1}{g(\bar{f}(x))} = \frac{1-\sqrt{1-4x}}{2x}$, the generating function of the Catalan numbers.

9.3 Let $(g, f) = \left(\frac{1}{(1+x)^2}, \frac{x}{(1+x)^2}\right)$. Again, we obtain $\bar{f}(x) = \frac{1-2x-\sqrt{1-4x}}{2x}$, and so the first column of the inverse matrix has generating function $\frac{1}{g(\bar{f}(x))} = \left(\frac{1-2x-\sqrt{1-4x}}{2x}\right)^2$, or $\frac{1-2x-\sqrt{1-4x}}{2x^2}$, which is the generating function of C_{n+1}.

9.4 Let $(g, f) = \left(\frac{1}{1+x+x^2}, \frac{x}{1+x+x^2}\right)$. Solving the equation $\frac{u}{1+u+u^2} = x$ gives us $\bar{f}(x) = \frac{1-x-\sqrt{1-2x-3x^2}}{2x}$. As this is a Bell type matrix, the generating function of the first column of the inverse is then $\frac{1-x-\sqrt{1-2x-3x^2}}{2x^2}$. This is the generating function of the Motzkin numbers.

9.5 Let $(g, f) = \left(\frac{1+x}{1+x+x^2}, \frac{x}{1+x+x^2}\right)$. As in the previous question, we find that $\bar{f}(x) = \frac{1-x-\sqrt{1-2x-3x^2}}{2x}$. Then we have that $\frac{1}{g(\bar{f}(x))} = \frac{2}{1+x+\sqrt{1-2x-3x^2}}$, which is the generating function of the Riordan numbers.

9.6 Let $(g, f) = \left(\frac{1}{1+x}, \frac{x}{1+3x+2x^2}\right)$. Solving the equation $\frac{u}{1+3u+2u^2} = x$, we obtain that $\bar{f}(x) = \frac{1-3x-\sqrt{1-6x+x^2}}{4x}$. With $g(x) = \frac{1}{1+x}$, we find that $\frac{1}{g(\bar{f}(x))} = \frac{1+x-\sqrt{1-6x+x^2}}{4x}$, which is the generating function of the little Schröder numbers.

9.7 Let $(g, f) = \left(\frac{1}{1+2x}, \frac{x}{1+3x+2x^2}\right)$. We solve the equation $\frac{u}{1+3u+2u^2} = x$ to get $\bar{f}(x) = u(x) = \frac{1-3x-\sqrt{1-6x+x^2}}{4x}$. With $g(x) = \frac{1}{1+2x}$, we find that $\frac{1}{g(\bar{f}(x))} = \frac{1-x-\sqrt{1-6x+x^2}}{2x}$. This is the generating function of the large Schröder numbers.

9.8 Let $(g, f) = \left(\frac{1-x}{1+x}, \frac{x}{(1+x)^2}\right)$. Then $\bar{f} = \frac{1-2x-\sqrt{1-4x}}{2x}$. With $g(x) = \frac{1-x}{1+x}$, we find that $\frac{1}{g(\bar{f}(x))} = \frac{1}{\sqrt{1-4x}}$. This is the generating function of central binomial coefficients $\binom{2n}{n}$.

9.9 Let $(g, f) = \left(\frac{1-x}{1+x^2}, \frac{x}{1+x^2}\right)$. Solving the equation $\frac{u}{1+u^2} = x$, we find that $\bar{f}(x) = \frac{1-\sqrt{1-4x^2}}{2x}$. With $g(x) = \frac{1-x}{1+x^2}$, we find that $\frac{1}{g(\bar{f}(x))} = \frac{2}{1-2x+\sqrt{1-4x^2}}$. This is the generating function of $\binom{n}{\lfloor \frac{n}{2} \rfloor}$.

9.10 Let $(g, f) = \left(\frac{1-x}{(1+x)^2}, \frac{x}{(1+x)^2}\right)$. We find that $\bar{f}(x) == \frac{1-2x-\sqrt{1-4x}}{2x}$. With $g(x) = \frac{1-x}{(1+x)^2}$, we obtain that $\frac{1}{g(\bar{f}(x))} = \frac{\sqrt{1-4x}+4x-1}{2x(1-4x)}$. This is the generating function of $\binom{2n+1}{n+1}$.

9.11 The generating function of the Chebyshev polynomials of the first kind $T_n(x)$ is given by $\sum_{n=0}^{\infty} T_n(x)t^n = \frac{1-tx}{1-2xt+t^2}$. Using the fundamental theorem of almost Riordan arrays, we have

$$\left(\frac{1}{1+t^2}\middle| \frac{1-t^2}{(1+t^2)^2}, \frac{2t}{1+t^2}\right) \cdot \frac{1}{1-tx} = \frac{1}{1+t^2} + \frac{t(1-t^2)}{(1+t^2)^2}\tilde{h}\left(\frac{2t}{1+t^2}\right) = \frac{1-tx}{1-2xt+t^2}.$$

Here, we use

$$h(x) = \frac{2t}{1+t^2} \Longrightarrow \tilde{h}(x) = \frac{x}{1-tx}.$$

9.12 Solving $\frac{u}{1+u^2} = x$ for $\bar{f}$, and using $g(x) = \frac{1-x-x^2}{1+x^2}$, we obtain $\mu(x) = \frac{1}{\sqrt{1-4x^2}-x}$. Thus $\frac{x}{\mu(x)} = x\left(\sqrt{1-4x^2} - x\right)$. We find that $\frac{1}{x}\text{Rev}\left(\frac{x}{\mu(x)}\right) = \frac{1}{x}\sqrt{\frac{1-2x-\sqrt{1-4x-16x^2}}{10}}$, which is the generating function for $F_{n+1}C_n$.

9.13 We have $a = 1, b = 2$ and $c = 2, d = 1$. We find that

$$g(x) = \frac{1 - 3x + x^2 - \sqrt{1 - 6x + 7x^2 + 6x^3 - 7x^4}}{2x^2(1 - x)}.$$

Forming the Riordan array $(g(x), xg(x))$ we find that

$$\begin{bmatrix} 1 & 0 & 0 & 0 & 0 & 0 \\ 1 & 1 & 0 & 0 & 0 & 0 \\ 3 & 2 & 1 & 0 & 0 & 0 \\ 9 & 7 & 3 & 1 & 0 & 0 \\ 29 & 24 & 12 & 4 & 1 & 0 \\ 95 & 85 & 46 & 18 & 5 & 1 \end{bmatrix} \begin{bmatrix} 1 \\ 2 \\ 4 \\ 8 \\ 16 \\ 32 \end{bmatrix} = \begin{bmatrix} 1 \\ 3 \\ 11 \\ 43 \\ 173 \\ 705 \end{bmatrix}.$$

9.14 We form the Riordan array $\left(\frac{1-3x}{1+2x}, \frac{x(1-3x)}{1+2x}\right)^{-1}$. We then calculate

$$\begin{bmatrix} 1 & 0 & 0 & 0 & 0 & 0 \\ 5 & 1 & 0 & 0 & 0 & 0 \\ 40 & 10 & 1 & 0 & 0 & 0 \\ 395 & 105 & 15 & 1 & 0 & 0 \\ 4360 & 1190 & 195 & 20 & 1 & 0 \\ 51530 & 14270 & 2510 & 310 & 25 & 1 \end{bmatrix} \begin{bmatrix} 1 \\ 3 \\ 9 \\ 27 \\ 81 \\ 243 \end{bmatrix} = \begin{bmatrix} 1 \\ 8 \\ 79 \\ 872 \\ 10306 \\ 127568 \end{bmatrix}.$$

References

1. R. Graham, D. Knuth, O. Patashnik, *Concrete Mathematics*, 2nd edn. (Addison-Wesley, 1994)
2. D. Merlini, M. Nocentini, Algebraic generating functions for languages avoiding Riordan patterns. J. Integer Seq. **21**(1), 1–25 (2018)
3. D. Merlini, R. Sprugnoli, Algebraic aspects of some Riordan arrays related to binary words avoiding a pattern. Theor. Comput. Sci. **412**(27), 2988–3001 (2011)

Index

© Springer Nature Switzerland AG 2022

L. Shapiro et al., *The Riordan Group and Applications*, Springer Monographs in Mathematics, https://doi.org/10.1007/978-3-030-94151-2